gcse geography

AQA A

Catherine Hurst

Jane Holroyd

Steve Rickerby

Jack Gillett

Meg Gillett

OxBox

OXFORD UNIVERSITY PRESS

Great Clarendon Street, Oxford OX2 6DP

Oxford University Press is a department of the University of Oxford. It furthers the University's objective of excellence in research, scholarship, and education by publishing worldwide in

Oxford New York

Auckland Cape Town Dar es Salaam Hong Kong Karachi Kuala Lumpur Madrid Melbourne Mexico City Nairobi New Delhi Shanghai Taipei Toronto

With offices in

Argentina Austria Brazil Chile Czech Republic France Greece Guatemala Hungary Italy Japan Poland Portugal Singapore South Korea Switzerland Thailand Turkey Ukraine Vietnam

Authors: Catherine Hurst, Jane Holroyd, Steve Rickerby, Jack Gillett, Meg Gillet

First published 2011

British Library Cataloguing in Publication Data

Data available

ISBN 978-0-19-913549-3

10 9 8 7 6 5 4 3 2

Printed in Singapore by KHL Printing Co Pte Ltd

Paper used in the production of this book is a natural, recyclable product made from wood grown in sustainable forests. The manufacturing process conforms to the environmental regulations of the country of origin

Acknowledgements

The publisher and authors would like the thank the following for permission to use photographs and other copyright material:

Demotix Demotix/Photolibary **P6**; Kevin West/Getty Images **P8**; Barcroft Media/Getty Images **P10**; National Geographic/Barry Bishop/Getty Images **P14(t)**; AFP/Getty Images **P14(b)**; National Geographic/Skip Brown/Getty Images **P15**; Reuters Photo Agency **P16**; Reportage/Getty Images **P17**; Image Bank/Getty Images **P18**; United States Geographical Survey **P19**; STR/AP/Press Association **P15(t)**; Reuters Photo Agency **P25(b)**; Cristobel Fuentes/Press Association **P24(t)**; MC2 Justin Stumberg/ Press Association **P24(b)**; David Rydevik **P26**; Tarmizy Harva/Reuters Photo Agency **P27**; Shutterstock **P28**; Graeme Peacock/Alamy **P30(t)**; STR/AFP/Getty Images **P31**; www.edupic.net/geo **P32(t)**; Geophotos/Tony Waltham **P30, P32(b), P33(t), P34(b), [illegible]34(c), P35, P37, P39(tl), P39(tc), P39(bc), P39(br), P40, P44, P45, P118(t), [illegible]27(t), P127(b), P128(b), P136 (t&b), P142**; Neale Haynes/Rex Features **P34(t)**; A R[illegible]n With A View/Alamy **P36**; Robert Harding Picture Library/Alamy **P39(tr)**; Peter Cr[illegible]ton/Alamy **P39(bl)**; B&J Photos/Alamy **P41(t)**; Jane Hallin/Alamy **P41(b)**; Alan Cur[illegible]lamy **P42(t)**; Simon King/Nature Picture Library P42(b); Watermark Holiday Hom[illegible]46; Adrian Davies/Nature Picture Library **P47(t)**; Joan Gravell **P47(b)**; Paul J Hudso[illegible]tty Images **P48**; Robert Harding Picture Library/Photolibrary **P53**; Apex News P[illegible]Getty Images **P56**; Anna Gowthorpe/Press Association **P57**; Getty Images **P58**; AFP[illegible]y Images **P60**; NASA **P63(t)**; Shutterstock **P63(b)**; Cartoon Stock **P64(t)**; An[illegible] Drysdale/Photolibrary **P64(b)**; Nigel Howard/Rex Features **P65(t)**; 10:10 Camp[illegible] **P65(b)**; NASA **P66**; Getty Images/AFP **P68**; National Hurricane Center **P69**; [illegible]ew/Reuters Photo Agency **P71**; Arco Images GmbH/Alamy **P72**; Getty Images [illegible]Paul Oomen/Getty Images **P77**; Alastair Shay/Photolibrary **P80(t)**; Peter Lewis/Ph[illegible]ary **P80(b)**; Britain On View/Photolibrary **P81(b)**; Gary Roebuck/Alamy [illegible]; Neil Setchfield/Alamy **P82(t)**; Getty Images **P82(b)**; John McKenna/Alamy [illegible]hotolibrary **P84(t)**; Peter Arnold Images/Photolibrary **P84(b)**, **P86(t), P87(b)**; Re[illegible]hoto Agency **P86(b)**; Mark Edwards/Still Pictures **P87(t)**; Reserva Ecologica d[illegible]acu **P88(t), P88(b)**; Slow Images/Getty Images **P89(t)**; Imagebroker/Photoli[illegible]89(b); Joern/Shutterstock **P90(l)**; Susanna Bennett/ Alamy **P90(r)**; Spectru[illegible]ofile Inc/Photographers Direct **P91**; Roger Ressmeyer/ Corbis **P92**; Mark Edwards/Still Pictures **P93**; Photoshot Holdings Ltd/Alamy **P94**; Bob Digby **P96(t), P96(b), P98**; Webb Aviation **P100(t)**; Neil Holmes Freelance Digital/Alamy **P100(b)**; Webb Aviation **P101**; Chris Clark/Press Association **P102**; Global Warming Images/Alamy **P104**; Mostafizur Rahmann/Press Association **P105**; Andrew Yates/Getty Images **P106(t)**; Christopher Furlong/Getty Images **P106(b)**; Arif Ali/AFP/Getty Images **P108, P109(t)**; Tim Wimbourne/Reuters Photo Agency **P109(b)**; Barry Morgan/Alamy **P100**; Michael Harvey/Alamy **P112**; Robert Harding Picture Library/Getty Images **P115**; Monpix/Alamy **P116**; MM_Photo/Alamy **P120(t)**; The Print Collector/Alamy **P121(t)**; Steve Frost/Alamy **P121(b)**; Bert Kammerlander/ Getty Images **P122**; Jane Holroyd **P123(l&r), P137(t)**; Jo Katanigra/Alamy **P124**; John Lamb/Getty Images **P125**; John Schweider/Alamy **P126**; Julian Love/Corbis **P130(t)**; Mike Harrington/Getty Images **P130(b)**; Travel Ink/Getty Images **P131(t)**; Danita Delimont/Alamy **P131(b)**; PJ-Foto/Alamy **P132(t), P132(b)**; Hemis/Alamy **P133(t)**; Jon Arnold Images Ltd/Alamy **P133(b)**; Galen Rowell/Mountain Light/Alamy **P134**; STR/Reuters Photo Agency **P135**; Top Pics/Alamy **P137(b)**; Terra/Corbis **P138**; Jason Feast **P140**; Shutterstock **P142(t)**; Tradewinds/Alamy **P144**; Mervyn Rees/ Alamy **P145**; Pat Downing/Alamy **P146(t)**; Flickr/Getty Images **P146(b)**; Simmons Aerofilms/Jason Crossley **P147(t)**; Dan Burton/Alamy **P147(b)**; Richard Austin/Rex Features **P148**; Andrew Stacey **P149**; Melvin Bourne **P150**; Chris Brunnen **P151**; Mohammen Seenen/AP/Press Association **P152(t)**; Chad Ehlers/Alamy **P152(b)**; Getty Images/AFP **P153**; Leslie Garland Picture Library/Alamy **P154(r)**; Nigel Forrow **P155**; Photolibrary **P156(b)**; Geoffrey Swaine/Rex Features **P156(t)**; John Giles/Press Assocation **P157**; Tony Phelps/Nature Picture Library **P158(t)**; Hugh Clark/Frank Lane Picture Agency **P158(b)**; Skyscan/Corbis **P159**; Image Source/Photolibrary **P160**; Astock/Corbis **P162**; Worldmapper **P163**; Toru Yamanka/AFP/Getty Images **P164**; Picture Contact/Alamy **P169**; Tim Graham/Alamy **P170**: AFP/Getty Images **P171**; Sean Sprague **P172, P173**; Chris Ison/Press Association **P174**; Libby Welch/ Alamy **P176(t)**; Ian Miles/Flashpoint/Alamy **P176 (b)**; Getty Images **P177**; Shutterstock **P178**; Getty Images **P179(t)**; Bloomberg/Getty Images **P179(b)**, Alamy **P180**; AFP/Getty Images **P181**; Comstock/Getty Images **P182**; Fredrik Renander/ Alamy **P184(inset)**; Getty Images **P184**; Getty Images **P186(t)**; Images of Birmingham **P186(b)**; Stringer India/Reuters Photo Agency **P187(t)**; Getty Images **P187(r)**; Shutterstock **P188(t)**; Derek Askill **P188(b)**; Images of Birmingham **P189**; Art Directors and Trip **P190(t)**; Images of Birmingham **P190(b)**; David Stowell **P191(t)**; Andrew Fox/Alamy **P192(t)**; Art Directors & Trip **P192(b)**; Images of Birmingham **P193(t)**; David Jones/Alamy **P193(b)**; Still Pictures **P194(t)**; Sean Sprague **P194(b)**; Adrian Fisk **P195(t)**; Reuters Photo Agency **P195(b)**; Fly Fernandez/Zefa/Corbis **P196**; Graeme Williams/Getty Images **P197**; AFP/Getty Images **P198**; AFP/Getty Images **P199(t)**; Bloomberg/Getty Images **P196(b)**; Nic Hamilton Photographic **P201**; Matt Cardy/Getty Images **P202(t)**; Skyscan **P202(b)**; View Pictures Ltd/Alamy **P203**; Shutterstock **P204**; Anne Gilbert **P206**; Snoasis Concern Ltd **P209(t)**; Webb Aviation **P209(b)**; Peter Evarard-Smith **P210**; David Wootton/Alamy **P211(t)**; Rob Atherton **P211(t), P212**; Julia Gavin **P231(t)**; Robert Crook **P213(b)**; Mike Page **P214**; Aidan Semmens Photography **P215**; Sid Frisby **P216(t)**; David Hobard Photography **P216(b)**; Geophotos/Alamy **P217**; Kathy Wright/Alamy **P218**; Justin Kase/Alamy **P219**; Jim Laws/Alamy **P220**; Keith Morris/Alamy **P221(t)**; Arco Images GmbH/Alamy **P221(b)**; Sami Sarkis **P222(t)**; Yann Arthus-Bertrand/Corbis **P222(c)**; Jeremy Horner/ Alamy **P222(b)**; Victor Englebert **P223(t)**; EcoImages/Universal Images Group/Getty Images **P223(b)**; Age/Fotostock/Robert Harding Picture Library **P224(t)**;Yves Gellie/ Corbis **P224(b)**; Sean Sprague **P225**; Danita Delimont/Alamy **P226**; Still Pictures/ Christian Aid **P228**; (c) SASI (University of Sheffield) & Mark Newman (University of Michigan) **P229**; Sylvia Cordaiy/Alamy **P232**; Still Pictures **P233**; Caroline Irby/Water Aid **P235**; NASA/Goddard Space Flight Centre/Science Photo Library **P236**; AFP/Getty Images **P237(t)**; Fabienne Fossez/Alamy **P238(b)**; Image Bank/Getty Images **P238**; Simon Rawles/Getty Images **P241**; Getty Images **P242**; Robert Harding Picture/Getty Images **P243**; Karel Prinsloo/AP/Press Association **P244(t)**; Bob Digby **P244(b)**; Christopher Furlong/Getty Images **P245**; Peter MacKinven/Photolibrary **P246**; Noel Hapgood/Getty Images **P247(t)**; Design Pictures/Getty Images **P247(b)**; Tyrone Siu/ Reuters Photo Agency **P248**; AFP/Getty Images **P250**; Middle East/Alamy **P251**; Sipa Press/Rex Features **P252(t)**; Stringer India/Reuters Photo Agency **P252(b)**; Richard Vogel/Press Association **P254**; Bloomberg/Getty Images **P255**; Worldmapper **P256**; Digital Globe **P258**; Eightfish/Reportage/Getty Images **P259(t)**; Eightfish/Getty Images **P259(b)**; Gerald Herbert/Press Association **P260**; Sami Sarkis/Alamy **P261(t)**; Wilfredo Lee/Press Association **P261(b)**; Jantsen Taleb **P262**; Kawal Kishore/Reuters Photo Agency **P263**; Martin Bond/Still Pictures **P265(t)**; Ashley Cooper Pictures/ Alamy **P265(b)**; Press Association **P266(t)**; Suzanne Porter/Alamy **P266(b)**; Image State Media Partners Ltd-Impact Photos/Alamy **P269(t)**; Nobel Images/Alamy **P269(b)**; Shutterstock **P270**; Reuters/Corbis **P272(l)**; Aurora Photos/Alamy **P272(r)**; Lonely Planet **P273(t)**; IML Images Group Ltd/Alamy **P273(b)**; Dave Pratt/Press Association **P274(t)**; Photodisc/Getty Images **P274(b)**; Moodboard/Photolibrary **P275(t)**; Taxi/Getty Images **P275(b)**; Axiom Photographic Agency/Getty Images **P277**; Photographers Choice **P276**; Ashley Cooper/Alamy **P279(l)**; Visuals Unlimited/ Getty Images **P279(r)**; Hemis/Alamy **P280(t)**; Danny Lehman/Corbis **P280(b)**; Robert Harding Picture Library/Photolibrary **P282**; Rolf Richardson/Alamy **P283**; Robert Harding Picture Library/Getty Images **P284**; Ron Watts/Photolibrary **P284(inset)**; OSF/Doug Allan/Photolibrary **P285(t)**; Workbook Stock/Getty Images **P285(b)**; Press Association **P286(t)**; Peter Arnold/Martin Harvey/Photolibary **P286(b)**; John Warburton Photography/Photolibrary **P287**; Shutterstock **P288**; Yachana Ecolodge **P289(t), P289(b), P290(t), P290(b), P291**; Thomas Mueller/Alimdi.net **P290(b)**; Bob Digby **P292**; OUP **P295**; Geophotos/Tony Waltham **P301**; Derek Askill **P308**; Webb Aviation **P309**; Ashley Cooper/Alamy **P312**.

Ordnance Survey maps reproduced by permission of Ordnance Survey on behalf of HMSO © Crown copyright. All rights reserved – OS License number 00000249.

Front cover: Shutterstock/Mike_Expert

Illustrations are by: Barking Dog Art, Q2A, Steve Evans and Hart McLeod.

The sections about Buckden Beck (pages 96 – 101) are based on material written by Bob Digby for GCSE Geography Edexcel B (Digby et al., OUP, 2009).

Every effort has been made to contact copyright holders of material reproduced in this book. Any omissions will be rectified in subsequent printings if notice is given to the publisher.

Contents

Unit 1 Physical geography

Section A

Section B

Contents

Unit 2 Human geography

Section A

Section B

Unit 3 Local fieldwork investigation

Exam skills and practice

1 » The restless Earth

What do you have to know?

This chapter is from **Unit 1 Physical Geography Section A** of the AQA A GCSE specification. It is about the Earth's plates, earthquakes and volcanoes; the impacts that earthquakes and eruptions have, and our responses to them. The table shows how the pages in this chapter match the content in the specification.

Specification content	Pages in this chapter
Different types of plates and plate margins, and where you find them.	p8-11
The location and formation of fold mountains, ocean trenches and two types of volcano.	p12-13
Fold mountains, and how people use them.	Case study of the Himalayas p14-15
Different types of volcano, their characteristics, and volcanic eruptions – causes, effects and responses.	p13, case study of the volcanic eruption on Montserrat p16-17
Supervolcanoes – characteristics, and the likely effects of an eruption.	p18-19
Earthquakes – location, causes and features, and how we measure them.	p20-21
The effects and responses to earthquakes in different parts of the world.	Case study of earthquakes in Chile and Haiti p22-25
Tsunamis – causes, effects and responses.	Case study of the Boxing Day tsunami 2004 p26-27

Your key words

Plates

Plate margins
(constructive, conservative and destructive)

Collision zones

Oceanic crust, continental crust

Subduction

Fold mountains

Ocean trenches

Composite volcanoes, shield volcanoes

Pyroclastic flows

Supervolcano

Caldera

Volcanic Explosivity Index (VEI)

Shock waves

Focus

Epicentre

Richter Scale

Mercalli Scale

Tsunami

Exam help ...

Advice See pages 297-299 for information on how to be successful in your exams.

Practice See page 300 for exam questions on this chapter.

What if ...

- you were caught up in a volcanic eruption?
- your home was destroyed by a lava flow?
- you lived in an earthquake zone?
- the UK was hit by a tsunami?

1.1 » Unstable Earth

On this spread you'll find out about the structure of the Earth, different types of crust and plate margins.

Imagine a vision of Hell – hot ash clogging up your eyes and ears, and making it hard to breathe; a stinking smell of sulphur; a red glow colouring the whole sky; and molten lava raining down on you. It's also roasting hot. Luckily, you just have to imagine a scene like that, but many people who live near the world's most active volcanoes actually experience your vision of Hell – and the lucky ones survive.

The structure of the Earth

Inside the Earth it's hot – really hot. We know that because of the molten lava that spews from volcanoes like the one on the right. Also, in some places, you get hot springs and geysers.

The Earth isn't a solid ball of rock. It's made up of layers – a bit like a chocolate-covered peanut. The Earth's core is the peanut, the mantle is the chocolate surrounding it, and the crust is the sugar shell.

your planet
Geysers are hot springs that shoot out jets of steam and hot water. They're named after a word in Old Norse, geysa, which means to gush.

The Earth's crust is divided into large pieces, called **plates**. It's unstable because these plates move – very slowly. The different plates meet at **plate margins**, or boundaries. There are three main types of plate margins (see the map opposite):

- **Constructive** – where the plates are moving apart.
- **Conservative** – where the plates slide past each other.
- **Destructive** – where the plates collide.

(There are also **collision zones**, which are a type of destructive margin.)

▼ *The structure of the Earth*

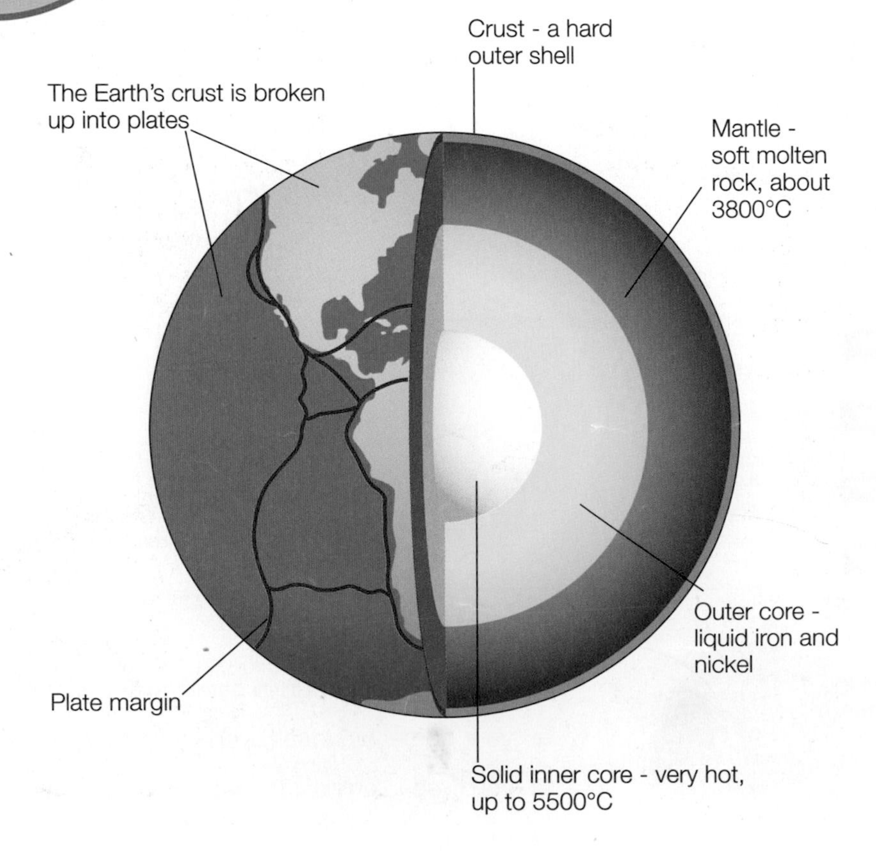

Two types of crust

The Earth's crust is made up of:

- **oceanic crust** – what you get under the oceans
- **continental crust** – what you get under the continents, or land masses.

Oceanic crust	Continontal crust
• 5-10 km thick • Dense • Sinks into the mantle when oceanic and continental crust meet • The oldest is only about 180 million years old • Forms constantly at constructive plate margins • Destroyed at destructive plate margins	• 25-100 km thick • Less dense • Doesn't sink • Very old – 3-4 billion years old • New crust isn't formed • Cannot be destroyed

Plates – where they are and how they move

▼ *Different types of plate margins*

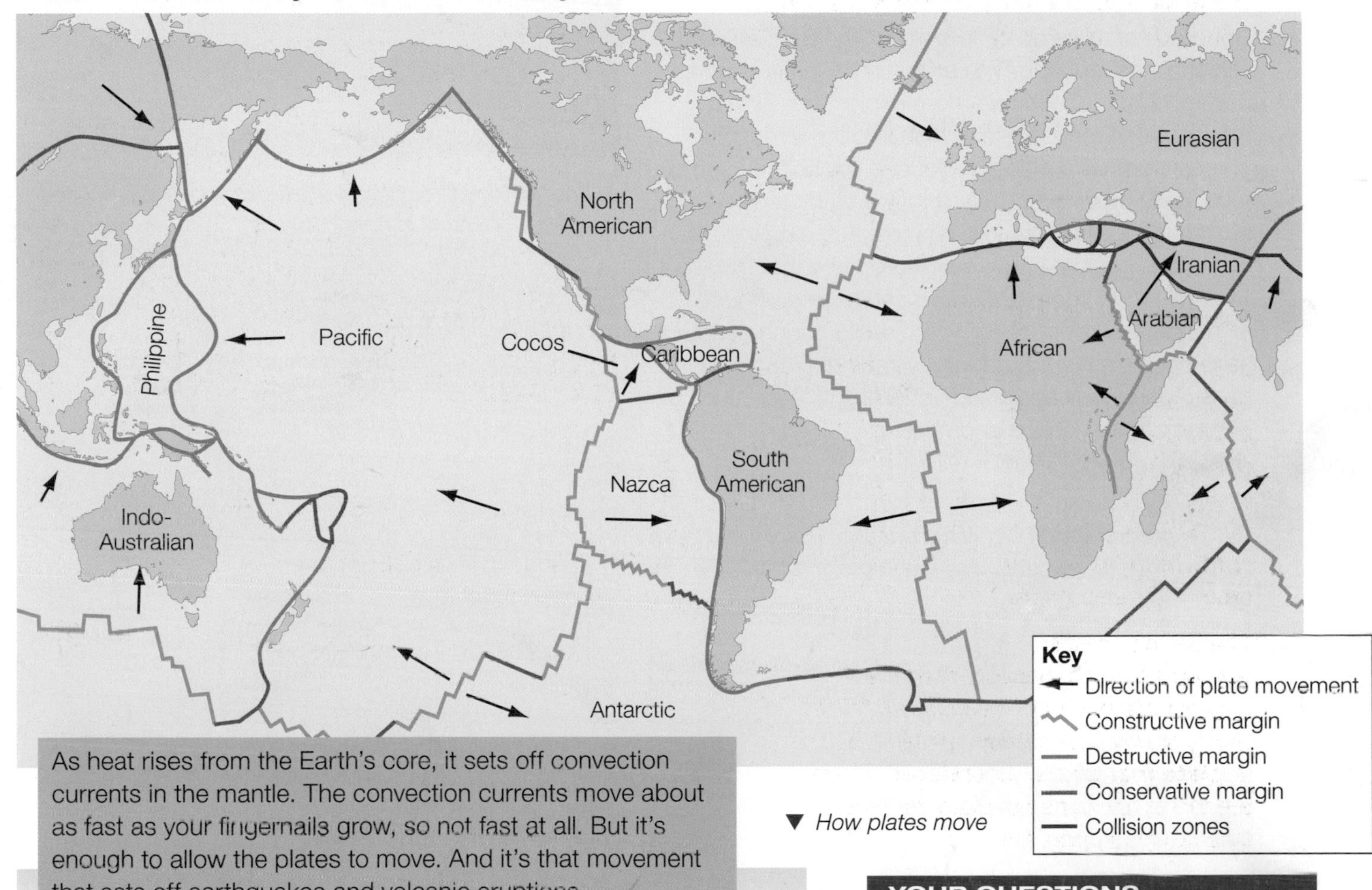

As heat rises from the Earth's core, it sets off convection currents in the mantle. The convection currents move about as fast as your fingernails grow, so not fast at all. But it's enough to allow the plates to move. And it's that movement that sets off earthquakes and volcanic eruptions.

▼ *How plates move*

Continental crust
Oceanic crust
Sea
Convection currents in Earth's mantle

YOUR QUESTIONS

1 a Explain what these terms mean: plates, plate margins.

b Name the two types of crust and list three differences between them.

2 Describe the differences between constructive, conservative and destructive plate margins.

3 Look at the map above.

a Which plate is the UK on?

b Name a country which is being split by two plates.

c Name two plates that are moving apart.

d Name two plates that are colliding.

On this spread you'll learn about different types of plate margins.

The day the sky went quiet

For decades, they'd been deafened by the roar of aircraft flying into and out of nearby Heathrow Airport. But, on 15 April 2010 – for the first time in years – people living in southwest London experienced the wonderful sound of silence. Why?

Hundreds of miles away, in Iceland, the Eyjafjallajokull volcano was erupting. The eruption began in March, but on 14 April it became more explosive – throwing thousands of tonnes of ash high into the atmosphere. The ash formed a massive cloud at the height used by most passenger aircraft, and they weren't allowed to fly through it. Over the following days, the growing ash cloud was pushed south over Europe by the prevailing wind direction. More and more countries grounded all flights. The residents of southwest London might have been pleased, but for hundreds of thousands of passengers, the cancellation of all flights over much of Europe brought chaos and misery.

▲ *The ash cloud from the Eyjafjallajokull volcano grounded thousands of aircraft in April 2010*

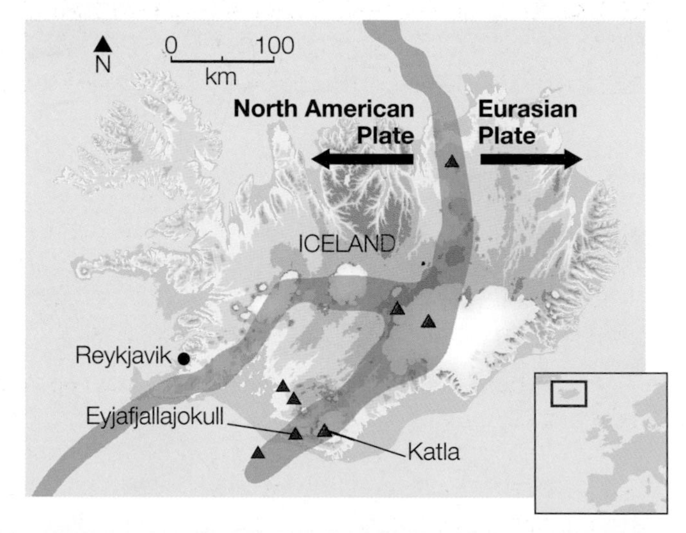

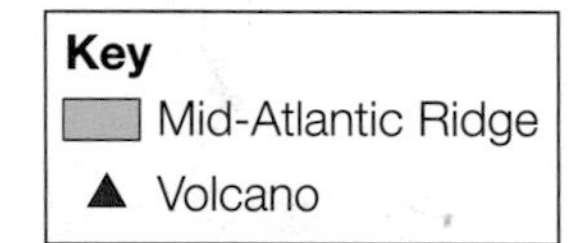

▶ *Iceland sits on a constructive plate margin, and has a number of active volcanoes*

Constructive plate margins

Iceland sits on top of two tectonic plates, at a constructive plate margin. At constructive margins, the two plates are moving apart – like the North American and Eurasian Plates in Iceland. This allows magma from the mantle to rise up to the Earth's surface and make – or construct – new crust. This is happening under the Atlantic Ocean, along the Mid-Atlantic Ridge.

The plate movement at constructive margins can cause volcanic eruptions, like Eyjafjallajokull in Iceland. Earthquakes can also be caused by the friction of the plates as they move over the mantle.

The crust on either side of a constructive plate margin often has big cracks or faults in it – caused by the massive pressure of the moving plates.

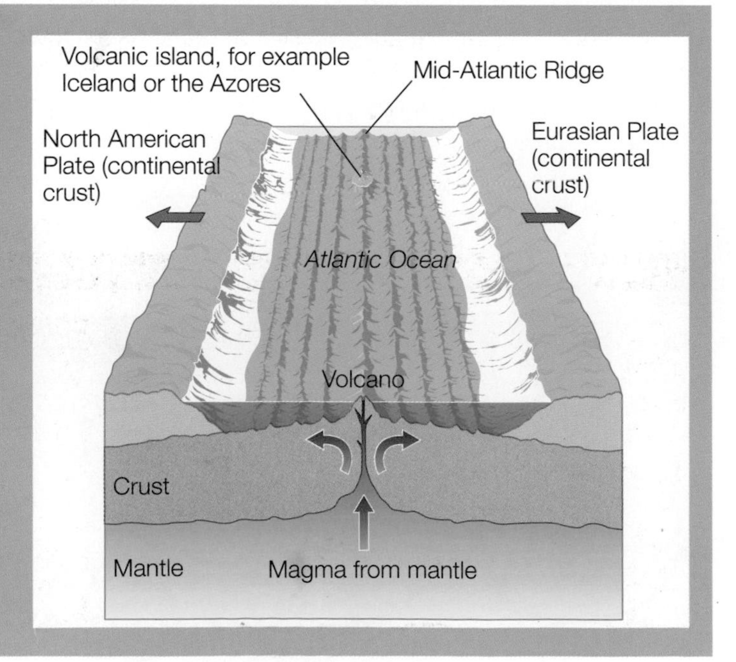

Destructive plate margins

At destructive plate margins, an oceanic plate and a continental plate collide. The oceanic plate is denser than the continental plate, so it sinks beneath the continental plate (also known as **subduction**). As it sinks down into the mantle, it melts – forming magma. It also takes sea water down with it – making it less dense than the mantle. This means that it will eventually rise up through the continental crust and explode at the surface as a volcano. The trapped sea water then turns into steam and this makes the volcano very explosive. There's not much lava, but there's a lot of ash, steam and gas.

As the oceanic plate sinks below the continental plate, it creates friction. This friction then causes earthquakes.

As the continental plate moves towards the oceanic plate, land on the edge of the plate gets crumpled up to form fold mountains, like the Andes.

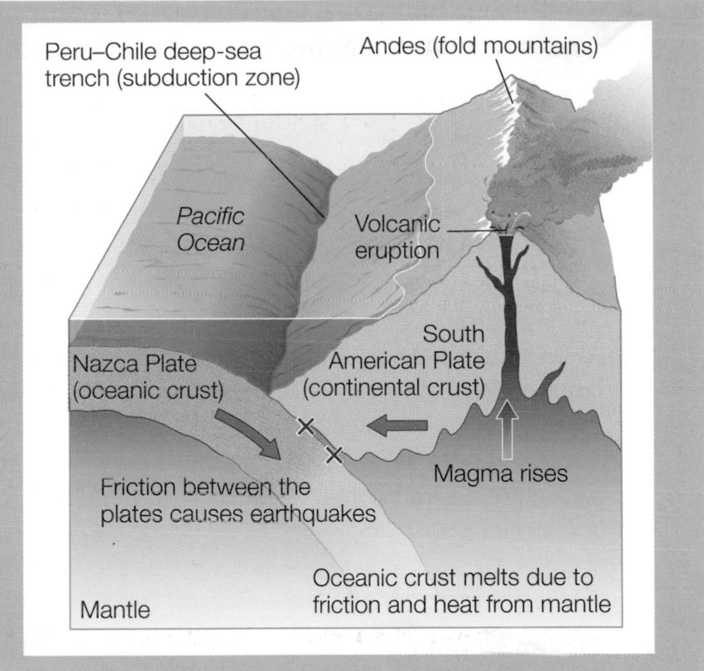

Conservative plate margins

At conservative plate margins, the two plates are moving past each other. Friction between the plates then causes earthquakes. These are rare but can be very destructive, because they occur close to the Earth's surface.

The San Andreas Fault in California is a famous example of a conservative plate margin. You can actually see the fault on the surface of the Earth. Here, the North American Plate and the Pacific Plate are both moving in the same direction – but at different speeds.

There are no volcanoes at conservative plate margins.

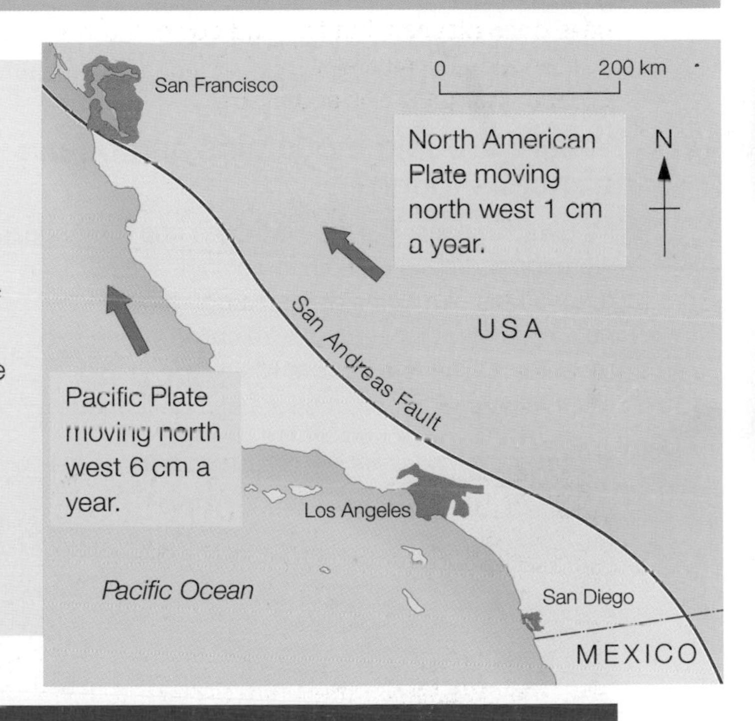

YOUR QUESTIONS

1 What does subduction mean?

2 Copy the table on the right. Then complete your table for the three types of plate margin discussed on this spread:

3 a Make a simple outline copy of the diagram of the constructive plate margin.

b Annotate it to show why Eyjafjallajokull erupted in 2010.

Hint: Annotate means add explaining labels.

Plate margin	Example	Earthquakes?	Volcanoes?

1.3 » Landforms at plate margins

On this spread you'll learn how fold mountains, ocean trenches, and volcanoes are formed.

The world's highest mountain (Mount Everest) and deepest ocean trench (The Mariana Trench) are both located at plate margins. Why is that?

Fold mountains

Mount Everest is in the Himalayas, which – like the Andes and the Alps – is a range of **fold mountains**. The Himalayas lie on a collision zone (a type of destructive plate margin). As the top diagram on the right shows, two continental plates are pushing into each other. Because they're the same density, neither sinks, so they're pushed upwards – forming fold mountains.

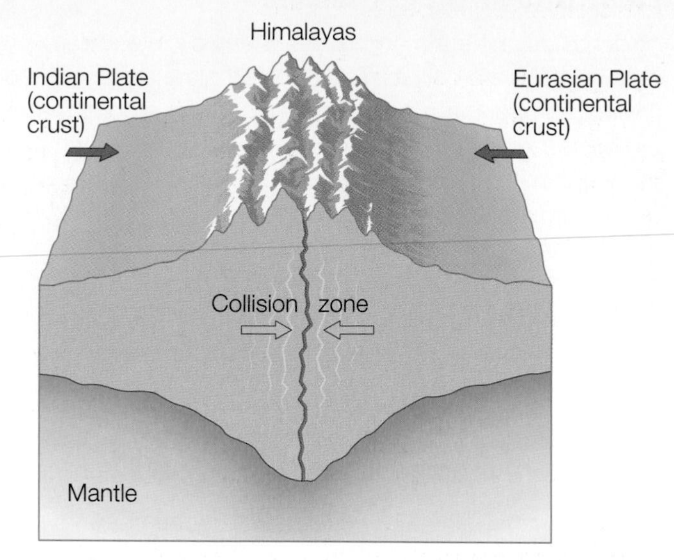

▲ ▼ *How fold mountains form. The pressure of the plates pushing together can cause strong earthquakes at collision zones.*

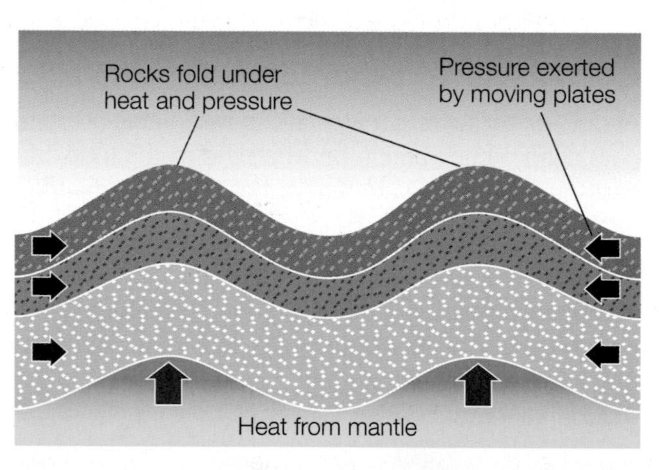

Ocean trenches

Ocean trenches are also found at destructive plate margins. Look at the diagram of a destructive plate margin on page 11. As the Nazca Plate sinks below the South American Plate, it creates a deep-sea trench.

Where are fold mountains and ocean trenches found?

Mountain ranges, like the Himalayas and the Andes, were formed quite recently in geological time. That means tens of millions of years ago, rather than hundreds of millions! So they're called young fold mountains. The map shows where young fold mountains and ocean trenches can be found.

▼ *Young fold mountains and ocean trenches*

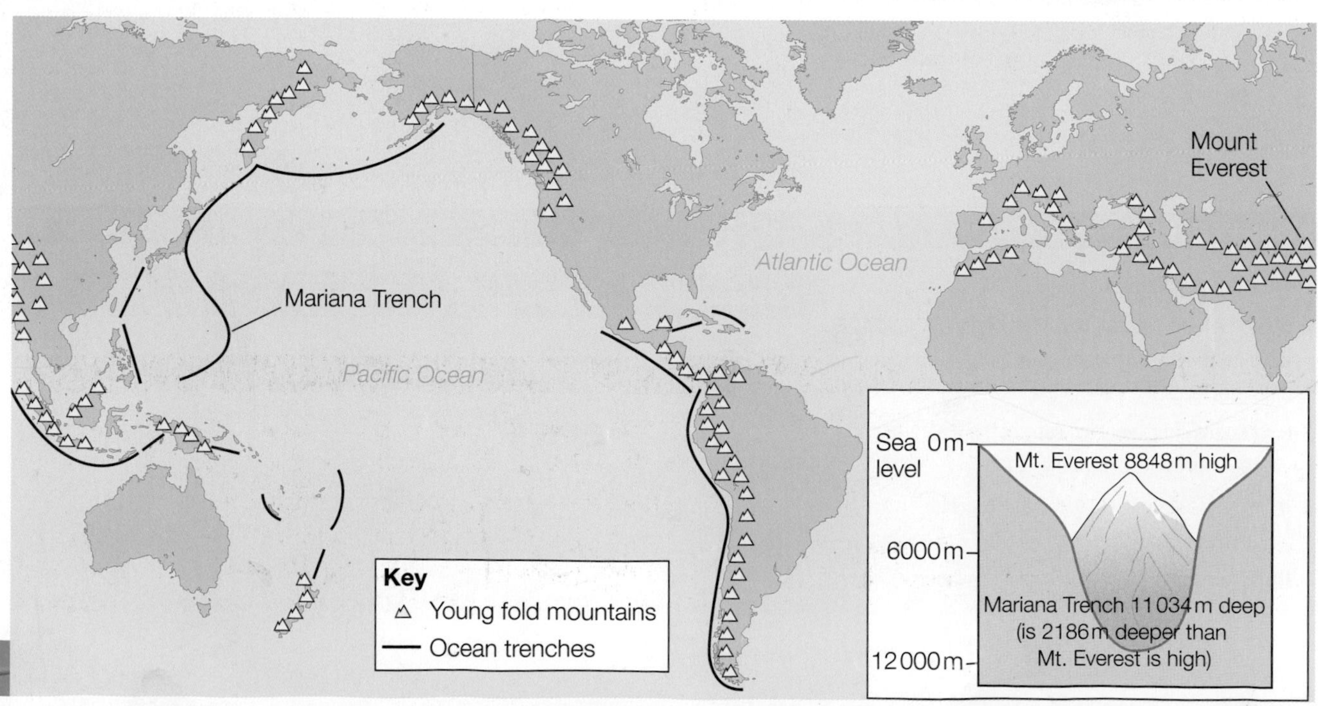

Volcanoes

There are two main types of volcano.

Composite volcanoes

- They're found at destructive plate margins. When the oceanic plate sinks into the mantle and melts, it forms magma. Magma mixed with sea water then rises up through cracks in the Earth's crust and erupts at the surface – forming volcanoes (page 11).
- Composite volcanoes have steep sides, and are made up of alternate layers of ash and lava.
- The lava is sticky, so it doesn't flow far. It's also acidic.
- Eruptions can be violent – expelling steam, ash, lava and rock – but they don't happen very often.

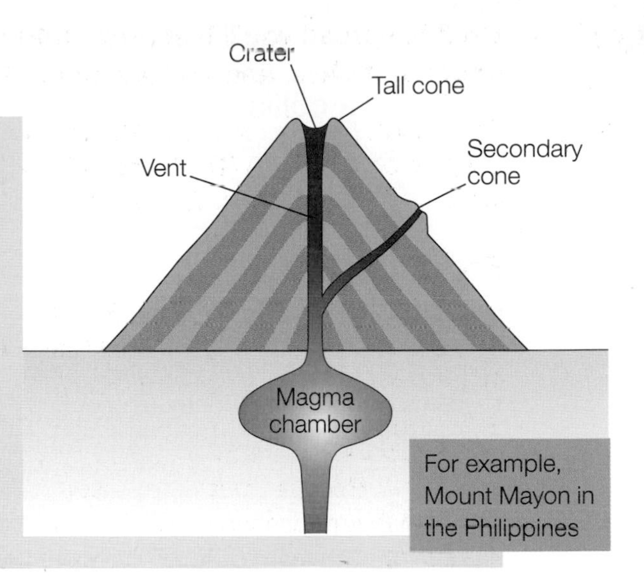

For example, Mount Mayon in the Philippines

Shield volcanoes

- They're found at constructive plate margins. As the two plates move apart, magma rises up from the mantle. Some magma is forced to the surface through a vent – forming a volcano.
- Shield volcanoes have a wide base and gently sloping sides.
- The lava is runny and flows a long way. It's also basic (that's the opposite of acidic).
- There can be frequent eruptions, but they're not violent.

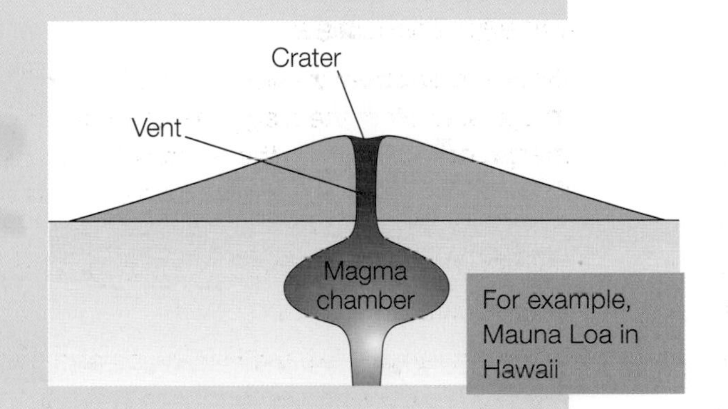

For example, Mauna Loa in Hawaii

Where are volcanoes found?

You tend to find volcanoes in long belts, as the map shows. The biggest one wraps around the Pacific Ocean. It's called 'The Pacific Ring of Fire'. There's another big one running down the middle of the Atlantic Ocean (at the Mid-Atlantic Ridge (see page 10).

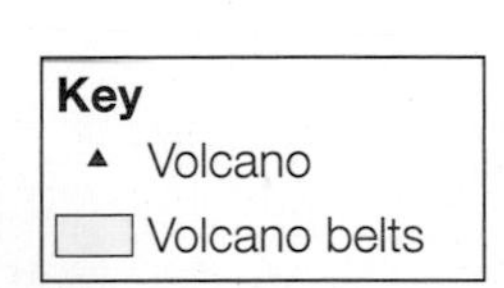

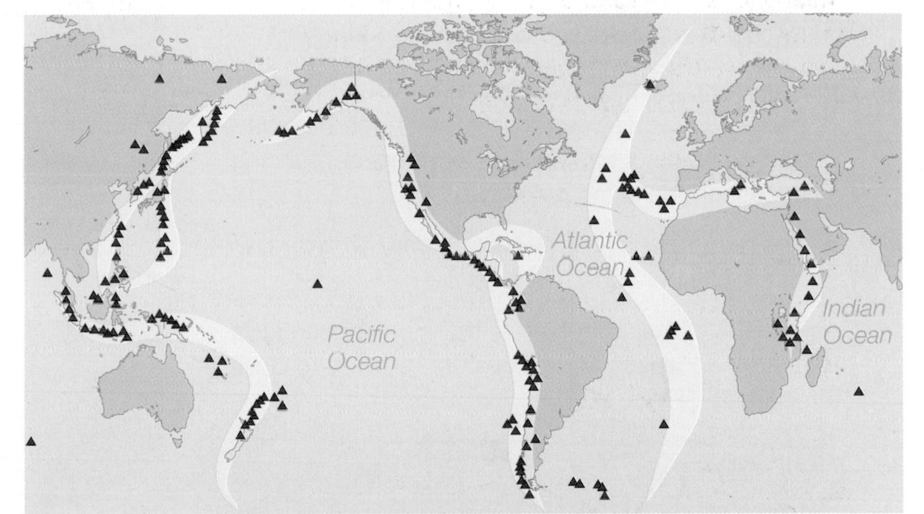

YOUR QUESTIONS

1 a What are fold mountains and ocean trenches?

b Draw two simple diagrams, with labels, to explain how both are formed.

2 Make a table with two columns headed 'Composite volcanoes' and 'Shield volcanoes'. Complete your table with as much relevant information as you can from your work so far on Spreads 1.1 – 1.3. Think about location, structure, cause, type of eruption, etc.

3 Compare the map opposite showing the distribution of young fold mountains and ocean trenches, with the map on page 9 showing different types of plate margins. What do you notice?

1.4 » Using fold mountains – the Himalayas

On this spread you'll explore how people use fold mountains.

Living in a collision zone

The Himalayas are more than just a huge range of fold mountains. People also live there, like the 29 million Nepalese. Nepal's mountain environment is used in many different ways.

Tourism

Nepal is one of the world's poorest countries, and tourism brings in vital money. The country has a rich culture and religious tradition, but the best-paying and longest-staying tourists travel there because of Everest.

Climbing Everest isn't cheap. To even set foot on the mountain, each team of seven climbers has to pay £50 000 to the Nepalese government. The climbers also provide much-needed income for local Nepalese villagers. Ang Dawa, a Sherpa guide, explains: 'There are tens of thousands of people who depend on trekkers and mountaineers for their income. A Sherpa who gets to the summit of Everest can make at least £1600 for 60 days' work. That's a lot of money in Nepal. It can support an entire village.'

But tourism isn't all good news:

- It's destroying the environment. Everest has been described as the highest rubbish dump in the world – covered with discarded equipment and general rubbish left behind by climbers and trekkers.
- Providing precious electricity and water for local people becomes very difficult when there are tens of thousands of tourists and climbers demanding access to the same resources.
- The mountain is now becoming very crowded, with all the climbers and other visitors.

As a result of these problems, there have been calls to restrict access to Mount Everest, or even to close it temporarily.

Mining

Despite having important mineral reserves, mining and quarrying aren't big industries in Nepal at the moment. They only account for 0.5% of the country's GDP. But the Nepalese realise that extracting these minerals will help their country to develop. So work is being done to build up these industries. Small-scale mining is being encouraged and more mineral surveys are being undertaken to locate the best reserves.

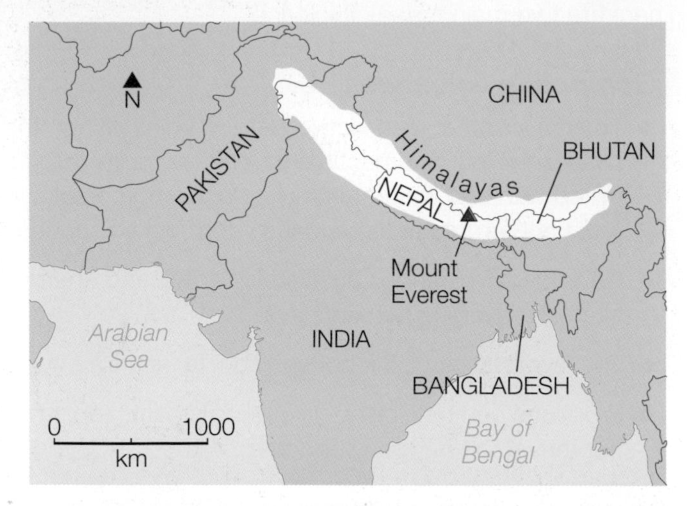

▲ *The view from the top of Everest – the roof of the world*

▼ *What the climbers leave behind on Everest*

Farming

Many farmers are ***subsistence farmers*** *(only growing enough food for their own families). And many farms are very small and fragmented.*

Increasing the amount of land used for farming has led to major deforestation. This has caused soil erosion and flooding – threatening the livelihoods of farmers throughout Nepal.

76% of Nepalese work in farming. It makes up 35% of the country's GDP. Most exports go to India, because in southern Nepal it's easier to transport goods across the border to India than to mountainous regions in Nepal itself.

Himalayan slopes are steep, and soils are thin, so crops are often grown on terraces.

Farmers grow rice and vegetables, and also keep cattle, goats and poultry.

The Nepalese government is now setting up projects to encourage the production of ***cash crops*** *like tea, ginger and mangoes for export.*

Hydroelectric power

Hydroelectric power (HEP) is clean and renewable – and Nepal has the potential to generate a huge amount of it. But only 30% of rural Nepalese have electricity (90% in urban areas). So, exploiting this potential – and generating enough electricity to eventually export it to India – are seen as important ways of helping Nepal to develop.

The diagram shows the things needed to generate HEP on a large scale. With the help of India and China, Nepal has now developed HEP projects on the Kosi and Gandak rivers.

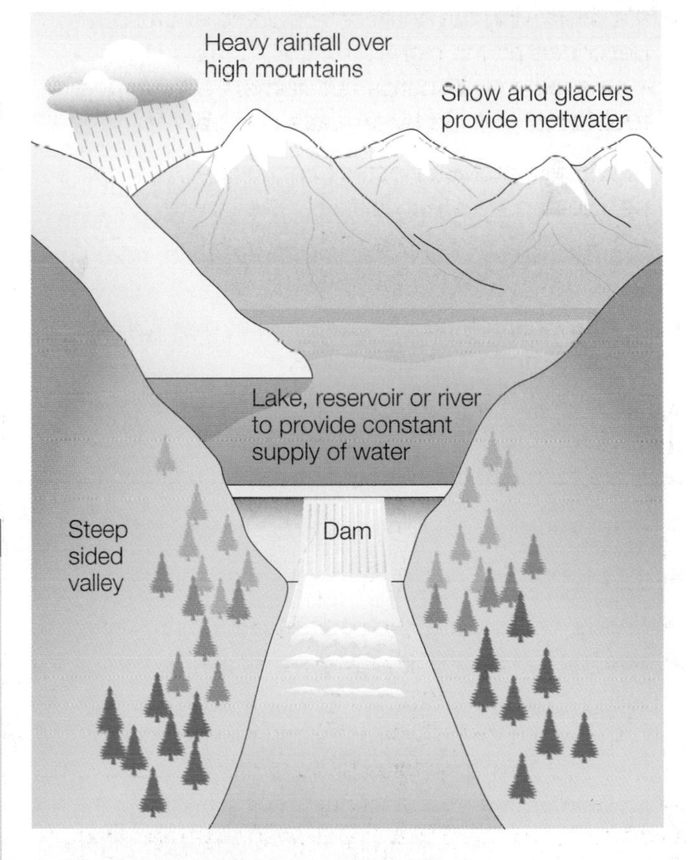

▲ *The things needed to generate HEP*

YOUR QUESTIONS

1. What are subsistence farmers?
2. Produce a poster – or prepare a PowerPoint presentation – about farming or tourism in the Himalayas. Include information on:
   - how people use the mountains
   - any problems they face, and how they could be overcome.
3. Nepal could provide more electricity in rural areas by developing micro-hydro projects. Find out what these are and how they could help people in Nepal.

1.5 » Volcanic eruption – Montserrat

On this spread you'll learn about the big volcanic eruption on Montserrat – the causes, the effects and the responses to it.

Montserrat remembered

When Montserrat's Soufrière Hills volcano suddenly began erupting in August 1995, it came as a big surprise! Before that, the volcano had been dormant for hundreds of years. Over the next two years, there was a series of small eruptions. Then, in June 1997, the big one came – killing 19 people (see right).

Rosemond Brown's father, Joseph, was one of them. She recalls: 'He didn't leave when he was told to, because he's not a person that would like to live in shelters and stuff like that. On the morning of the eruption he went back in, said he was going to take a nap, and that was the last of him'.

Adapted from an article on the BBC News website in 2005

Why did the volcano erupt?

Montserrat is part of a volcanic island arc in the Caribbean, which has developed at a destructive plate margin (see page 11 and the diagram on the right). At this plate margin, the Atlantic Plate is sinking, or subducting, below the Caribbean Plate – causing magma to rise to the surface and erupt as a series of volcanoes (see the map).

The first eruption, in August 1995, smothered Plymouth (Montserrat's capital) in dense clouds of volcanic ash. Two years of 'gentle' eruptions then followed. But the big eruption, in June 1997, sent massive **pyroclastic flows** – mixtures of volcanic fragments, ash, mud and toxic gases (at temperatures of over 500°C) – flowing down the sides of the Soufrière Hills at over 130 km/h, covering everything in their path.

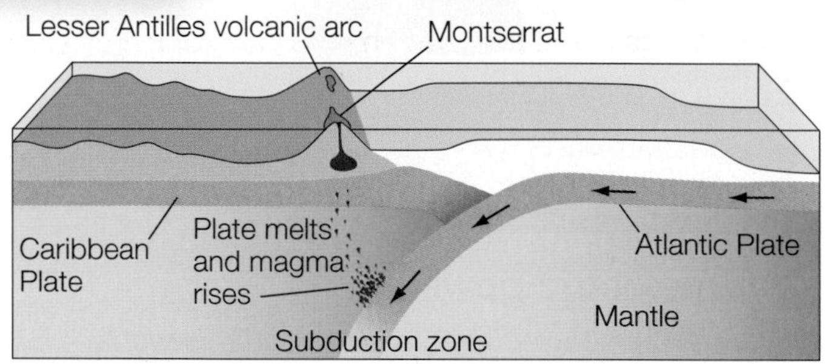

▲ ▼ *Montserrat is part of a volcanic island arc in the Caribbean – each red triangle on the map represents a volcano! The arc sits above a subduction zone at a destructive plate margin.*

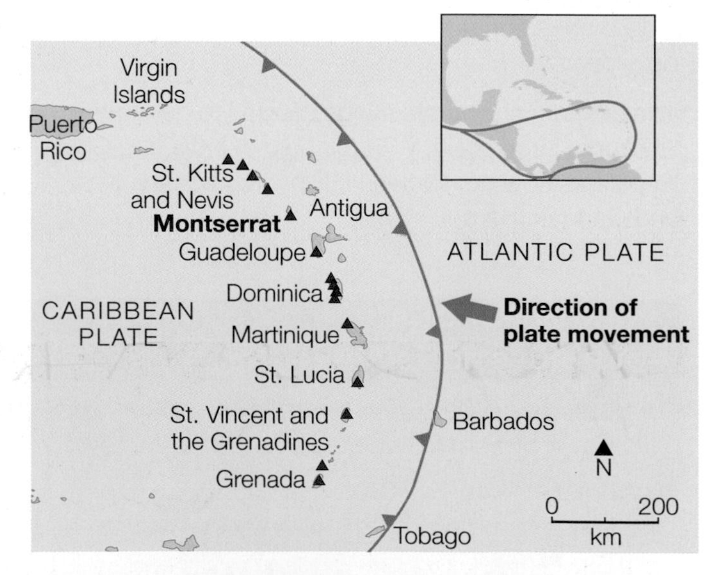

The effects of the big eruption

- More than half of Montserrat became uninhabitable.
- Infrastructure, including the airport, was destroyed.
- 19 people died and many people fled the island.
- Farms, homes, and the capital were destroyed and abandoned.
- Tourists stopped visiting the island.
- Montserrat's economy was devastated.
- There were long-term health effects, because the volcanic ash contains a type of quartz that causes a lung disease called silicosis.

But the good thing is that lessons learned from the Montserrat eruption will help protect people around the world from future disasters.

Responses to the eruptions

Immediate responses	Long-term responses
• In August 1995, many residents were evacuated to the north of the island. In April 1996, Plymouth was evacuated. An 'exclusion zone' was set up in the south of the island before the big eruption, which saved many lives. • Many people left the island completely. By November 1997, Montserrat's population had fallen from 12 000 to 3500. • Montserrat is a British Overseas Territory, so the British government spent millions of pounds on aid – including temporary buildings and water purification. • Charities set up temporary schools, and sent emergency food for farm animals.	• Some people returned to the island. By 2010, the population had risen to nearly 5000. • The island's population structure changed. Many younger people left and didn't return. Many older people either never left, or came back. • The British government spent over £200 million helping Montserrat to restore electricity and water, build a new harbour in the north of the island at Little Bay, a new airport, and new roads.

Monitoring and predicting volcanic eruptions

The Montserrat Volcano Observatory was set up in 1996. Almost everything that scientists can think of has been used to monitor the volcano, including:

- checking changes in its shape, using electronic tilt meters and global positioning systems
- using seismometers to listen to the rumbling of the volcano, as magma flows towards the surface
- creating a seismology network to collect information on earthquake activity
- measuring sulphur dioxide emissions.

The monitoring is ongoing, and scientists use the information collected to predict what type of eruption could happen, and when. Daily reports and advice are passed to the local radio station, so residents can be kept informed.

▲ *Montserrat's capital, Plymouth, had to be abandoned*

Montserrat's future

The future for Montserrat is looking up. In 2005, tourists began returning to the island. Little Bay is being developed as Montserrat's new capital, as well as a tourist destination and a harbour for trade. It's hoped that Little Bay will stimulate Montserrat's economic growth. But, in the meantime, the volcano rumbles on with occasional eruptions and pyroclastic flows.

A volcanic risk map for Montserrat in September 2007 – based on data from the Montserrat Volcano Observatory ▼

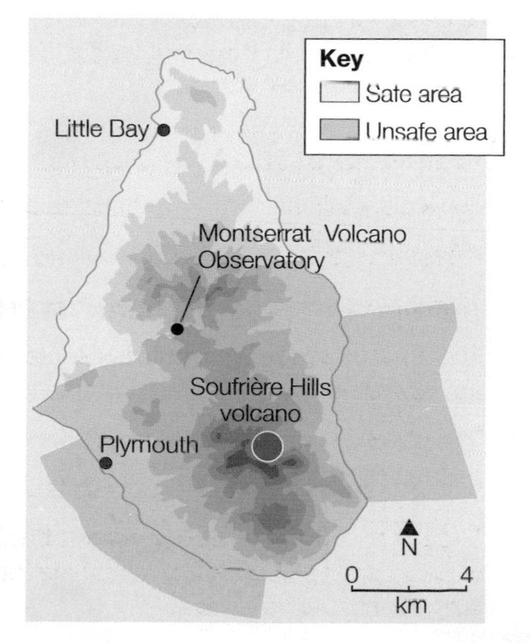

YOUR QUESTIONS

1. Describe a pyroclastic flow.
2. Sort the effects of the 1997 eruption into **primary** and **secondary**.

   Hint: Primary effects are those that happened straight away. Secondary effects are those that happened later, as a result of the eruption.
3. Write a newspaper article (no more than 300 words) about the eruptions on Montserrat. You need to tell people why they happened, and what the effects and responses were. Add some photos or diagrams to illustrate your article.

1.6 » Supervolcanoes

On this spread you'll find out about supervolcanoes – what they are, and what the effects of an eruption might be.

▲ *The Old Faithful geyser, in the USA's Yellowstone National Park, is powered by one of the world's supervolcanoes. Nowhere else on Earth has as many steam vents, hot springs and active geysers as Yellowstone.*

Supervolcano!

Volcanic eruptions aren't usually too much of a problem, if you live in a country a long way away from a plate margin. But beware the **supervolcano**. It's one of nature's greatest killers, and capable of setting off a super-eruption. 'A super-eruption is the world's biggest bang' says Professor Bill McGuire of the University of London. 'It's a volcanic explosion big enough to dwarf anything else – and affect everyone on the planet.'

What's a supervolcano?

Put simply, they're volcanoes capable of erupting on a much bigger scale than a normal volcano. A supervolcano also looks different to a normal volcano, because it often doesn't have a peak or cone. In a normal volcano, magma flows up from deep inside the Earth and pours out of the cone. But in a supervolcano:

- the magma is blocked from reaching the Earth's surface
- the pressure begins to build up, more rock melts and more magma is formed – creating a vast magma chamber
- when the pressure becomes too much, the entire surface above the magma chamber is blown away by a huge explosion, forming a **caldera**.

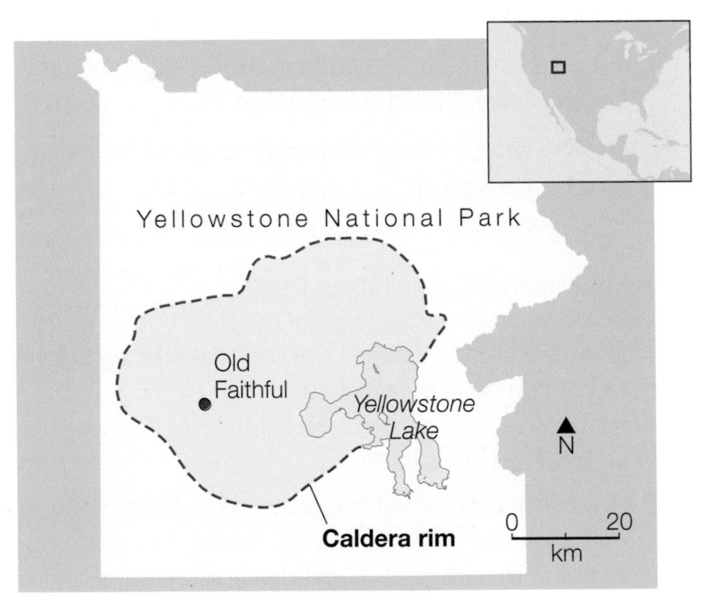

▲ *The Yellowstone caldera*

Scientists estimate that there are six or seven supervolcanoes, including:

- Yellowstone, USA
- Long Valley, USA
- Taupo, New Zealand
- Toba, Indonesia

At Yellowstone, the caldera is so big that scientists had to use photos from space to check that it ***was*** a caldera. About 6.5 km below the Yellowstone National Park is a 64 km-wide magma chamber. This is what powers Yellowstone's hot springs and geysers.

How big are supervolcanic eruptions?

In 1980, Mount St. Helens (in the USA) erupted. The blast was so powerful that it blew 400 metres off the top of the volcano! It also:

- created a crater 3 km long and 500 metres deep (pictured on the right)
- destroyed and flattened every single tree in a blast zone 25 km north of the volcano
- emitted 0.25 km^3 of material.

your planet

In June 2010, 131 earthquakes were recorded in the Yellowstone region. The largest measured 2.9 on the Richter Scale.

Impressive? Maybe, but Mount St. Helens is just an ***ordinary*** volcano. A supervolcano could emit over 11 000 times more material than Mount St. Helens – and create a global catastrophe. For instance, the Toba supervolcano emitted about 2800 km³ of material when it erupted 74 000 years ago!

One way of measuring eruptions is to use the **Volcanic Explosivity Index (VEI)**. This measures the volume of material erupted. The index is logarithmic. That means that each point on the scale represents an eruption 10 times bigger than the one below. So, a VEI 5 is 10 times bigger than a VEI 4, and 100 times bigger than a VEI 3, etc. The diagram on the right shows the relative sizes of some eruptions.

What would the impacts of a super-eruption be?

Utter devastation. The impacts would be felt around the world. According to Bill McGuire, if Yellowstone erupted again:

- magma would be flung 50 km into the atmosphere
- virtually all life up to 1000 km away would be killed by falling ash, lava flows and the force of the explosion
- 1000 km³ of lava would pour out of the volcano – enough to cover the whole of the USA with a layer 12.5 cm thick.

The amount of ash and gas thrown up into the atmosphere by the explosion would dramatically reduce the level of radiation from the sun reaching the Earth's surface. This would trigger a freezing cold volcanic winter across the planet.

- Crops wouldn't grow and people would starve.
- Economies would collapse.
- Society would not survive.

Should we worry?

A super-eruption is certainly possible, and we can't stop it. But the Yellowstone supervolcano is constantly monitored to check for signs of activity, so it's likely that the threat of a catastrophic eruption there would be picked up. Anyway, super-eruptions don't happen that often – the last one was at Toba, and that was 74 000 years ago!

▲ *Mount St. Helens after the 1980 eruption – minus its top and north side*

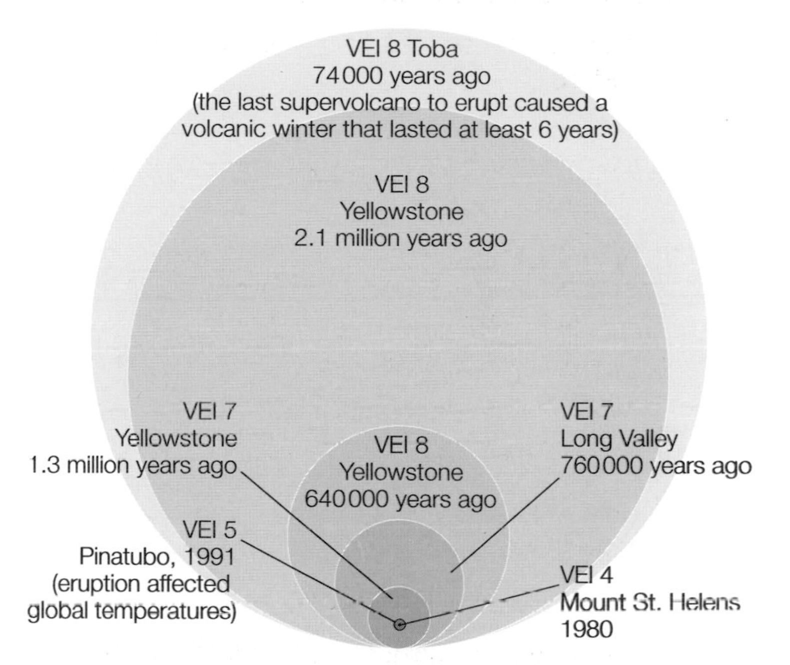

▲ *Some supervolcanic eruptions, measured according to the VEI, with two 'normal' volcanic eruptions for comparison*

YOUR QUESTIONS

1. Explain the terms: caldera, VEI.
2. Describe the differences between a volcano and a supervolcano.
3. Imagine that Yellowstone has just erupted. Work in pairs to write a radio bulletin to say what's happened, and what the impacts might be – on the USA and on the whole world.

1.7 » Earthquake!

On this spread you'll discover where and why earthquakes occur, and how we measure them.

Where do earthquakes occur?

The world map on the right shows where earthquakes occurred in the first week of July 2010. There were 206 – in just one week! Many were relatively small, but it proves that the Earth ***is*** restless and constantly on the move.

The world map below shows the global pattern of earthquake activity. Earthquakes tend to occur in long belts. The biggest one goes round the whole Pacific Ocean, and another big one runs down the middle of the Atlantic Ocean. These earthquake belts follow the plate margins (see page 9).

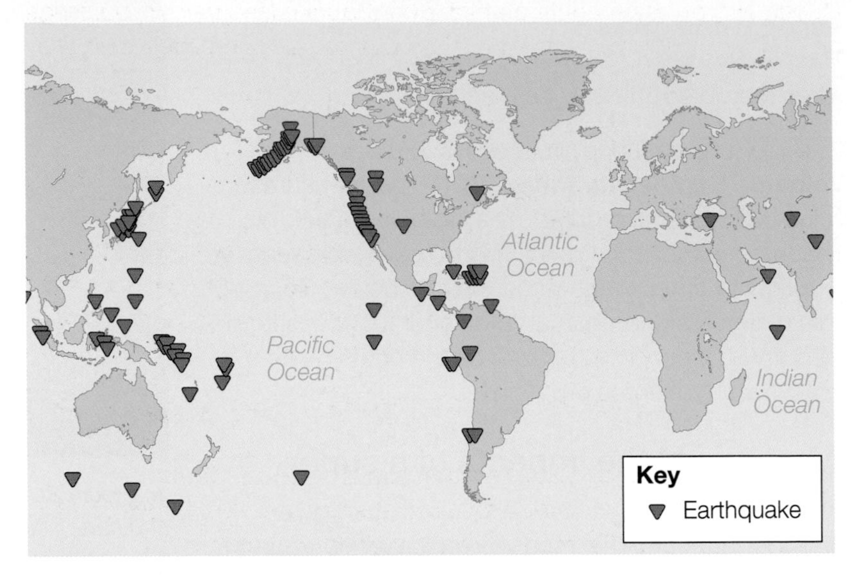

▲ *Earthquakes occurring over a seven-day period in July 2010*

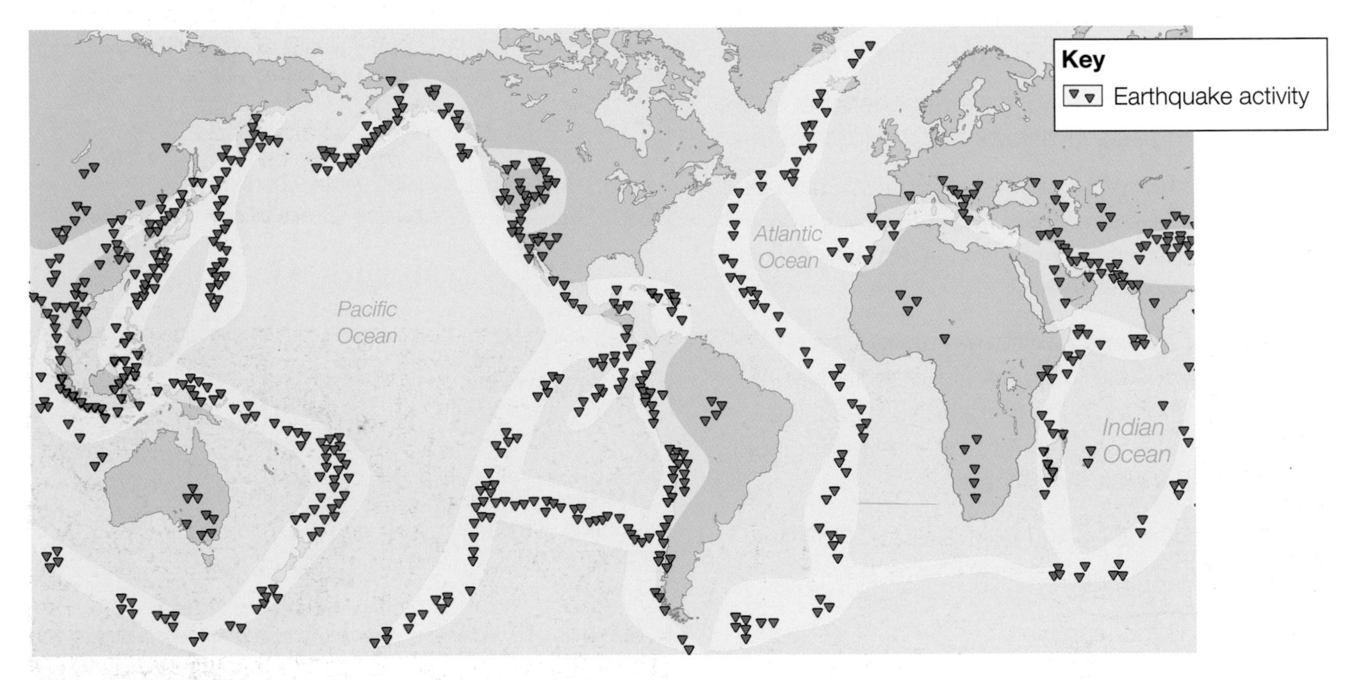

▲ *Pattern of earthquake activity*

What causes earthquakes?

Earthquakes occur at plate margins because of friction – the plates try to move and get stuck. Pressure builds up as the plates keep trying to move. If the pressure is released slowly, we hardly notice the movement. But if the pressure is released suddenly, it can send out huge pulses of energy – causing the Earth's surface to move violently. And that's an earthquake. Earthquakes occur at all four types of plate margin. They occur at:

- destructive margins, when one plate tries to sink below the other
- collision zones, where the plates are pushing together
- constructive margins, when both plates try to move over the mantle
- conservative margins, when plates moving alongside each other get stuck.

Features of earthquakes

When an earthquake strikes, the ground shakes really violently. Depending on the magnitude of the earthquake (see below), buildings and bridges and other man-made structures can completely collapse.

1 September 1923 began like any other day for the residents of Tokyo and Yokohama in Japan. But, for many, it was their last. Just before noon, a low deep rumbling sound grew to a monstrous roar, as a fault under Sagami Bay ripped itself apart and sent shock waves tearing north towards the two cities. The earthquake measured a massive 8.3 on the Richter Scale. The ground shook so hard that it was impossible to stand. Within seconds, thousands of buildings had collapsed – killing those inside.

Adapted from *Global Catastrophes, A Very Short Introduction* by Bill McGuire

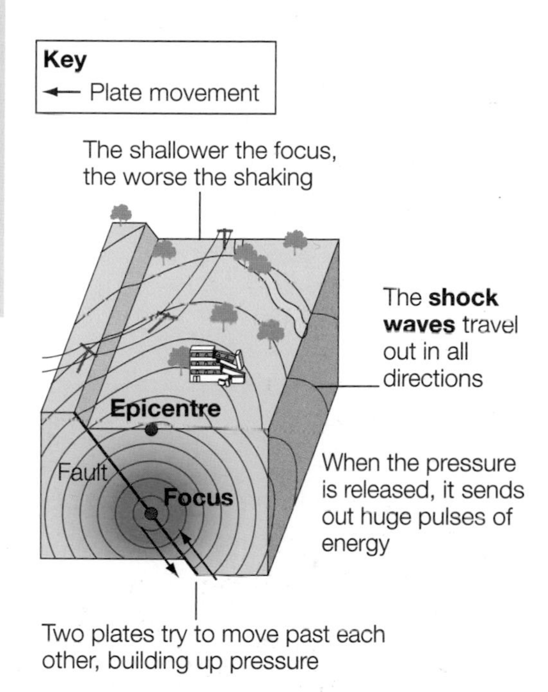

▲ *The basic features of any earthquake*

The Japanese call the above earthquake 'The Great Kanto Earthquake'. It was a major disaster that killed 143 000 people. But it had the same basic features as any other earthquake:

- When plates that have been stuck eventually move, the built-up pressure is released along faults (cracks). This sends out huge pulses of energy.
- The **shock waves** travel out from the **focus** (where the earthquake starts).
- The **epicentre** is the point on the Earth's surface directly above the focus.

Measuring earthquakes

The Richter Scale

The strength, or magnitude, of an earthquake is measured using the **Richter Scale**. This is a logarithmic scale. So, for example, a magnitude 6 earthquake is 10 times more powerful than a magnitude 5 earthquake, and 100 times more powerful than one measuring 4. There is no upper limit to this scale.

The Mercalli Scale

This scale goes from 1-12 (but uses Roman numerals I-XII). It measures the power and effects of earthquakes, based on what is observed. So, for example:

I Most people wouldn't even feel the earthquake.

V Most people would feel it and windows would break.

XII Everything would be destroyed – buildings, bridges, railways. The ground would move in waves.

YOUR QUESTIONS

1. Describe these earthquake features: shock waves, focus, epicentre.
2. What is the difference between the Richter Scale and the Mercalli Scale?
3. Using the lower map opposite and the map on page 9, describe the pattern of earthquake activity around the world.
4. **a** Find out where earthquakes have occurred on this day in the past. Use this website: http://earthquakes.usgs.gov/learn/today/
   b Plot your earthquakes on a blank world map.
   c Find out what type of plate margin the earthquakes were on.
   d Add notes to explain why the earthquakes happened (use information on this spread and Spread 1.2).

On this spread you'll learn about the effects of two earthquakes in two different countries.

Chile and Haiti

Chile might not be the richest country in the world, but its GDP (one way of measuring how wealthy it is) comes in at 46th out of 227 countries. It has one of the best-run economies in South America. By contrast, Haiti is described as the poorest country in the Western Hemisphere. Its GDP is 143rd (out of 227 countries).

In early 2010, Chile and Haiti were both struck by powerful earthquakes. This spread and Spread 1.9 compare the effects of, and responses to, those earthquakes.

Chile, 2010

- On 27 February 2010, an earthquake measuring 8.8 on the Richter Scale struck just off the coast of central Chile.
- The earthquake occurred at a destructive plate margin, where the Nazca Plate is sinking underneath the South American Plate (see page 11).
- Because the earthquake's focus was just off the coast, it distorted the seabed – causing a tsunami (huge waves, see pages 26-27), which raced across the Pacific Ocean. At least 59 countries received tsunami warnings.
- The huge initial earthquake was followed by a number of strong aftershocks (smaller earthquakes which follow the main earthquake).
- The earthquake's focus was 34 km underground, and the epicentre was 115 km from Concepcion (Chile's second-largest city).

▼ *Chile's earthquake in February 2010*

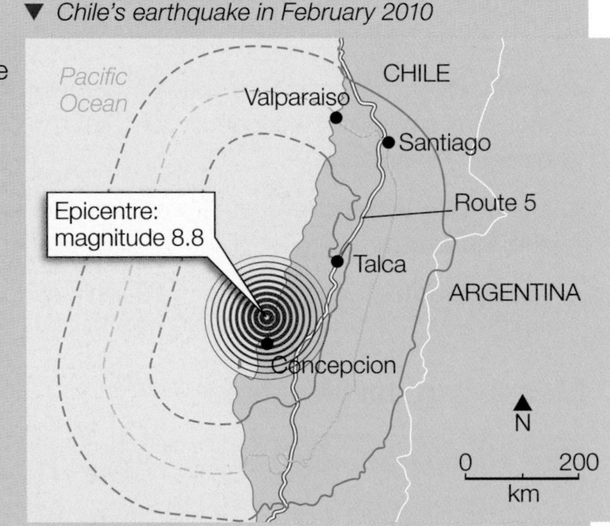

Haiti, 2010

- On 12 January 2010, an earthquake measuring 7.0 on the Richter Scale struck close to Haiti's capital, Port-au-Prince.
- The earthquake occurred at a destructive plate margin between the Caribbean and North American Plates, along a major fault line (see the map).
- The earthquake's focus was 13 km underground, and the epicentre was just 25 km from Port-au-Prince.
- Haiti suffered a large number of serious aftershocks after the main earthquake.

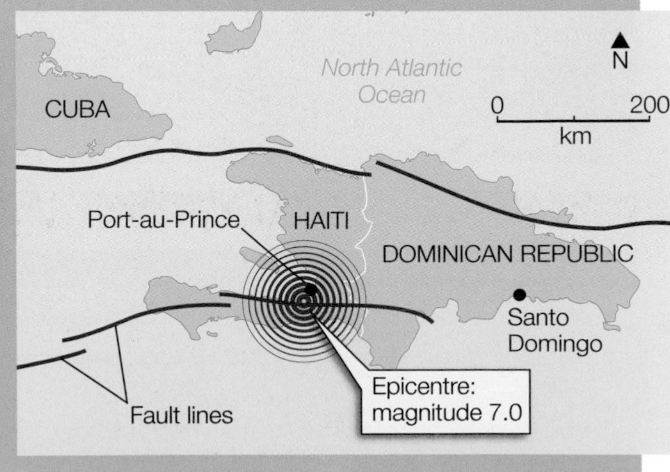

▲ *Haiti's earthquake in January 2010*

What were the effects of the earthquakes?

Chile

Primary effects

Primary effects are things that happen immediately as a result of an earthquake or other disaster.

- About 500 people were killed and 12 000 injured.
- Some buildings were completely destroyed, while many others stayed standing. At least 500 000 homes were damaged.
- Several bridges and roads were destroyed, along with much of the hospital in Talca.
- Santiago's airport was slightly damaged.

Secondary effects

Secondary effects happen in the hours, days and weeks after the initial earthquake.

- Much of Chile lost power, water supplies and communications.
- Several Pacific countries were hit by the tsunami set off by the earthquake.
- A fire at a chemical plant on the outskirts of Santiago meant that the area had to be evacuated.
- Chile's copper mines (crucial to its economy) suffered little damage.

Haiti

Primary effects

- About 220 000 people were killed and 300 000 injured.
- The main port was badly damaged, along with many roads that were blocked by fallen buildings and smashed vehicles.
- Eight hospitals or health centres in Port-au-Prince collapsed, or were badly damaged. Many government buildings were also destroyed.
- About 100 000 houses were destroyed and 200 000 damaged in Port-au-Prince and the surrounding area. Around 1.3 million Haitians were displaced (left homeless).

Secondary effects

- Over 2 million Haitians were left without food and water. Looting became a serious problem.
- The destruction of many government buildings hindered the government's efforts to control Haiti, and the police force collapsed.
- The damage to the port and main roads meant that critical aid supplies for immediate help and longer-term reconstruction were prevented from arriving, or being distributed effectively.
- Displaced people moved into tents and temporary shelters, and there were concerns about outbreaks of disease. By November 2010, there were outbreaks of cholera.
- There were frequent power cuts.
- The many dead bodies in the streets, and under the rubble, created a health hazard in the heat. So many had to be buried in mass graves.

YOUR QUESTIONS

1. Explain the difference between primary and secondary effects.
2. Begin a table to compare the earthquakes in Chile and Haiti. You need two columns (one for each country). Include information on:
   - When the earthquake happened, and why
   - Location of focus and epicentre
   - Primary effects
   - Secondary effects

   You will finish this table after working on Spread 1.9.
3. How does the depth of the earthquake's focus, and the location of its epicentre, help to explain some of the effects seen in Chile and Haiti?

   Hint: Look back at page 21 if you need help with this.

On this spread you'll find out about the responses to the earthquakes in Chile and Haiti, and about preparing for earthquakes in advance.

Responding to the earthquakes

Chile

Immediate responses

- Chile responded quickly. The President insisted on a rapid analysis of the situation, and within hours was asking for specific help from other countries, e.g. field hospitals, satellite phones, and floating bridges.
- The important Route 5 north-south highway was temporarily repaired the day after the earthquake. Thick metal plates were used to cover the big cracks. That meant that aid could be driven quickly from Santiago to the worst hit areas.
- Ten days after the earthquake, over 90% of homes had had their power and water restored.
- A national telethon raised $60 million – enough to build small emergency shelters for those people whose homes had been destroyed.

Longer-term responses

- A month after the earthquake, Chile's government launched a housing reconstruction plan – to help about 196 000 households affected by the earthquake.
- Chile has a strong economy, so it can recover and rebuild without relying on foreign aid. The country's huge copper reserves earn plenty of income to help pay for the reconstruction efforts.

Haiti

Immediate responses

- Because the main port and roads were badly damaged, crucial aid (such as medical supplies and food) was slow to arrive and be distributed. The airport couldn't handle the number of planes trying to fly in and unload aid.
- American engineers and diving teams were used to clear the worst debris and get the port working again, so that waiting ships could unload aid.
- The USA sent ships, helicopters, 10 000 troops, search and rescue teams (pictured) and $100 million in aid.
- The UN sent troops and police, and set up a 'Food Aid Cluster' to feed 2 million people.
- Bottled water and water-purification tablets were supplied to survivors.
- Field hospitals were set up, and helicopters flew wounded people to nearby countries.
- The Haitian government moved 235 000 people from Port-au-Prince to less-damaged cities.

Longer-term responses

This chapter was written about six months after the earthquake struck. At that stage, the likelihood was that:

- Haiti would be dependent on overseas aid to help it recover
- new homes would need to be built to a higher standard, costing billions of dollars
- large-scale investment would be needed to bring Haiti's road, electricity, water and telephone systems up to standard, and to rebuild the port.

Preparing for earthquakes

Prediction

A lot is known about where earthquakes are ***likely*** to happen, but there is no known way of predicting ***when*** they will happen. So, all we can do is prepare for them and protect people from their effects as much as possible.

Chile – preparation and protection

The Chilean government was well prepared for the February 2010 earthquake. Following a devastating earthquake in 1960 (which measured 9.5 on the Richter Scale and was the largest earthquake ever recorded), they weren't taking any chances:

- After 1960, new buildings were built to withstand earthquakes – with reinforced concrete columns, strengthened by a steel frame.
- In 2002, the government introduced a plan that set out the responsibilities of central and local government in the event of a disaster. This plan worked well in 2010.
- The government also organises regular anti-disaster drills. Every few months, children practise earthquake drills in Operacion Deyse.
- All Chileans know that if an earthquake strikes, they must 'Drop, cover, hold' (drop to the ground, take cover under a heavy table or doorframe, and hold on until the shaking stops).

▲ *In Chile, many buildings survived the earthquake*

Haiti – preparation and protection

- Haiti wasn't prepared in January 2010, because there hadn't been any earthquakes there in living memory.
- It also had a weak government and very little money.
- Port-au-Prince was home to more than 2.5 million people. It was overcrowded and had poorly built homes that weren't designed to withstand earthquakes.

▶ *The buildings in Port-au-Prince didn't stand a chance when the earthquake hit*

YOUR QUESTIONS

1 Complete the table you began on Spread 1.8. Add information about the immediate and longer-term responses to the two earthquakes. Also consider the preparation for, and protection against, earthquakes in Chile and Haiti.

2 How far do you think the differences in the effects of the two earthquakes, and the responses to them, are due to the fact that Haiti is a poor country and Chile is relatively wealthy?

Hint: 'How far' means you need to look at whether the differences are due to differences in wealth, or not. You need to look at both sides of the argument.

On this spread you'll find out about the 2004 Boxing Day tsunami – what caused it, what effects it had, and the responses to it.

Tsunami horror

Imagine the biggest wall of water you can. Then double it. The waves that hit the Indonesian coast near Banda Aceh on Boxing Day 2004 were nearly 17 metres high. Surprised tourists and local people had watched the sea retreat far more than usual – leaving fish floundering on the sand. Then the sea came back as a massive wall of water – a **tsunami**.

A tsunami is a series of large waves that form when an earthquake occurs in a fault below the seabed. The earthquake permanently jolts the fault up or down by several metres. The movement of the fault displaces the volume of water above it, and large waves begin moving through the ocean. In deep water, the waves move at high speed. When they reach shallower water near the coast, they begin to slow down – but build in height.

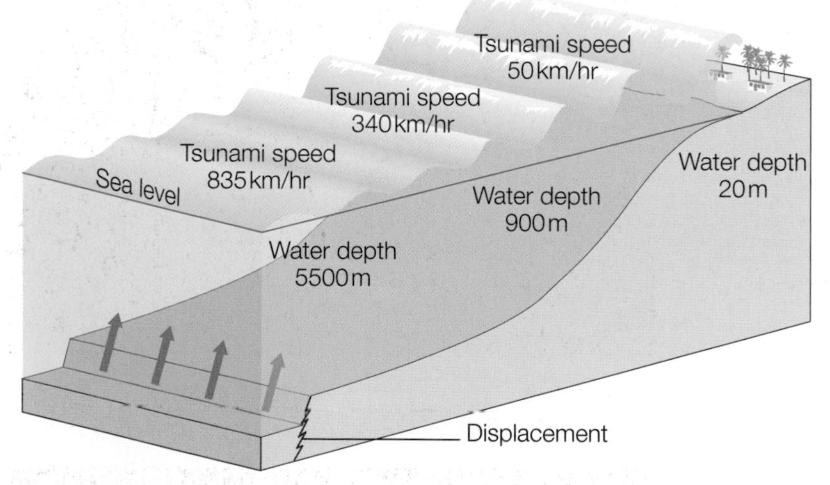

▲ *How tsunami form*

▼ *The victims of the tsunami*

	Country affected	Killed or missing
1	Indonesia	236 169
2	Sri Lanka	31 147
3	India*	16 513
4	Thailand	5395
5	Somalia	150
6	Burma	61
7	Maldives	82
8	Malaysia	68
9	Tanzania	10
10	Seychelles	3
11	Bangladesh	2
12	Kenya	1
	Total	**289 601**

* *Including the Andaman and Nicobar Islands*

The effects of the Boxing Day tsunami

The earthquake that caused the Boxing Day tsunami was estimated to be between 9.0 and 9.3 on the Richter Scale – one of the largest earthquakes ever recorded. The tsunami then moved as a series of 'ripples' across the Indian Ocean, which built up as massive waves when they approached land. It was one of the worst disasters in history – nearly 300 000 people were killed, or disappeared completely (see the table).

1 Indonesia

Western Sumatra was the closest inhabited area to the earthquake's epicentre, and was devastated. Up to 70% of some coastal populations were killed or missing.

2 Sri Lanka

The southern and eastern coastlines were ravaged, with homes, crops and fishing boats destroyed. 400 000 people lost their jobs.

4 Thailand

The west coast was severely hit, including islands and tourist resorts, so the dead there included 1700 foreigners from 36 countries.

3 India

The southeast coast of the mainland was the worst affected. In the Andaman and Nicobar Islands, salt water contaminated freshwater sources and destroyed arable land. Most of the islands' jetties were also destroyed.

▲ *Countries affected by the 2004 tsunami. The countries numbered on the map match those in the text boxes and the table opposite.*

Responses to the tsunami

Immediate responses

- Clean water, food, tents and plastic sheeting arrived as aid.
- $7 billion was donated worldwide for the affected countries.
- People in the UK donated £330 million (more than the government).
- The UN's World Food Programme provided food aid for more than 1.3 million people.

Longer term – tsunami warning system

The disastrous effects of the Boxing Day tsunami led to the setting up of a tsunami early warning system in the Indian Ocean. Formal warnings are now sent to countries throughout the region if there's a tsunami threat. These warnings are then passed on to individuals via radio, television and e-mail. Or by bells, megaphones and loudspeakers attached to mosques.

Longer term – restoring mangroves

The tourist industry in Thailand and Sri Lanka had grown rapidly before the tsunami. Many coastal areas had been cleared of mangrove swamps to make way for hotels. But mangroves act as a natural barrier – absorbing wave power and helping to protect coastlines and inland areas from tsunami.

After 2004, some projects to restore mangroves were started, e.g. the Green Coast Project in Aceh, Indonesia. The new mangroves will help to increase the protection from future tsunami, and also help to provide a livelihood for people affected by the 2004 tsunami, because mangroves are a good breeding ground for fish.

▲ *Countries around the Indian Ocean hold tsunami drills to check their responses to the tsunami early warning system*

YOUR QUESTIONS

1 What is a tsunami?

2 a Mark and label the 12 countries hit by the 2004 tsunami on a blank world map.

b Now annotate your map to explain how tsunamis form, and why this one was so devastating.

3 Work in pairs to write a tsunami preparation plan. You should think about how people can be better protected, and how they can prepare for tsunami.

2 » Rocks, resources and scenery

What do you have to know?

This chapter is from **Unit 1 Physical Geography Section A** of the AQA A GCSE specification. It is about different types of rocks; the landforms, landscapes and scenery they create; and the way that rocks and the landscape are used. The table shows how the pages in this chapter match the content in the specification.

Specification content	Pages in this chapter
The geological timescale, and where granite, carboniferous limestone, and chalk and clay fit into the timescale.	p30-31
The characteristics and formation of igneous, sedimentary, and metamorphic rocks, and where you find them in the UK. The rock cycle.	p32-33 (plus p36, 37, 38)
Rocks and weathering – mechanical, chemical and biological.	p34-35
Different types of rock create contrasting landforms and landscapes.	
• Granite	p36
• Chalk and clay	p37
• Carboniferous limestone	p38-39
The different ways that people use rocks and landscapes.	p40-43 (case studies – farming on Dartmoor p40-41; tourism in the Yorkshire Dales p42-43; London aquifer p43)
Quarrying and the issues it creates.	Case study of quarrying in the Yorkshire Dales p44-45
How the impact of quarrying on the environment can be reduced	Case study of quarrying in the Cotswolds p46-47

Your key words

- Sedimentary, igneous and metamorphic rocks
- Permeable and impermeable
- Chemical weathering (carbonation, solution)
- Biological weathering
- Physical/mechanical weathering (freeze-thaw or frost shattering, exfoliation or onion weathering)
- Granite intrusions (dyke, trench, batholith, ridge, sill)
- Tor and blockfield
- Escarpments (or cuestas) and vales
- Dry valleys
- Pervious
- Clints and grykes
- Aquifer
- Aggregate
- Biodiversity, Biodiversity Action Plan (BAP)

Exam help ...

Advice See pages 297-299 for information on how to be successful in your exams.

Practice See page 301 for exam questions on this chapter.

What if ...

- the view outside your window looked like the rocks opposite?
- rocks could talk and tell us their past?
- fossils came back to life?

2.1 » A matter of time

On this spread you'll learn about the geological timescale, and find out that geological time is very different to human time.

Secret geology places

Everyone has heard of the book *Dracula*. The churchyard of St Mary's Church in Whitby inspired the author, Bram Stoker, to write his famous book. The churchyard is on top of Whitby's east cliff.

Whitby's cliffs hold many secrets – they're made up of Jurassic mudstones that formed over 150 million years ago, and contain the fossils of many strange sea creatures like ammonites.

▲ *Whitby, on the north-east coast of England, is one of geology's 'secret' places*

Geological time and human time

The Earth was formed around 4600 million (or 4.6 billion) years ago. That's a staggeringly long time ago, and a lot has happened since then!

- The world's oldest rocks formed around 3800 million years ago.
- Oxygen levels began to rise about 2300 million years ago.
- Life on Earth eventually started to get going around 542 million years ago. (That's taken as the starting point for the geological time scale of Eras and Periods, such as the Jurassic Period, which is used by geologists and scientists – see the table opposite.)

If all those millions look a bit daunting, you could simplify them in your mind. Instead of thinking of the Earth as a planet which is 4600 million years old, get rid of eight noughts and think of it as just 46 years old. On that more manageable scale, a year represents 100 million years of geological time, and a day around 275 000 years. Given that the modern human species only began around 190 000 years ago (see the table opposite), you can see that we've only been around for a tiny fraction of the Earth's existence – less than a single day in 46 years! Because geological time and human time are so different in scale, thinking of them together can be tricky!

▲ *Ammonites are often found as fossils. Fossils are formed when the remains of plants or animals are buried in sediment that eventually becomes rock over many millions of years.*

your planet
Ammonites became extinct at roughly the same time as the dinosaurs.

Four types of rock

You are going to study four different rock types in this chapter. They formed at different times under different conditions.

- Carboniferous limestone formed 359 – 299 million years ago.
- Granite formed about 280 million years ago.
- Chalk formed during the Cretaceous period (145 – 65 million years ago).
- Clay formed particularly during the Jurassic, Cretaceous and Tertiary periods (199 – 2 million years ago).

▲ *The fossilised bones of huge dinosaurs being excavated from rock in China – they're at least 65 million years old*

The geological time scale

Geologists divide time into Eras and Periods. They are listed in the table below, which also lists major events in the last 542 million years, like the extinction of the dinosaurs and the evolution of modern humans.

▼ *The geological time scale*

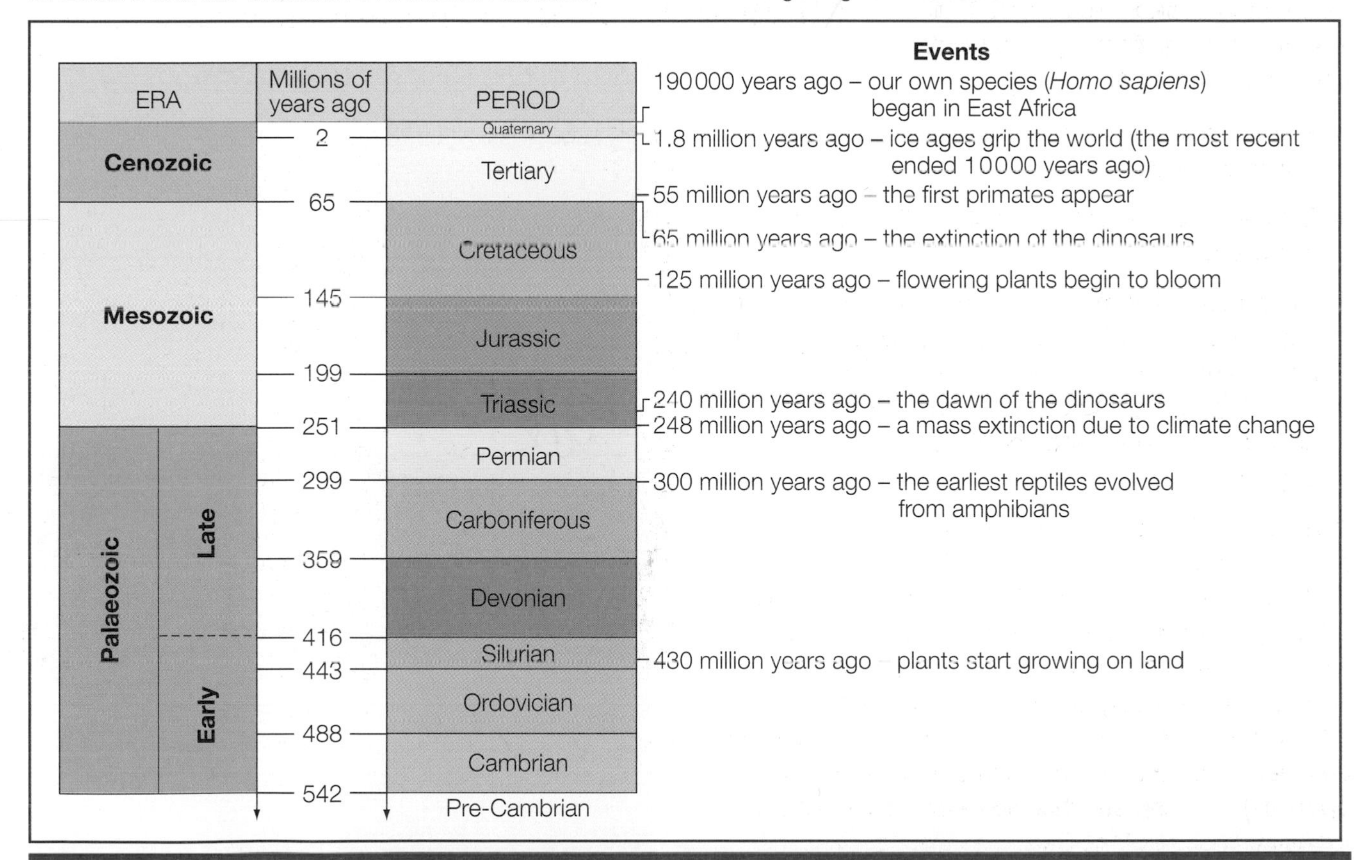

YOUR QUESTIONS

1. Explain in your own words how geological time is different to human time.
2. Use the information on this spread to draw up a poster about the geological time scale and major events in the last 542 million years (only some of them!).
3. Use this website www.bgs.ac.uk and follow the links for 'Popular Geology' and then 'Secret' geology places. Choose one of the places (but not Whitby) and write an e-mail to your partner explaining why your place is important.

2.2 » Rock groups

On this spread you'll learn about the three main groups or categories of rock, how they're formed, and what they're like (their characteristics).

Three categories of rock

There are lots of different types of rocks, but they all belong to one of three categories or groups: **sedimentary**, **igneous**, or **metamorphic**.

Sedimentary rocks

How are they formed?

Sedimentary rocks are formed when sediment (that's material carried by water) is deposited in layers on the bottom of a sea or lake. As more layers of sediment are deposited, the weight of the layers above compresses or squeezes the sediment below – to eventually form sedimentary rocks, like limestone (pictured) or sandstone.

What are their characteristics?

- Sedimentary rocks are laid down in layers. There are lines of weakness, called bedding planes, between the layers. There may also be vertical lines of weakness, called joints.
- Sedimentary rocks are **permeable** and allow water to pass through them.
- They are easily eroded.

Some examples

- Limestone and chalk consist of calcium carbonate, which comes from the remains of dead plants and animals – like the ammonites on page 30.
- Clay forms from the deposition of mud.
- Sandstone forms from the deposition of sand.

Igneous rocks

How are they formed?

Igneous rocks come from magma – that's molten rock below the Earth's crust. Some igneous rocks – like granite (pictured) – are formed from magma that cools inside the Earth. Others – like basalt – are formed from lava that erupts from a volcano and then cools on the Earth's surface.

What are their characteristics?

- Igneous rocks consist of crystals, which form as the magma or lava cools.
- They are hard and difficult to erode.
- They are **impermeable**, and won't allow water to pass through them.

Some examples

- Granite and dolerite (formed inside the Earth).
- Basalt and andesite (formed on the Earth's surface).

Metamorphic rocks

How are they formed?

Metamorphic rocks are formed when existing sedimentary or igneous rocks change (that's what 'metamorphosis' means) – because of either heat or pressure. This can happen along destructive plate boundaries and fault lines.

What are their characteristics?

- Metamorphic rocks are usually very hard and don't get eroded or weathered much.
- They consist of crystals.
- They are impermeable.

Some examples

- Marble is formed from limestone (a sedimentary rock) that has been subjected to heat and pressure.
- Schist (pictured) is formed from basalt (an igneous rock) or shale (a sedimentary rock) that has been put under pressure

The rock cycle

All rocks are part of one big rock cycle:

- Rock is weathered and eroded. It is then transported, deposited in layers, and compressed to form sedimentary rock.
- Sedimentary and igneous rocks can then be changed by heat or pressure to form metamorphic rocks.
- Metamorphic and sedimentary rocks may melt to form magma, and then eventually cool to form igneous rocks.
- ... and then the whole cycle begins again!

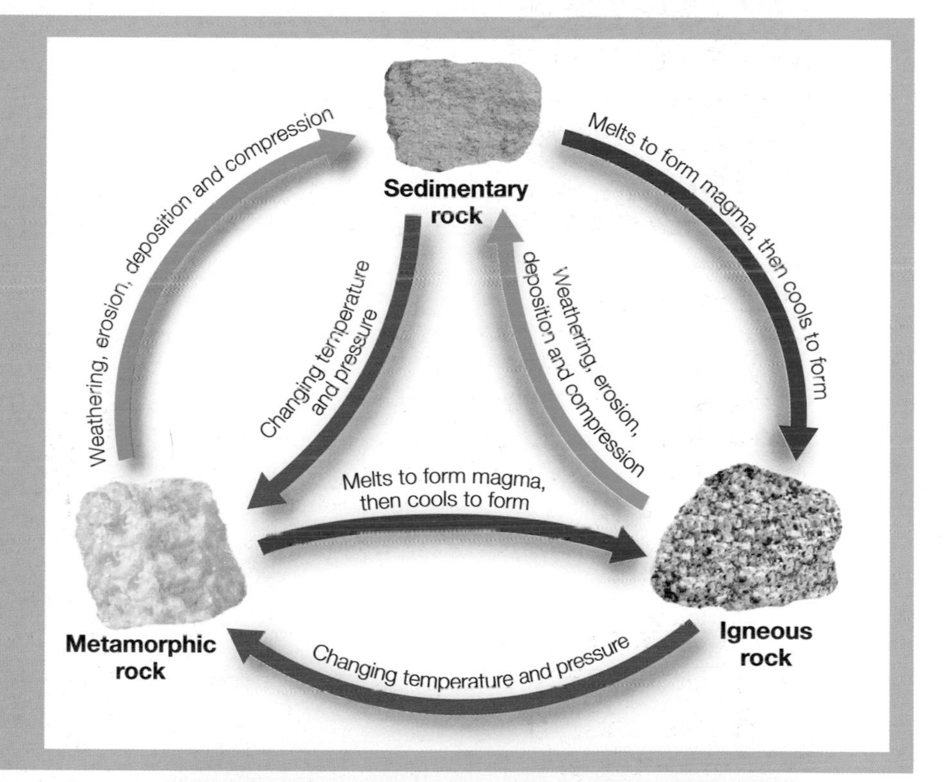

▶ *The rock cycle*

your planet

The word 'granite' comes from the Latin word 'granum', which means a grain and refers to the grainy structure of the rock.

YOUR QUESTIONS

1. Explain what these terms mean: impermeable, permeable.
2. Draw up a table with four columns, headed: Rock type, Formation, Characteristics, Examples. Complete the table for sedimentary, igneous and metamorphic rocks, using information from this spread.
3. Describe the journey of a piece of sediment around the rock cycle.

2.3 » Rocks and weathering

On this spread you'll find out about three types of weathering.

Scree running is a fun and fast way to get down a mountainside. You need a long stretch of loose, deep scree – that's all the loose bits of rubble on the mountainside – then you run and leap your way down. It's a bit like jumping down a sand dune – just a bit more painful if you fall! To understand where all that scree comes from, you have to know something about weathering.

Three types of weathering

Weathering is the process by which rocks are broken up into smaller pieces. It happens when they are exposed on the Earth's surface. Weathering happens *in situ* – that means it happens just where the rock is. There are three main types of weathering: mechanical, chemical and biological (and all three can happen at once – but *very slowly*). The main type of weathering that occurs depends on the climate and what kind of rock is involved.

Chemical weathering

This happens when rocks react with water and the air. Chemical changes in the rocks make them rot and decay. It occurs most when the climate is very warm and wet.

Carbonation

Carbonic acid is found naturally in rainwater. This acid reacts with rocks – like limestone – which contain a lot of calcium carbonate. The acid makes the limestone slowly dissolve, and it's then removed by running water. This particular weathering process can create very distinctive landforms, like the Guilin Hills in China.

Solution

Some minerals (like rock salt) and rocks simply dissolve in rainwater. This process is called solution.

▲ *The amazing limestone hills at Guilin in China*

Biological weathering

Plants and animals make this happen, e.g. tree roots grow into cracks in a rockface and widen them. This eventually forces bits of rock to fall off. Animals can also burrow into weak rocks, forcing them apart.

▶ *Biological weathering at work*

Mechanical weathering

This is also known as physical weathering. It is most likely to happen where the rock is bare and there's nothing to protect it from extreme temperatures.

Freeze-thaw, or frost shattering

This happens if a rock – like granite – contains cracks and joints, and the temperature regularly alternates above and below freezing point (perhaps above freezing point during the day and below it at night).

- When it rains, water trickles into cracks and joints in the rock.
- If the temperature then drops below freezing point (0°C), the water in the cracks freezes and turns to ice.
- Water expands when it freezes – which forces the cracks to widen.
- When the temperature warms up, the ice melts (or thaws) – releasing the pressure on the rock.
- Then, when it rains again, even more water trickles into the widened cracks, which then freezes – widening the cracks even further.
- Repeated freezing and thawing widens the cracks and weakens the rock so much that eventually bits of rock begin to break off.
- And all the broken bits then form scree.

Exfoliation, or onion weathering

This happens in very warm climates, where rocks alternately heat up and cool down. During the warm day, the outer layers of rock heat up and expand. In the cooler night, they cool down and contract. Repeated heating and cooling makes the outer layers peel off, just like an onion.

Cracks and joints in rock fill with water

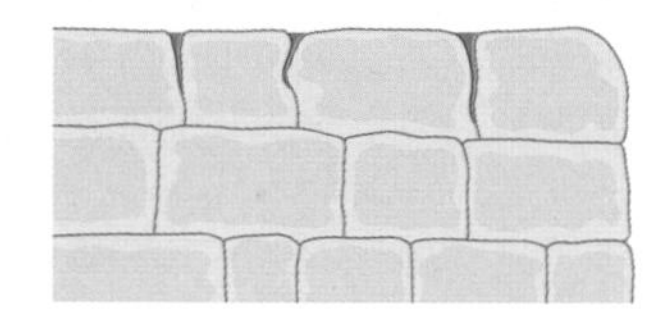

The water freezes and expands – pushing the rocks apart

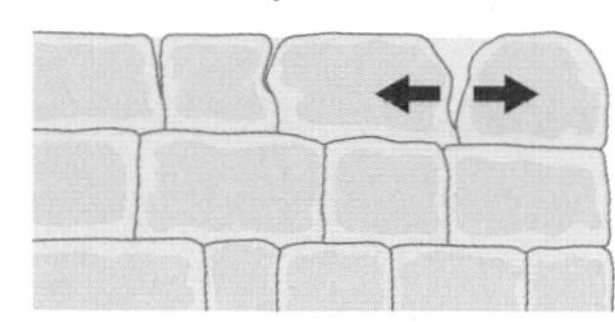

Repeated freezing and thawing widens the crack, and eventually bits of rock fall off

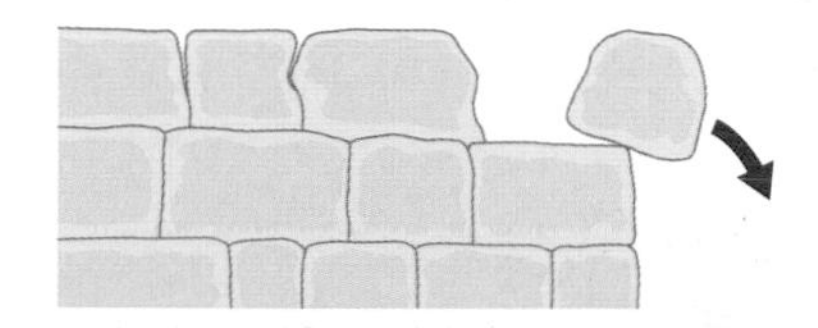

▲ *How freeze-thaw weathering works*

▶ *A great example of onion weathering*

YOUR QUESTIONS

1 What does the term weathering mean?

2 Divide up into pairs. Each partner should take turns to describe one type of chemical and one type of mechanical weathering to the other.

3 Draw a series of diagrams to show how exfoliation works. Add annotations to your diagrams.

Hint: Annotation means explaining labels.

4 Give an example to show how the type of weathering that happens depends on the climate and type of rock involved.

2.4 » Landforms and landscapes – 1

On this spread you'll find out about the different landforms and landscapes that granite – and chalk and clay – create.

Granite

What is granite?

Granite is an igneous rock, made from magma that cools inside the Earth. Where the magma intrudes – or goes into – the Earth's crust, it makes **granite intrusions**. These can create different landforms. The diagram below shows different types of intrusions.

Where do you find granite?

It's not that common. The map shows where you can find it in the British Isles.

What's a granite area like?

Granite creates distinctive landforms, with features like **tors** and moorlands, as the photo of Hay Tor on Dartmoor shows.

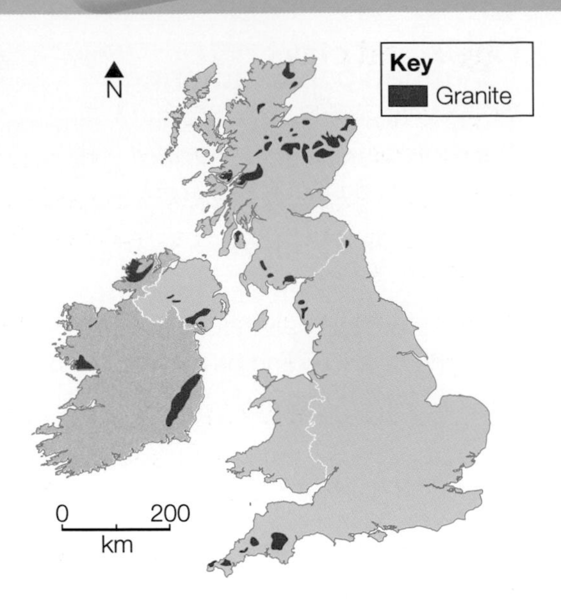

▲ *Granite areas in the British Isles*

A vertical intrusion is called a **dyke**. If it's softer than the rest of the rock, it makes a dip called a **trench**.

A big dome-shaped intrusion is called a **batholith**.

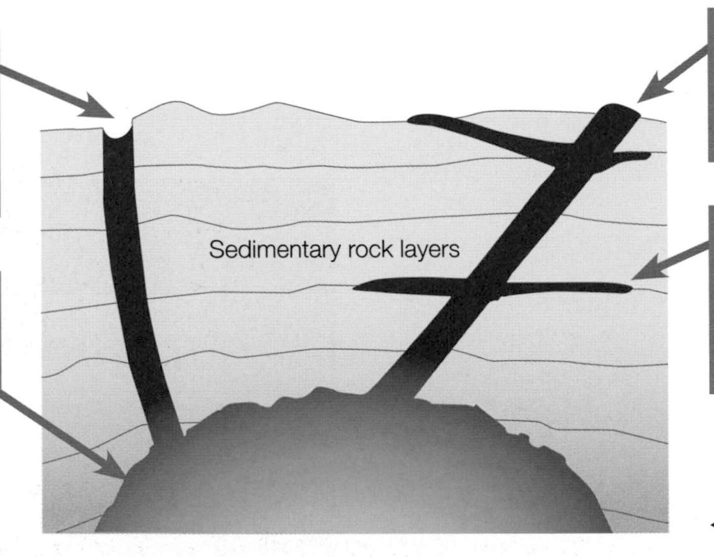

If the dyke is harder than the surrounding rock, it sticks up – leaving a **ridge**.

An intrusion that goes sideways along a bedding plane between layers of rock is called a **sill**.

◀ *Granite intrusions*

Dartmoor is the top of an exposed batholith. The granite there was more resistant than the rock above it, so that was eroded – eventually exposing the granite at the Earth's surface.

The moorland is covered with grass. In prehistoric times, it would have been trees.

Dartmoor has over 160 tors like this one. They are isolated outcrops of rock.

The joints in the granite are weathered by freeze-thaw. The bits of rock that fall off the tor create a **blockfield**.

There are deep river valleys on Dartmoor, where rivers have slowly eroded the jointed rock.

◀ *Hay Tor on Dartmoor*

Chalk and clay

What are chalk and clay?

They are both sedimentary rocks.

- Chalk is a type of limestone, and consists of calcium carbonate. It's made from the shells of dead sea creatures falling to the seabed and becoming compacted over time.
- Clay is very fine-grained. It is formed from the chemical weathering of other rocks and minerals.

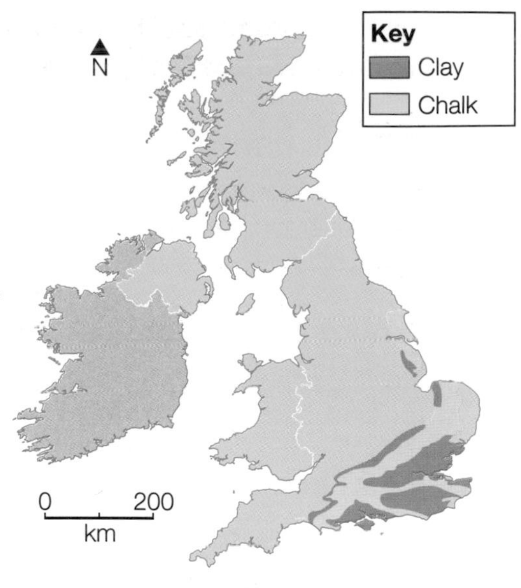

▲ *Chalk and clay areas in the British Isles*

Where do you find chalk and clay together?

Clay is very common in the British Isles, but you mostly find chalk and clay together in south and east England, as the map shows.

What are chalk and clay areas like?

- Chalk and clay are very different.
- Chalk is permeable and lets water pass through it – so there aren't many rivers on chalk.
- Chalk is quite resistant to erosion, so it tends to stand above other more easily eroded rocks to form hills and cliffs.
- Clay is impermeable, so it doesn't allow water through.
- Clay is easily eroded.

Where clay and chalk are found together – and where the rock is tilted – they form a particular type of landscape, consisting of **escarpments** (also known as **cuestas**) and **vales**. You can see this type of landscape in the photo.

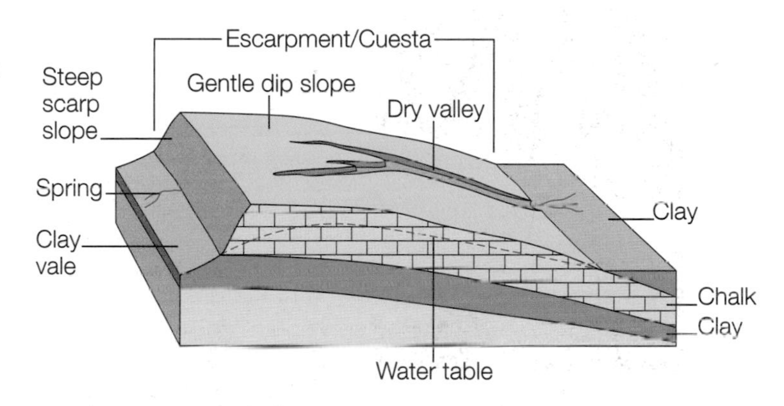

▲ *Features of a chalk and clay landscape*

Escarpments have a steep scarp slope and a gentle dip slope.

You don't usually get streams or rivers on the chalk. They occur in the clay vales.

Dry valleys are found on chalk. They were formed either when the water table was higher than it is today, or when the chalk was frozen during the Ice Age.

◀ *The North Downs in Kent*

YOUR QUESTIONS

1 Describe what these features are and what type of rock you find them on:

Either batholith, dyke, trench, sill, ridge, tor, blockfield

Or escarpment, vale, dry valley

2 Describe the physical landscape of Dartmoor. Use the photo opposite to help you.

3 a Use www.geograph.org to find a photo of a chalk and clay area in the British Isles, e.g. search for South Downs. It should show an escarpment or a dry valley. Put your photo in the middle of a large sheet of paper.

b Create a mind map around the photo to describe what chalk and clay landscapes are like, and where you find them.

On this spread you'll find out about the different landforms and landscapes that carboniferous limestone creates.

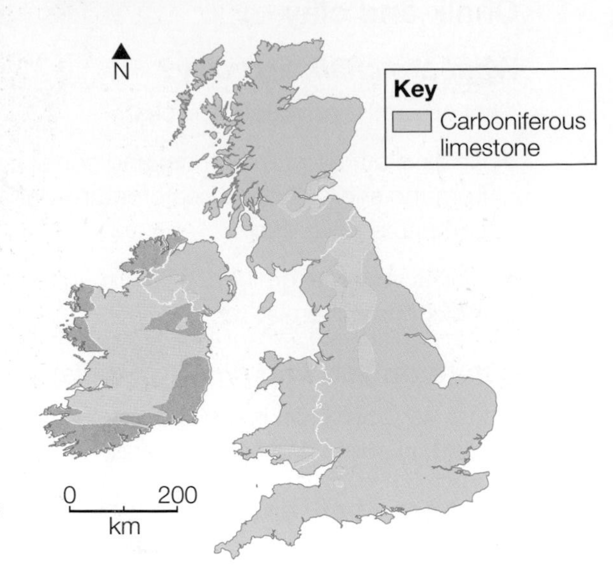

▲ *Carboniferous limestone areas in the British Isles*

What is carboniferous limestone?

Carboniferous limestone is a sedimentary rock formed during the Carboniferous period. It was formed on the seabed and contains the fossils of coral and shellfish.

Where do you find carboniferous limestone?

The largest areas in the British Isles occur in Ireland and in northern England, such as in the Yorkshire Dales.

What is a carboniferous limestone area like?

Carboniferous limestone produces distinctive landforms and features – for three main reasons:

- Limestone is made from calcium carbonate, which is dissolved by the carbonic acid in rainwater, i.e. chemical weathering (see page 34).
- The **structure** of this rock type means there are weak areas, which are attacked by chemical weathering. The different layers of carboniferous limestone are separated by bedding planes. There are also vertical joints within each layer. Both joints and bedding planes are vulnerable to chemical weathering.
- Carboniferous limestone is also **pervious** – water flows along the bedding planes and down the joints.

For these reasons, a lot of different features can be found in limestone areas. Some of them are on the surface and some are underground.

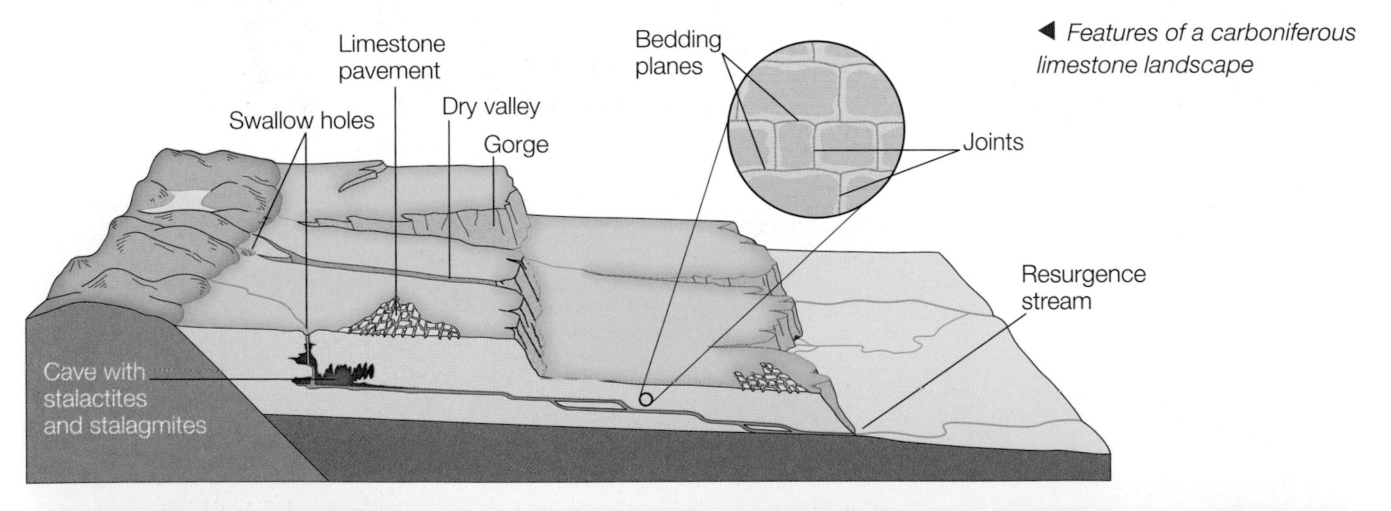

◀ *Features of a carboniferous limestone landscape*

YOUR QUESTIONS

1 Which is the odd one out in each trio below, and why?

- stalactite, stalagmite, gorge
- clints, pillar, grykes
- swallow holes, joints, bedding planes
- caves, swallow holes, dry valleys

2 Draw up a table with these headings: Carboniferous limestone feature, Example, Description, How it forms. Complete your table using the information on this spread.

3 Draw a spider diagram to show the reasons why carboniferous limestone produces distinctive landforms.

Limestone features in the Yorkshire Dales

Limestone pavements

These are areas where the limestone has been exposed on the surface. Rainwater has started to weather the vertical joints – leaving a pattern of enlarged joints (called **grykes**) and flat surfaces (called **clints**), which looks like a pavement! This one near Malham is a good example.

Swallow holes

These happen when a stream dissolves a joint in the limestone and then flows down it, rather than over the surface. The hole 'swallows' the stream! This photo shows a famous swallow hole – Gaping Gill near Ingleborough.

Dry valleys

These are old river valleys. They show that a river once flowed over the surface – possibly during the last Ice Age, when the ground was frozen, or afterwards, when the ground was saturated. Now water flows down swallow holes, so the valleys are dry. This one is above Malham Cove.

Resurgence streams

Streams that disappear down swallow holes continue to flow underground. If the water meets an impermeable rock, it flows over it until it reaches the ground surface again as a resurgence stream, like this one at the foot of Malham Cove.

Caves

These develop when water flows underground through joints or swallow holes and then weathers and erodes the limestone beneath. Eventually, large cave systems can form, like the ones leading from Gaping Gill. Inside the caves water drips from the roof. Some of the water evaporates and calcium carbonate is deposited to form stalactites (which hold 'tite' to the roof of the cave), and stalagmites (which grow up from the floor). Occasionally (and it's pretty rare) a stalactite and a stalagmite will link up to form a pillar.

Gorges

If the roof of a cave system collapses, it can leave a steep-sided narrow gorge, like this one at Trow Gill near Clapham.

your planet

The world's biggest stalactite is in the Jeita Grotto in Lebanon. It's nearly 27 feet long (or 8.2 metres)!

2.6 » Rocks and their uses – 1

On this spread you'll investigate the different ways in which people use rocks – and look in more detail at granite.

How do we use rocks?

Rocks are a major resource that are used in many different ways – from actually extracting them and turning them into something else, to finding varied ways of using the landscapes they create. The table lists some of the many uses of rocks.

Rock	Uses after extraction	Opportunities for farming	Opportunities for tourism
Granite	Used as building stone. Weathered granite is mined as china clay and used to make ceramics, and by the paper industry. Granite can also contain layers (veins) of copper and tin, which have many uses.	Granite soil is poor and acidic, so it tends to be used for grazing animals, rather than for crops, e.g. sheep are farmed on granite moorlands like Dartmoor.	Granite moorlands, such as Dartmoor, attract tourists who enjoy outdoor activities, e.g. walking, camping, mountain biking, bird watching.
Chalk	Used to make plaster, putty, cement and mortar (all used for building). Also used to produce lime for farming and industry.	Chalk soil is thin, but wheat and barley can grow on it. It's also suitable for grazing animals like sheep and sometimes cows.	Chalk can form rolling hills (like the North and South Downs), and steep cliffs (like the Seven Sisters). Both types of scenery attract tourists. Areas like the Downs also have unique habitats, with rare insects and flowers that attract people.
Clay	Used to make pottery, and also by companies making paper, chemical filters and the tips of spark plugs. It can also be used for making bricks and for waterproofing landfill sites.	Clay soil is high in nutrients, so it's used for arable farming (crops). It can also produce rich grass that is good for cows.	A clay landscape tends to be flat and featureless, and can get quite waterlogged, so it's not very attractive for tourism.
Carboniferous limestone	Used for making cement. Used as stone for dry-stone walls. Also used to produce lime for farming and industry.	Limestone soil is thin and not good enough for crops. The grass is also not good enough for cows, but it's fine for sheep.	Tourists are attracted by the scenery – both above and below ground. It's good for outdoor activities like walking, mountain biking, climbing. The caves attract potholers.

Farming on Dartmoor – a granite area

A farmer's tale

It's January now and I hope we've seen the worst of the cold and snow. It's been a harsh winter, even for Dartmoor.

I've had to dig sheep out of snowdrifts for the first time in my life. The old boys saying 'winter's not like it used to be' were wrong this time. We've been checking the sheep as often as possible, and moved them to avoid them getting stuck in the snow again.

We've started scanning to see how many lambs we'll get this year. At the moment the price for lamb is strong. Let's hope it stays that way.

(Adapted from 'A Farmer's Tale' on www.dartmoorfarmers.co.uk)

Dartmoor's farmers face problems

Dartmoor's moorland looks natural enough, but it only looks that way because it's been maintained by grazing for at least 3500 years. Without the farmers and their animals, Dartmoor's landscape would change – affecting the plants, birds and animals that live there.

Dartmoor's farmers are hill farmers. And it's a tough life. Most of the farmers keep sheep and cows to produce lambs and calves to sell for meat. All farmers need their farms to be profitable in order to keep going. But things are looking grim:

- Farming alone isn't providing enough income.
- The costs of feed, fuel, and bedding for the animals have all gone up.
- Some farmers have **diversified** into other areas, but not everyone can. And if they do, the moorland won't be maintained as it has been for thousands of years by grazing farm animals.
- Changes to payments from the EU mean that farmers now get less money.

What's being done to help?

The Dartmoor Hill Farm Project is aiming to make hill farming more profitable. It has:

- provided support to farmers selling meat directly to the public
- helped to organise specialist sheep sales
- set up an apprenticeship scheme to train workers for hill farming on Dartmoor.

As well as that, Dartmoor's farmers would like to be recognised and rewarded for their role in maintaining the landscape.

YOUR QUESTIONS

1 What does diversification mean?

2 Draw a spider diagram to show the different ways in which granite (or its landscape) is used.

3 **a** Go to www.dartmoorfarmers.co.uk. Follow the link for Dartmoor Farmers.

b Find out who set up the Dartmoor Farmers Association and why.

4 Hold a class discussion. The topic is: 'Should farmers be paid to look after the landscape?'

Hint: You need to think about the work farmers do, where the money would come from to pay them to look after the landscape, and what would happen if they gave up farming.

Diversification

Farmers can diversify, which means doing other things as well as farming. Examples include opening campsites, turning barns into holiday accommodation, and running farm shops.

▲ *Not a sheep in sight*

▲ *If farmers can't afford to carry on farming, the moorland landscape will change*

On this spread you'll find out how areas of carboniferous limestone are used, as well as areas of chalk and clay.

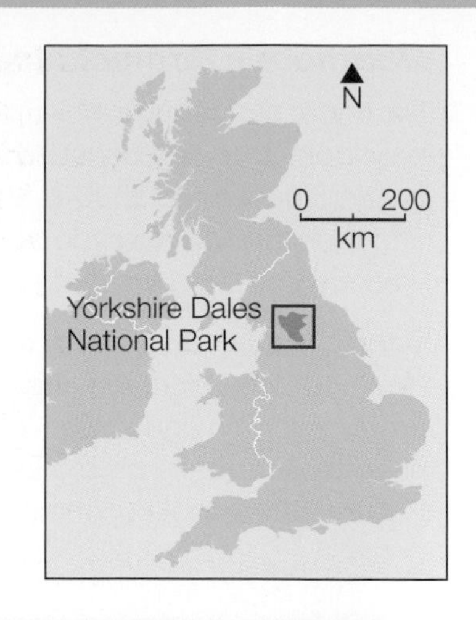

Tourism in the Yorkshire Dales – a carboniferous limestone area

The Yorkshire Dales National Park (YDNP) is in the north of England. It has stunning scenery – thanks to its carboniferous limestone environment (see page 39) – and it attracts loads of tourists, with over 8 million visitor-days every year. That means fewer than 8 million visitors in total, because some of them will only visit for a day, while others will stay for longer. There's so much to do there ...

If you're energetic, you can:

- walk the 1458 km of public footpaths
- ride the 623 km of public bridleways
- cycle, climb, or go caving.

Those less energetic can:

- visit the picturesque villages, or the 203 ancient monuments
- visit tea shops and pubs
- go bat watching, or on a fungal foray!

Tourism – the good and the bad

Tourism is very important to the UK's economy:

- More than 1.3 million people (about 5% of the working population) work in tourism, and over 8% of the UK's GDP depends on it.
- In the Yorkshire Dales, 15 000 people work in tourism (13% of the working population).
- Tourists spend £478 million a year in that area alone.

But the bad news is that:

- 15% of homes in the YDNP are either second homes or rented out as holiday cottages. This pushes up house prices, which means that local people often can't afford to buy their own homes.
- only 6% of the homes there are classed as social housing (that's housing provided by councils and housing associations). So there is also a limited supply of cheap housing to rent, as well as to buy.
- tourism jobs are often seasonal, low-paid and part-time.
- the large numbers of tourists who tend to visit the area by car cause congestion and pollution.

▲ *Malham Cove (above) is the place to see peregrine falcons (pictured), owls and woodpeckers*

What's being done to help?

The Yorkshire Dales National Park needs special care to conserve it for future generations. The National Park Authority also has to think about the economic and social well-being of the Park's almost 20 000 residents. The National Park Management Plan aims to tackle some of the problems listed opposite:

- In 2009, 19 affordable homes were built (and more sites for new homes were identified).
- The Plan hopes to increase the use of public transport within the YDNP, but by 2010 little progress had been made in this area.
- The Plan also aims to widen the range of adult training and learning opportunities, in order to increase job prospects within the YDNP.

The London aquifer – a chalk and clay area

What is an aquifer?

An **aquifer** is an underground reservoir of water. Areas of chalk and clay make great aquifers, because:

- chalk is permeable – water passes through its joints
- clay is impermeable – water can't pass through.

Where the two rocks occur together, they trap water to form an aquifer – as the diagram illustrates.

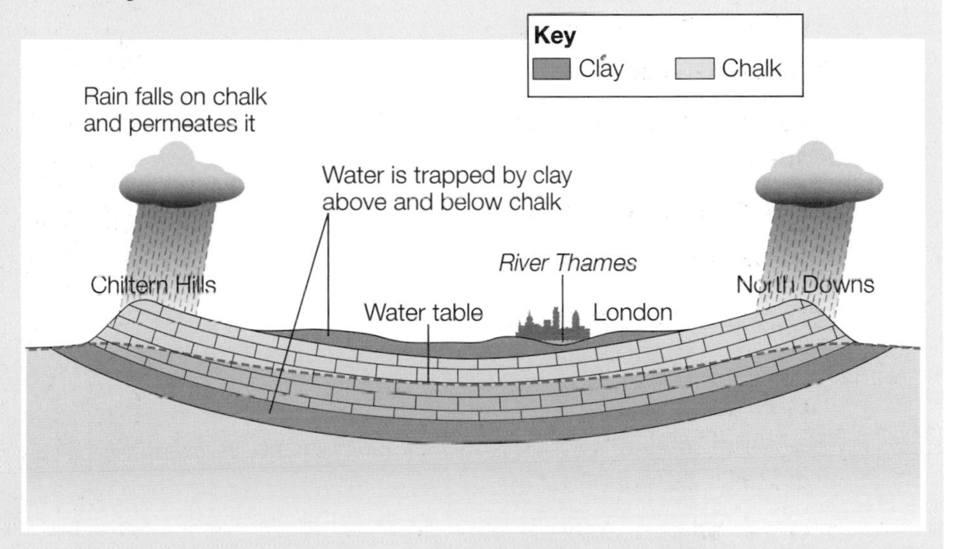

▲ *How the London aquifer works*

Why is it useful?

- It provides a clean water source, because the water is naturally filtered as it permeates its way through the chalk down to the aquifer.
- The water can be reached easily via boreholes.
- The aquifer can be used to supply extra emergency water during periods of dry weather. For example, in 2005, Thames Water pumped 200 million litres of water a day from its emergency aquifer under North London, which provided London with 10% of its water needs.

YOUR QUESTIONS

1 a What is an aquifer?

b Why do areas of chalk and clay make good aquifers?

Hint: You can either explain this in words, or draw an annotated diagram.

2 Design a poster to encourage people to visit the YDNP. You can either use the information on this spread, or if you want to find out more, go to www.yorkshiredales.org.uk

3 Work in pairs. One of you should list the benefits of tourism to the Yorkshire Dales, and the other should list the costs (i.e. the disadvantages or problems). Compare your lists and decide whether the benefits outweigh the costs.

2.8 » Quarrying in the Yorkshire Dales

On this spread you'll investigate quarrying in the Yorkshire Dales, and the issues that it raises.

Ingleton Quarry

Ingleton Quarry is in the Yorkshire Dales National Park – an area very popular with tourists (see pages 42-43). The quarry produces gritstone, which is used for road surfaces. It is one of a number of quarries within the National Park (see the map opposite), and different people have different views about it.

The quarry manager's view

This is a rural area and there's not many jobs. We employ 15 people at the quarry. That doesn't sound many, but on top of that we hire people to drive the trucks, contractors to do maintenance, electricians and so on. And we always use local people if we can.

Dust used to be a big problem, but we spent nearly £7 million on a new plant, and part of that was spent reducing the amount of dust we produced. It used to be all over the fields, roads and trees, but not any more.

The National Park Authority was keen to make sure that when we finish quarrying we reduce the impact we have on the environment. So, we've done a lot of tree planting – around 28-29 000 trees so far – a mixture of trees, including silver birch. By the time this place closes, it'll have less of an impact than it does now.

The tourist's view

The quarry ruins the view. It's noisy too, and there are those huge trucks on the narrow country roads. I might not come here again. I'll go somewhere else instead, where there are no noisy quarries to spoil the landscape and the peace and quiet.

The environmentalist's view

The rock is a non-renewable resource – once it's gone, it's gone for good. And if they take it out, will the rare plants that live here ever come back? Is it sustainable to use the land like this? Is it being used in a way that doesn't harm the ecosystem and environment in the long term?

Quarrying – the good and the bad

Ingleton Quarry is one of five large working quarries in the Yorkshire Dales National Park, which together produce about 4 million tonnes of **aggregate** a year. Aggregate is crushed stone and most of the aggregate from the Yorkshire Dales is used in road building and repair, and for making concrete.

What are the main issues to do with quarrying?

- Look at the view of Ingleton Quarry from the hillside above it. Some people think that that's enough of an issue on its own.
- About 75% of the quarried stone is transported by road in big lorries, which can create problems with congestion and pollution. Only 25% goes by rail.
- As well as transport, quarrying itself creates dust and noise.
- However, quarrying does provide jobs for local people. In 2006, 88 people were employed directly in the quarries, but – as the quarry manager opposite said – other people were also employed in industries and jobs related to quarrying.
- The quarries are also working hard to reduce their environmental impact by landscaping the area around them and planting trees.

What's being done to manage these issues?

The National Park Authority wants to protect the Yorkshire Dales, so it:

- monitors any plans to open new quarries in the National Park
- monitors any plans to make the existing quarries bigger – considering their possible effects on wildlife, the landscape, and the lives of local people
- works with quarry owners on plans for nature conservation management and the restoration of quarry sites once quarrying ends
- ensures that quarry companies monitor the effects of quarrying on the surrounding habitats, and take steps to prevent environmental damage and disruption to local people's lives.

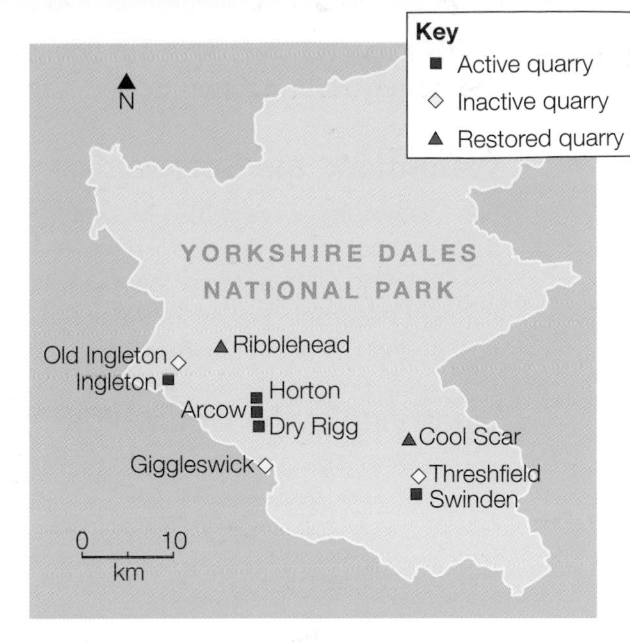

▲ *Aggregate quarries in the Yorkshire Dales National Park*

▲ *65% of the Yorkshire Dales aggregate is sold within the Yorkshire and Humber region, and most of it goes by road*

YOUR QUESTIONS

1 What is aggregate and what is it used for?

2 Create a mind map about quarrying in the Yorkshire Dales. Use all of the information on this spread, and sort it into Economic, Social and Environmental issues. Highlight the positive issues in one colour and the negative issues in a second colour.

3 **a** Create a conflict matrix like the one on the right.

b Complete the matrix using the key.

c Describe what it shows about who is likely to agree or disagree with quarrying in the Yorkshire Dales.

d Should quarrying be allowed to continue, or should the quarries be closed? Explain your answer.

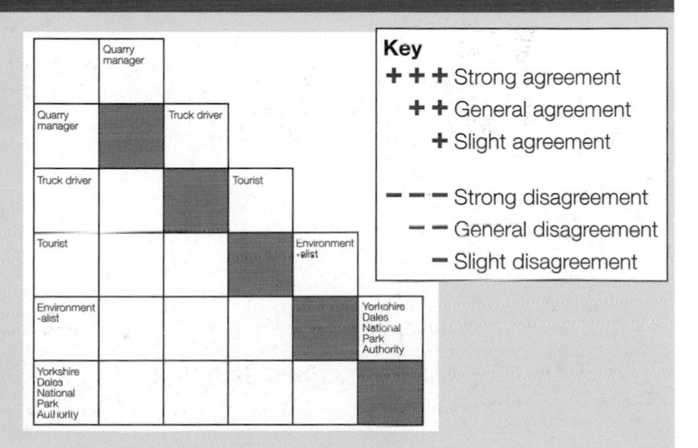

On this spread you'll find out how the impact of quarrying on the environment can be reduced.

Watermark, the Cotswolds

Fancy a holiday home in the Cotswolds? Not many of us can afford a second home, but houses like those on the right at Watermark, in the Cotswolds, are aimed at people who would like something a bit different to a traditional country cottage. And what's more, they're built on the site of old gravel pits, or quarries.

▲ *Idyllic – life next to an old gravel pit*

Quarrying in the Cotswolds

You'll find Watermark in the Cotswold Water Park. It's on the Gloucestershire-Wiltshire border and covers 40 square miles. It includes 150 lakes – all of which are old quarries. There are also eight working quarries there, with 50 years' worth of gravel still to be extracted.

The first lakes were created in the early twentieth century, when small fields were sold for sand and gravel extraction. The area has high groundwater levels, so any holes deeper than 1 metre quickly fill up with water – forming lakes.

Many of the lakes, or old quarries, have been restored and are now used for different purposes, including:

- fishing
- nature reserves
- an inland beach
- sailing, windsurfing and water skiing.

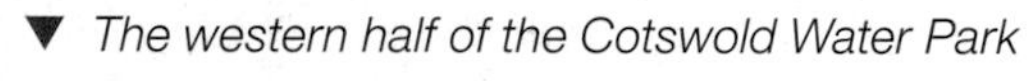

▼ *The western half of the Cotswold Water Park*

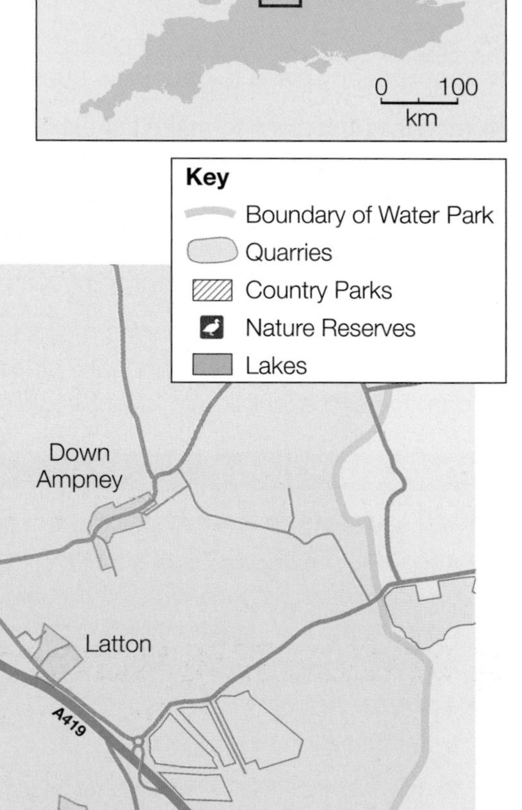

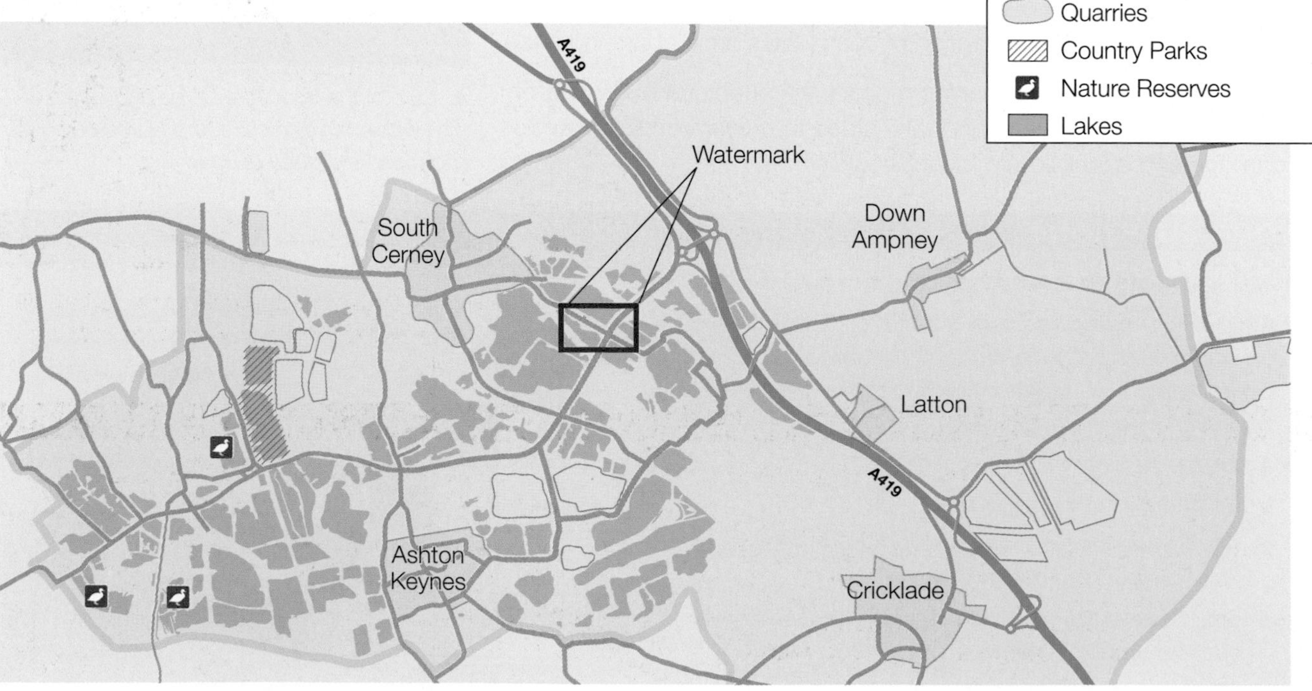

How do quarrying companies reduce their impact?

The eight working quarries are operated by several different companies, including: Hanson, Hill's Quarry Products, and the Cullimore Group. They extract just under 2 million tonnes of sand and gravel a year. Most of it is used in construction, and a lot of it goes locally – to places like Swindon and Gloucester.

If you look at their websites, you'll see what some of the companies say about how they care for the environment.

Hanson says: 'Caring for the environment is an important part of our business. We have invested heavily to make sure that we leave behind something as good as, or better than, the environment that existed before quarrying began. That process starts before we begin quarrying, and the restoration plan is often one of the most important aspects of a new quarrying scheme.

We have restored habitats and created nature reserves – and some former sites have become Sites of Special Scientific Interest. We have won awards for our restoration programmes.'

(Adapted from www.heidelbergcement.com)

What else is being done?

The Cotswold Water Park **Biodiversity Action Plan** covers the period from 2007-2016. The Plan's aim is for the Water Park to be an important site for nature conservation, which integrates the needs of industry, leisure, people and wildlife – both now and into the future. So, it is an attempt to help manage the Water Park in a sustainable way.

The Cotswold Water Park Society has the job of implementing the Biodiversity Action Plan. It:

- works with companies to enhance the biodiversity of working quarries and on the restoration of old quarries
- works with building developers on the creation of habitats
- advises landowners.

Biodiversity Action Plans (BAPs for short) set out a framework for nature conservation. They work at a local level.

Biodiversity is short for biological diversity. It is the variety of all forms of life on Earth – plants, animals and microorganisms.

▲ *Thanks to the BAP, the number of otters in the Cotswold Water Park has increased*

▲ *Shallow reed beds encourage a greater variety of wildlife than just having deep water, but more are needed in the Cotswold Water Park*

YOUR QUESTIONS

1. Explain these terms in your own words: biodiversity, Biodiversity Action Plan.
2. Describe how quarrying companies, such as Hanson, try to reduce their impact on the environment.
3. Write the words 'Cotswold Water Park and sustainable management' in the middle of an A4 piece of paper. Create a spider diagram to show how the Water Park is managed in a sustainable way to reduce the impact of quarrying.
4. Go to www.waterpark.org and search for the Aggregates Levy Sustainability Fund. Write a paragraph to explain what it is, and what it aims to do.

3 » Challenge of weather and climate

What do you have to know?

This chapter is from **Unit 1 Physical Geography Section A** of the AQA A GCSE specification. It is about weather and climate, and the challenges we face from extreme weather events and climate change. The table shows how the pages in this chapter match the content in the specification.

Specification content	Pages in this chapter
The UK climate – its characteristics, reasons for the climate, and why there are differences within the UK.	p50-51
The importance of depressions and anticyclones in the UK, and the weather that they bring.	p52-55
Weather in the UK is becoming more extreme. The evidence for this, and the impact it has on people.	p56-57 p58-59 Case study of Gloucestershire flooding
The evidence for and against global climate change, and the possible causes of global warming.	p60-61
The consequences of global climate change for the world and the UK.	p62-63
Responses to the threat of global climate change from international to local levels	p64-65
Tropical revolving storms (hurricanes): how they form, the impacts they have, and responses to them in different parts of the world.	p66-67 p68-69 Case study Hurricane Katrina p70-71 Case study Cyclone Nargis

Your key words

- Climate ,weather
- Temperate maritime climate
- Continental climate
- Prevailing wind
- Relief rainfall, convectional rainfall
- Anticyclone, depression
- Isobars
- Synoptic chart
- Warm front, cold front, occluded front
- Climate change
- Global warming
- Greenhouse effect
- Enhanced greenhouse effect
- Carbon credits
- Hurricanes and tropical revolving storms
- Coriolis force
- Storm surge
- Saffir-Simpson scale

Exam help ...

Advice See pages 297-299 for information on how to be successful in your exams.

Practice See page 302 for exam questions on this chapter.

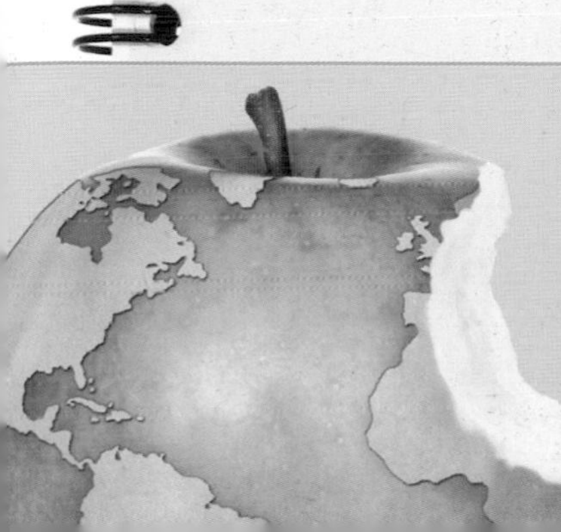

What if ...

- all our rain came at once?
- your town was hit by a hurricane?
- we had a heat wave every year?

On this spread you'll learn about the climate of the UK – what it's like, and why.

What is climate?

Some say that climate is what you want, but weather is what you get! If you go on holiday to Majorca in August, you expect it to be hot and sunny – that's **climate**. If it rains, you've got **weather**!

Weather is short-term day-to-day changes in things like temperature, wind, and sunshine. Climate is the average of those weather conditions, measured over thirty years.

Mild and wet

Our climate in the UK is generally mild and wet. We have a **temperate maritime climate**, which is influenced by the sea surrounding the UK. In summer the sea cools the climate, and in winter the sea insulates us – keeping us warmer than other places at the same latitude.

But the UK's climate isn't the same everywhere, as the diagram shows.

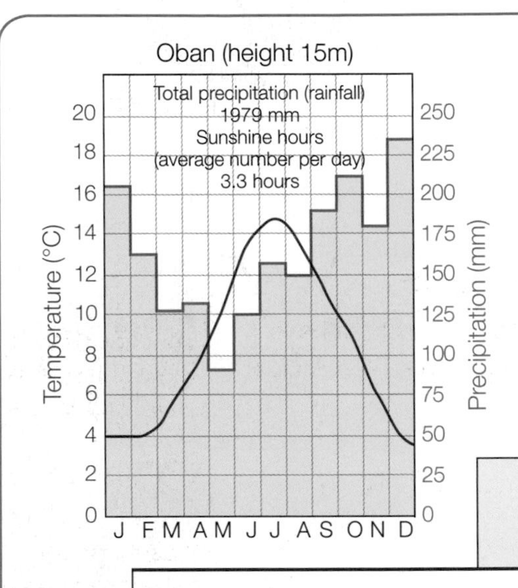

Why is it colder in the north?

Colder temperatures in the north are due to both the higher latitude and the mountains (air cools by 1°C for every 100 metres in height). So in winter the climate can be quite harsh. Arctic air from the North Pole can also blast Scotland, reducing temperatures further.

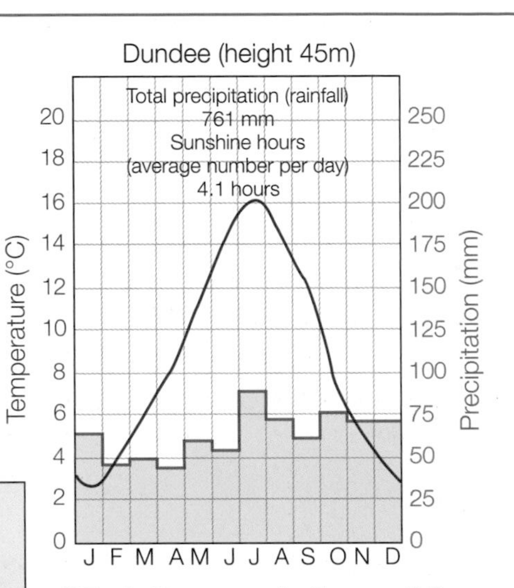

Key to map

15 °C July isotherm

4 °C January isotherm

isotherms are lines joining places of equal temperature

Why is it warmer in the south?

Warm air from Europe helps to lift summer temperatures. Southerly continental winds often produce a heat wave. And as they're at a lower latitude, southern areas also get more concentrated heat energy from the sun.

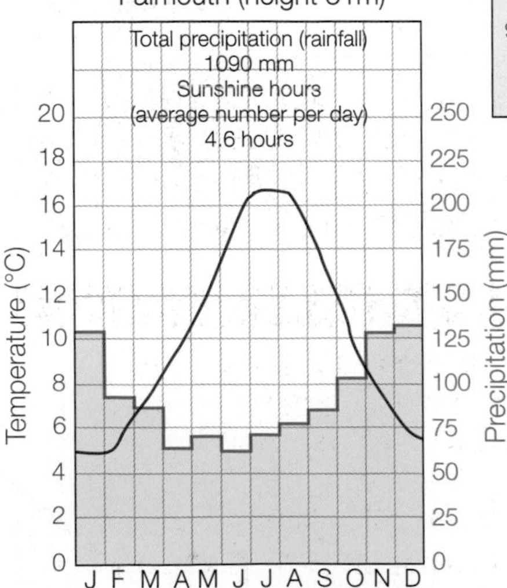

Why is it milder in the west?

The North Atlantic Drift is a very warm current that starts in the tropical Gulf of Mexico. The warmth from this water helps keep the west of the UK milder during the winter months – the prevailing south-westerly winds are warmed by this water.

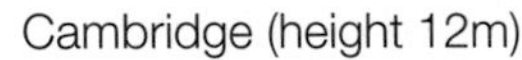

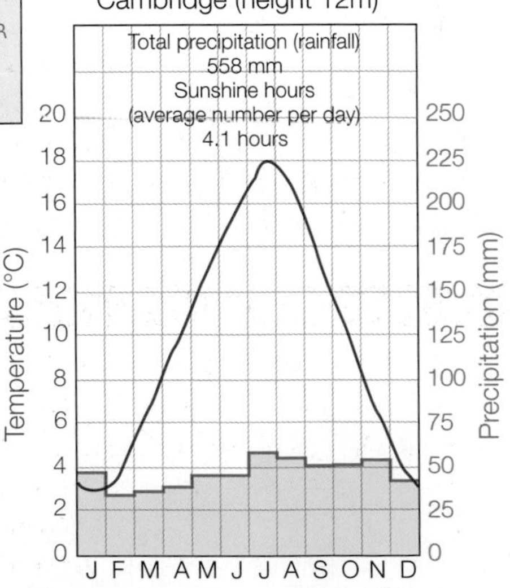

Explaining climate

Why is the UK's climate mild and wet? There are a number of factors that help to explain it, but the most important one is latitude.

Latitude

The further you go from the Equator, the cooler it gets. Why? Because the Earth's surface is curved, which means that the sun's energy isn't evenly distributed. The diagram shows how the sun's energy at the Equator is much more direct and concentrated than at the Poles. This means that the Earth is hottest here.

At the North and South Poles, the greater curvature of the Earth means that the sun's energy is spread over a larger area. This means colder temperatures. Ice forms there and reflects heat back into space, making it even colder.

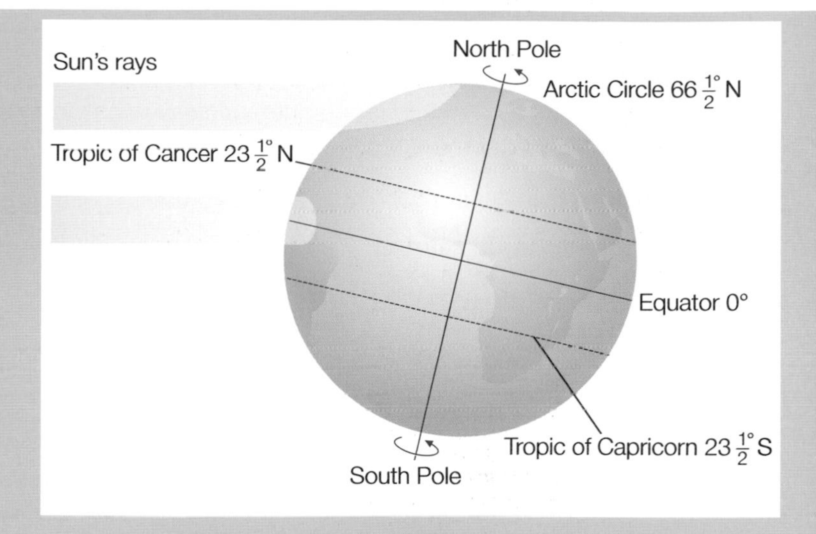

Distance from the sea

The sea is cooler than the land in summer – but warmer in winter – because it takes longer to heat up than the land, but is slower to cool down. The sea moderates our climate in the UK, making temperatures more even throughout the year.

The interior of countries with large land masses, like the USA, have a **continental** climate. It is too far from the sea to be influenced by it, and so they have very cold snowy winters, and hot dry summers.

Altitude

Temperatures decrease by 1°C for every 100 metres of altitude (height above sea level). So, mountainous areas are always cooler. These areas also tend to get more rain (because of relief rainfall – see page 52), so they are wetter too.

Pressure and winds

Around the world some areas have high **pressure** and some areas have low pressure.

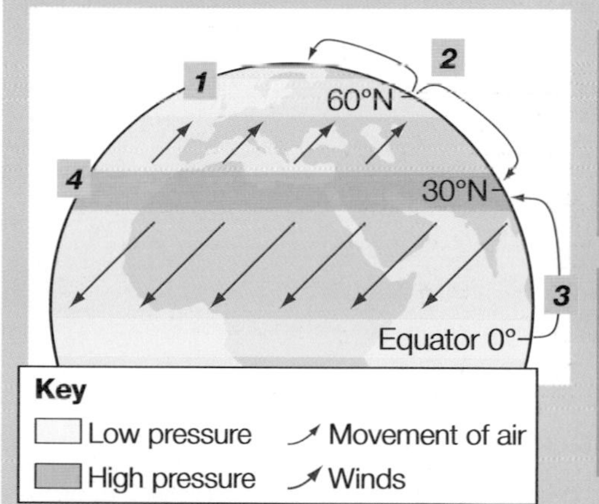

1 The UK is in a low-pressure zone. Air is rising here.

2 Air cools, condenses and forms clouds – then it rains and we get our wet weather.

3 Hot air rises at the Equator, and then sinks to create areas of high pressure.

*4 Winds (moving air) blow from areas of high pressure, to areas of low pressure – like the UK. Our **prevailing** winds (the ones we usually get) are from the south-west. They have blown across the Atlantic Ocean, so they often bring rain.*

YOUR QUESTIONS

1. Define these words from the text – climate, weather, temperate maritime climate, continental climate, prevailing wind.
2. **a** Draw your own sketch map of the UK and divide it into the four climate areas shown on the map opposite.
   b Using the climate graphs, annotate your map with information about the climate in each area. Remember to include figures, and use the following terms: maximum temperature, minimum temperature, temperature range (the difference between the maximum and minimum temperatures), total precipitation and sunshine hours.
   Hint: 'Annotate' means to add explaining labels.
3. Explain how either latitude or distance from the sea affects the climate of the UK.

3.2 » Rain and weather systems – 1

On this spread you'll find out about two types of rainfall, and also all about anticyclones.

Wet, wet, wet

It's always raining in the UK – or at least that's how it feels. The map on the right shows the average annual rainfall across the UK – much of it brought by the prevailing south-westerly winds from the Atlantic. In the UK we get three types of rainfall:

- **relief rainfall** (sometimes called orographic)
- **convectional rainfall**
- frontal rainfall (you will learn about this on page 55)

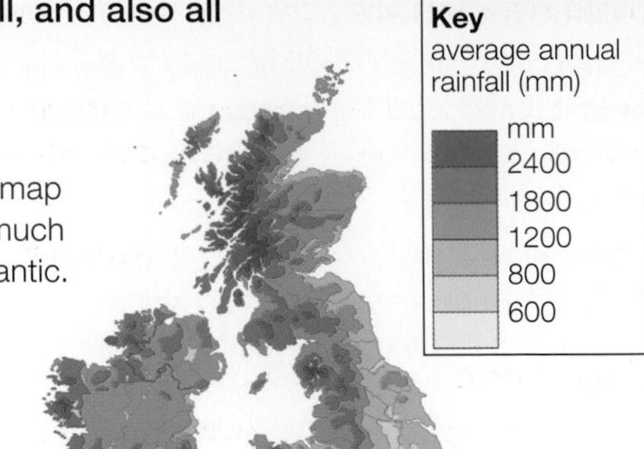

▶ *Average annual rainfall in the UK. Notice how the west is wetter.*

your planet
The wettest place in the UK is Crib Goch in Snowdonia, Wales, with 4470 mm of rain a year (or 176 inches).

Relief rainfall

Warm moist air arrives from the Atlantic Ocean and rises over the mountains on the western side of Britain – the Cambrians, Pennines and Grampians. When it rises, it cools, condenses into cloud and starts raining. The mountain peaks can receive a lot of rain – up to 2000 mm a year.

Once the air has passed over the mountains, it descends and gradually warms as it reaches lower ground. This creates drier conditions and is called the **rain shadow**. Therefore, mountains in the UK tend to be wetter on their windward, western sides and drier on their leeward, eastern sides.

Convectional rainfall

During the summer, strong sunshine causes the ground to heat up rapidly. This sets up rising pockets of warm air, known as **convection currents**. The warm air rises rapidly to a high altitude, where it cools and its water vapour condenses to form clouds. With time – particularly by late afternoon – thick cumulonimbus clouds can form. These can produce heavy rainfall and sometimes thunderstorms. Therefore, some places like Cambridge (see the climate graph on page 50) have their wettest months in July or August.

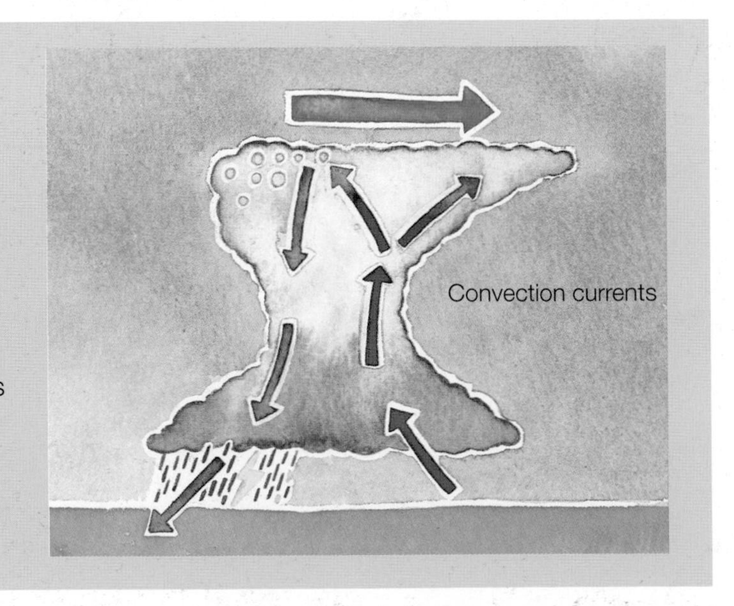

Two large-scale weather systems control the weather in the UK – anticyclones and depressions.

- **Anticyclones** have high pressure and result in clear and calm weather. However, there are differences between summer and winter anticyclones (see below).
- **Depressions** have low pressure and result in cloudy, wet and windy weather (see pages 54-55).

Anticyclones

The map on the right shows a **synoptic chart**, or weather map. It shows an anticyclone – an area of high pressure – over Western Europe (labelled **H**). Europe was in the grip of a heat wave at the time.

The symbols on a synoptic chart describe the weather conditions at a particular place at a particular time. This chart has the symbols for warm and cold fronts and pressure, as shown below.

Warm front

Cold front

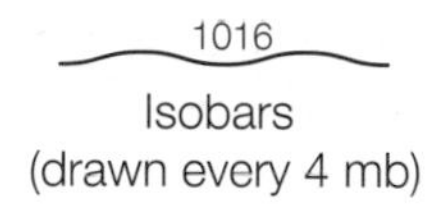

Isobars
(drawn every 4 mb)

Air pressure is shown by ***isobars*** *(lines of equal pressure). In an anticyclone they are far apart – air pressure doesn't change much over a long distance, so winds are light.*

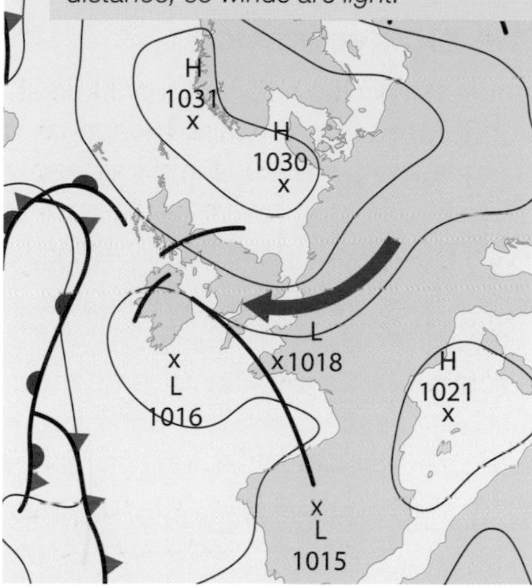

▲ *A synoptic chart for Western Europe at midday on 5 August 2003*

Summer anticyclones

The air descends in an anticyclone. As it descends it warms up, and any water vapour evaporates. This prevents clouds from forming. Cloudless skies mean that the sun is strong and the days are hot. If the anticyclone stays still, it can even lead to a heat wave – as in Europe in 2003. However, because there's no cloud, the evenings can be cool. When the ground cools at night, water vapour condenses to form dew.

No clouds means no rain – and that can lead to drought. But, on very hot days, the hot air can rise quickly, cool and form thick clouds – leading to thunderstorms.

Winter anticyclones

In winter, the cloudless skies we often get with anticyclones allow heat to escape into the atmosphere. This means that the ground cools quickly at night and water vapour condenses and freezes on cold surfaces – forming frost. It also condenses on dust and other particles in the air to form fog, which can linger into the day until the sun's heat evaporates it. Still water in ponds and puddles can freeze. The days are often clear, cold and bright.

YOUR QUESTIONS

1 Define and explain these terms: relief rainfall, convectional rainfall, rain shadow, anticyclone, depressions, isobars, synoptic chart.

2 a Research two photos to illustrate common weather in summer and winter anticyclones.

b Label the photos to show the type of weather associated with each type of anticyclone.

3 a Find out how the following are shown on synoptic charts: wind speed and direction, cloud and present weather (that's things like rain, snow, etc.). Geography books are a good place to start looking.

b Draw a chart to show these weather symbols.

3.3 » Rain and weather systems – 2

On this spread you'll find out what depressions are, and about the weather they bring.

One big depression

August 2004. The height of the summer, and Boscastle in Cornwall was busy. But a large **depression** was heading for the town. The result was heavy rain and flooding on a massive scale – as the photo opposite shows.

your planet

On 16 August 2004, an estimated 440 million gallons of water poured through Boscastle. That's roughly the equivalent of 880 Olympic-sized swimming pools!

Depressions, and how they form

Depressions are low-pressure weather systems. They form over the Atlantic Ocean when warm tropical air meets cold polar air. They move across the UK from west to east. Most of the UK's rainfall and most of our dramatic storms – like the one that hit Boscastle – are the result of depressions. In depressions, the isobars on synoptic charts are close together, indicating stronger winds.

The diagrams below show how depressions develop. They're in 3D.

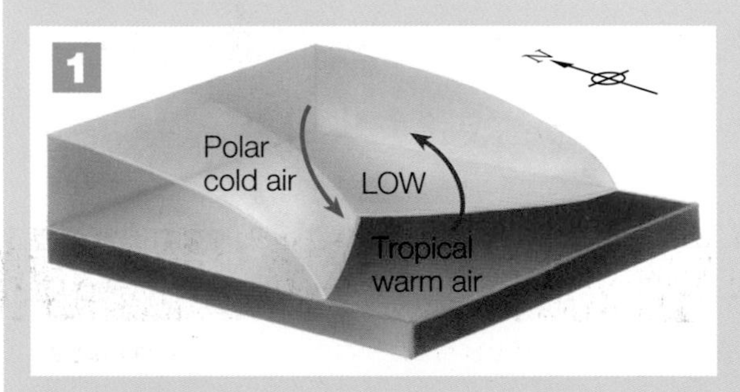

- Cold air flowing down from polar regions meets warm air flowing up from tropical regions.
- The warm tropical air is less dense and so starts to rise over the cold polar air, creating low pressure.

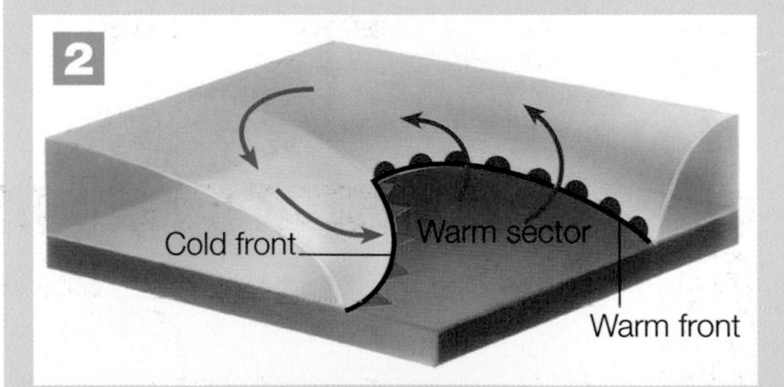

- Warm air is 'sucked' into the low-pressure area, creating a warm sector.
- Cold air is 'sucked' in behind and the air mass starts to spiral.
- Where warm air rises over cold air, it is called a **warm front**. As air rises, it cools and condenses – creating clouds and rain.
- Where cold air pushes in behind, it is called a **cold front**.

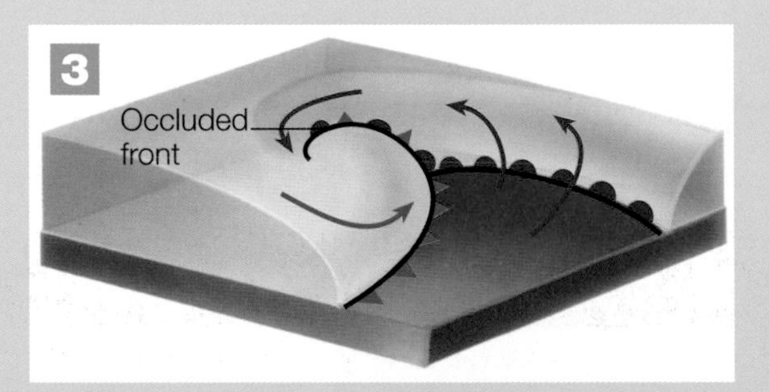

- The warm sector starts to 'climb' above the cold air.
- Below the rising air the cold front catches up with the warm front. It lifts the warm sector off the ground.
- Where the warm sector is lifted, it is called an **occluded front**.

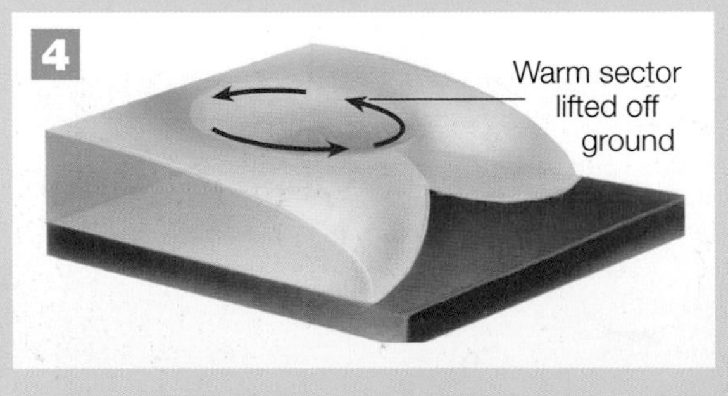

- Cold air has replaced the warm sector on the ground.
- Temperatures even out.
- The fronts disappear and the depression dies.

Depressions, rain and weather

Depressions bring rain – and often lots of it. The diagram below is a cross-section. It shows how rain forms along the warm and cold fronts of a depression.

▼ *Frontal rainfall in a depression*

When depressions move across the UK, there is a definite pattern to the weather – as the table below shows.

	Ahead of the warm front	Passage of the warm front	Warm sector	Passage of the cold front	Cold sector
Pressure	starts to fall steadily	continues to fall	steadies	starts to rise	continues to rise
Temperature	quite cold, starts to rise	continues to rise	quite mild	sudden drop	remains cold
Cloud cover	cloud base drops and thickens	cloud base is low and thick	cloud may thin and break	clouds thicken	clouds thin
Wind speed	speeds increase	becomes blustery with strong gusts	remains steady	speeds increase sometimes to gale force	winds are squally
Precipitation	none at first, rain closer to front, sometimes snow on leading edge	continues, and sometimes heavy rainfall	rains turns to drizzle, or stops	heavy rain, sometimes with hail, thunder or sleet	showers

A large depression brought torrential rain and flooding to Boscastle in 2004

YOUR QUESTIONS

1. Define these words: depressions, warm front, cold front, occluded front.
2. Write a 20-second radio weather forecast for Boscastle on 16 August 2004. Include these words in your forecast: depression, pressure, temperature, wind speed, and rainfall (or precipitation).
3. Imagine that you are the helicopter pilot in the photo on the left. Write a 50-word account to describe the weather conditions you are flying through.

3.4 » Extreme weather in the UK

On this spread you'll find out what extreme weather is, and look at the evidence that weather is becoming more extreme.

What is extreme weather?

Usually severe, and often unexpected, extreme weather can bring chaos and misery. Floods, droughts, heat waves and fires are just a few of the problems caused by extreme weather. Too much or too little rain, temperatures that are higher or lower than normal, extreme winds and intense low- or high-pressure weather systems cause problems which can affect us all.

The text boxes below show just a few of the extreme weather events which have affected the UK in the recent past.

your planet

By 2030, heat waves – like the one which affected Europe in 2003 – could happen every three years.

▼ *Skiers and tobogganists – not in the Alps but in Henley on Thames – enjoy 'The Big Freeze' in January 2010*

2003 *A heat wave affected not just the UK, but the whole of Europe. It began in June and continued until mid-August. Tourism increased in parts of the UK, because many people stayed at home to enjoy the Mediterranean-style weather, rather than holidaying abroad. However, 2045 people died in the UK as a result of the heat.*

2004 *Boscastle in Cornwall, was hit by major flash flooding (see page 54). No-one died, but 1000 residents and tourists were affected.*

2005 *The south of England had its driest period since the drought of 1976. Hosepipes and sprinklers were banned.*

2007 *Flooding affected a number of areas in the UK. Gloucestershire was particularly badly hit (see pages 58-59). Boscastle also flooded again.*

2008 *A torrential hailstorm in the Ottery St Mary area of Devon brought 100mm of rain and hail in a few hours on 30 October, causing major flooding.*

2010 *January saw 'The Big Freeze' hit most of the UK (parts of Scotland and northern England had been experiencing it since December). Forecasters declared that it was the coldest winter weather since the winter of 1962-3 – in places the night-time temperature fell to -20°C. The UK's gas supplies came under pressure, because many people turned up their heating. Some large companies had their gas supplies rationed to preserve supplies for homes. Supplies of rock salt to treat roads and pavements also reached extremely low levels, and many minor roads had to be left untreated.*

What evidence is there that weather is becoming more extreme?

Scientists say that it isn't possible to link a particular extreme weather event with climate change. However, the planet does seem to be warming up, and, at the same time, the whole world – not just the UK – is experiencing more severe weather. For instance:

- records kept by the International Disaster Database, show that the number of floods has increased significantly since the 1960s. The floods are worse, last longer and affect more people.
- climate models also show an increase in the frequency and length of extreme events. For example, in the USA, rainfall increased by 5-10% from 1997-2007, and rain and snow are falling in fewer, more-extreme events.

The Intergovernmental Panel on Climate Change (IPCC) is becoming more confident that certain weather events will become more frequent and intense during the twenty-first century, as climate changes. These will have an impact on people's homes and lives, on farming, industry and transport.

▼ *More extreme weather. Flooding in Cockermouth in 2009 caused chaos, and meant that many people had to be evacuated from their homes.*

2009: one year's weather in the UK

January	*Most places had temperatures a little below average, but northern Scotland was very mild – with temperatures of 14°C.*
February	*Parts of England had the heaviest snowfall since 1991, which disrupted travel in the South East.*
March	*Temperatures were slightly above average.*
April	*It was warm and dry, with Kent reaching a maximum temperature of 22°C.*
May	*Parts of Scotland had 50% more sunshine than normal, but also very strong winds.*
June	*It was warm and stormy. Marble-sized hailstones fell in Essex, and golf ball-sized ones in Fife. The month ended with a heat wave.*
July	*The heat wave didn't last long, and July was a washout with rain and flooding across much of the UK.*
August	*The wet weather continued. Many areas in western Scotland, Northern Ireland and Cumbria had twice the normal August rainfall.*
September	*The remnants of Hurricane Danny brought wet and windy weather to parts of the UK.*
October	*The weather was pretty uneventful.*
November	*More rain and more flooding. It was the wettest November since 1914. Western Cumbria, particularly Cockermouth, was hit by devastating floods (see the photo on the left and pages 104-107).*
December	*Saw the beginning of the longest cold snap since 1981, which lasted well into 2010 and became known in the media as 'The Big Freeze'. Heavy rain and snow caused travel disruption, especially over Christmas.*

YOUR QUESTIONS

1 Explain what extreme weather is to your neighbour, and give them two examples of extreme weather events in the UK.

2 Study the photo on the left. Draw a spider diagram to show as many impacts as you can think of that an extreme weather event, such as flooding, might have on people in the UK.

3 Look at the weather events timeline for 2009. How far do you think the weather events were extreme? Or was this just an average year?

Hint: 'How far' means you have to put both sides of an argument.

On this spread you'll find out about the impacts of the floods which hit Gloucestershire in July 2007, and about how Cheltenham manages flooding.

Gloucestershire under water

Tewkesbury in Gloucestershire is no stranger to flooding – it's a regular event. Floodwater normally flows onto a large meadow called 'The Ham', and the town usually stays dry. But in July 2007 it didn't.

What were the impacts of the flooding?

In late July 2007, some areas of England and Wales received up to three times the average rainfall for the whole of July – in just 24 hours! The rain caused massive flooding. In Gloucestershire, the River Severn flooded, and the table below shows the impacts it had. One of the biggest surprises was the effect that it had on power and water supplies.

your planet

The average rainfall for July is 5-6cm. Pershore, in Worcestershire, had over twice this amount in just over 24 hours.

▼ *The impacts of flooding in Gloucestershire in 2007*

Social impacts	• Three people died. • 5000 homes and businesses were flooded. • About 2000 people had to be evacuated from their homes and were housed in temporary accommodation. • 10 000 drivers were stranded overnight in their cars on the M5 and other roads across the county. • The Mythe water treatment plant at Tewkesbury was flooded and shut down. 135 000 homes (over half the homes in Gloucestershire) were left without piped water for up to 17 days. Severn Trent Water (the local water company) supplied over 40 million bottles of drinking water, and put 1400 bowsers (large water tanks) on the streets, so that people had water to use. • Castle Meads electricity substation in Gloucester was flooded, leaving 48 000 homes without power for two days.
Economic impacts	• Damage to the county's roads was estimated at £25 million (Gloucestershire County Council's total annual budget for road repairs). • Severn Trent Water estimated that the flooding cost them £25-35 million.
Environmental impacts	• The River Severn in Gloucester peaked nearly 5 metres above its normal level.

The map below shows some of the places affected by the floods.

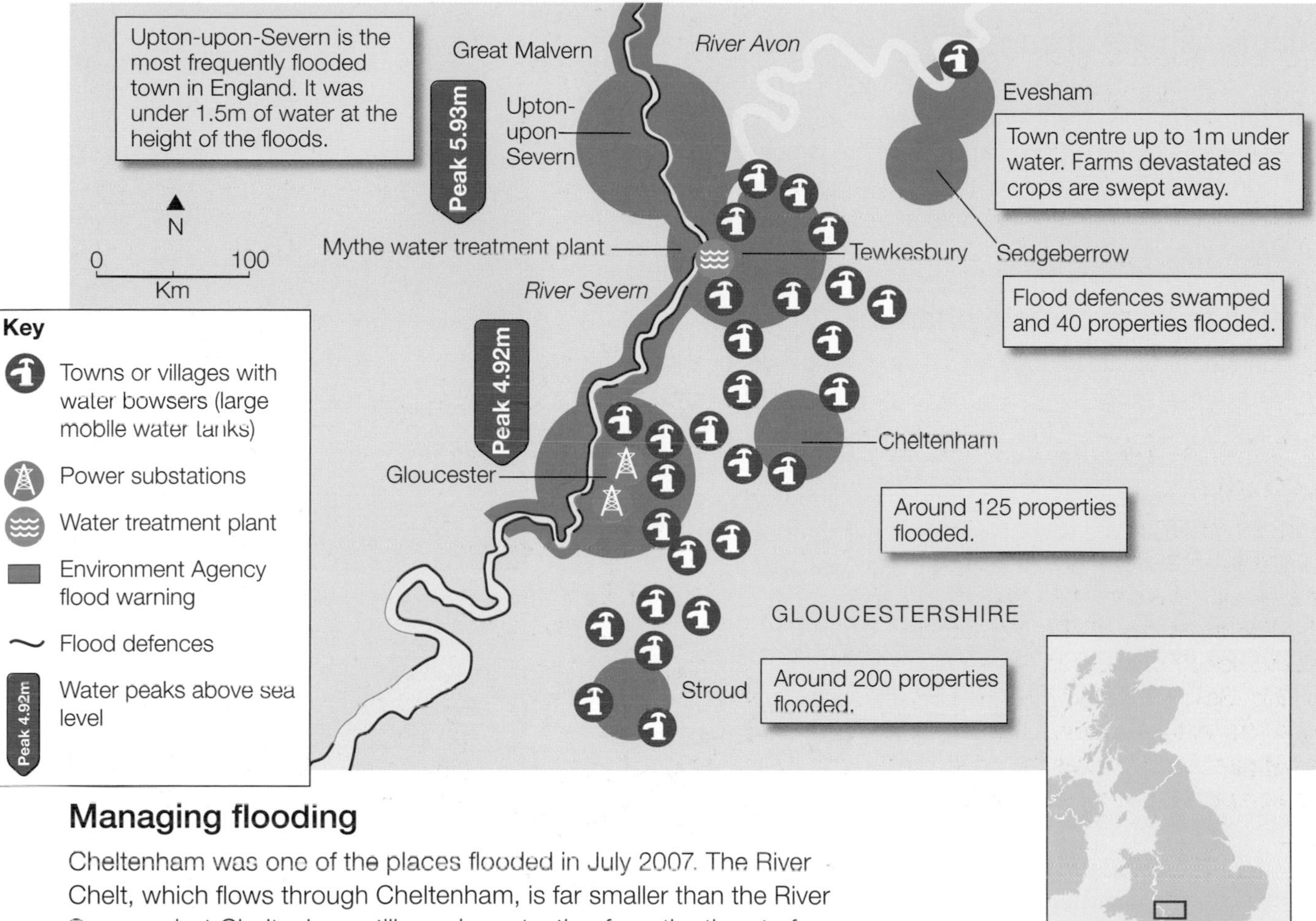

Managing flooding

Cheltenham was one of the places flooded in July 2007. The River Chelt, which flows through Cheltenham, is far smaller than the River Severn – but Cheltenham still needs protecting from the threat of flooding.

The flood defences on the River Chelt had been improved between 2000 and 2006, but couldn't protect the town from flooding in 2007. The improvements included:

- widening the river channel
- building flood walls
- lowering the river bed
- removing weirs
- moving culverts (pipes carrying water underground).

Flood forecasting and warning

The Environment Agency has the job of working out which places are at risk from flooding. They produce flood maps which show areas at risk (see page 107).

The Environment Agency also monitors rainfall and water levels in many main rivers 24 hours a day. They use this information, plus detailed weather forecasts from the Met Office, to predict flooding and issue warnings.

YOUR QUESTIONS

1 Write a report in your own words about the flooding in Gloucestershire. You should include sections on:

- Why the flooding happened
- The impacts of the flooding
- How flooding can be managed

You can use the information here, or do some research using www.gloucestershire.gov.uk Follow the links for flood information and summer 2007 floods.

Your report should include maps and photos, and should be no more than 500 words.

2 Imagine that you are in charge of the Mythe water treatment plant at Tewkesbury. Write a 30-second speech to explain to people why they had no clean running water.

3.6 » Climate change

On this spread you'll learn about climate change – what it is and the evidence for it.

What is climate change?

Climate change isn't something that's going to happen in the future – it's happening now! Disasters, like the severe droughts in Niger in sub-Saharan Africa, in 2005-6 and 2009, are wrecking people's lives more and more frequently. And it's going to get worse. So, what is climate change, and what is the difference between climate change and global warming?

What's causing climate change?

To understand climate change, you need to know about the **greenhouse effect**. This is a completely natural process where gases in the atmosphere trap heat from the sun. The gases act like the glass in a greenhouse – they let heat in but prevent some of it from getting back out. The diagram below shows how it works. Greenhouse gases are essential to keep the Earth warm. Without them, most of the planet would be a frozen wasteland.

But the problem is this. Human activities are releasing more and more greenhouse gases – such as carbon dioxide, methane and nitrous oxide – into the atmosphere. Activities like:

- burning fossil fuels for industry and transport, and to heat our homes
- clearing rainforests (which act as 'carbon sinks' to naturally absorb excess carbon dioxide from the atmosphere)
- farming (particularly cattle farming, which generates methane)

all contribute to the problem by releasing greenhouse gases.

As the greenhouse gases in the atmosphere increase, we are getting an **enhanced greenhouse effect** (the greenhouse effect is working more strongly). This is leading to an increase in average temperatures around the world – global warming. As a result, most people believe that the climate is changing because of human activities.

Climate change: The Earth's climate has always changed naturally over time. One reason for this is thought to be that the Earth's orbit varies around the sun, which has led to ice ages and warmer periods.

Global warming: Now most scientists think that the natural cycles of climate change have been overtaken by rapid global warming – the rise in average temperatures around the world.

▼ *The natural greenhouse effect*

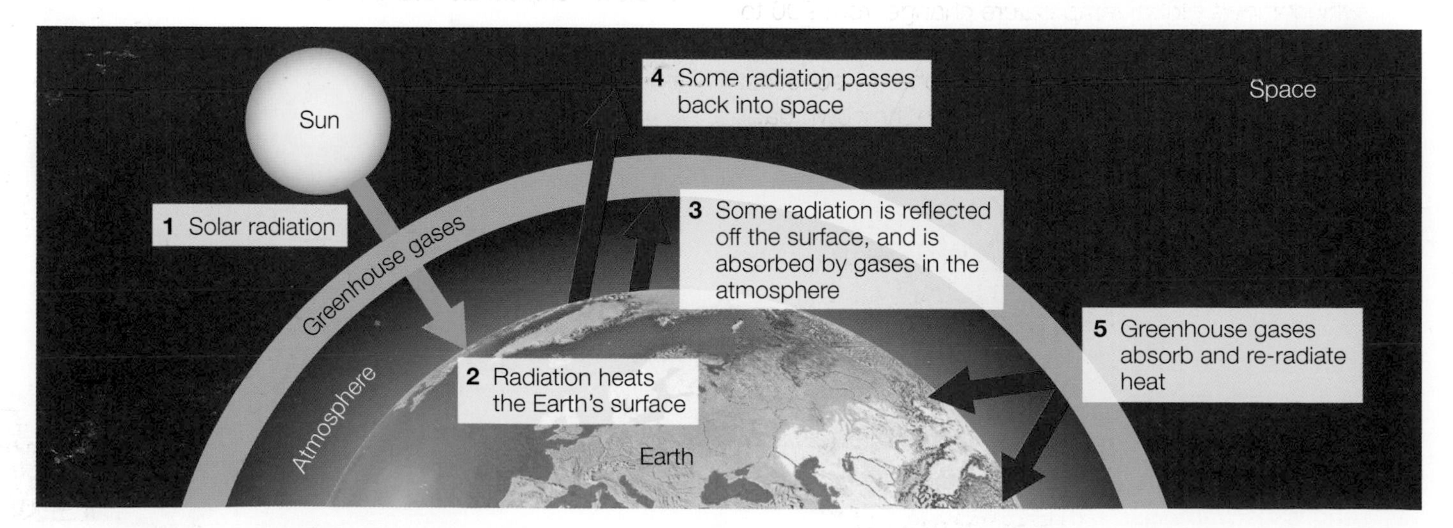

Is there evidence for global warming?

Yes, and plenty of it.

- The average global temperature has generally been on the rise since the early twentieth century (see the first graph).
- Since 1975, the area of Greenland's ice sheets melting every summer has increased by 30%.
- Research has shown that, since 1992, winter temperatures are 5°C higher on Greenland's ice caps, and spring and autumn temperatures are 3°C higher.
- Africa, as a whole, is 0.5°C warmer than it was in 1900. Droughts are becoming more common, rainy seasons are more unreliable, and overall rainfall is decreasing.

The list of evidence that global warming is happening goes on ... but there is some debate about the extent of global warming and its causes. The average global temperature has increased steeply in the last ten years but, if you look over a longer time period, you can see that the average global temperature has risen and fallen. For instance, the rate of increase in temperature between 1910 and 1920 was about the same as the rate of increase between 2000 and today, but there were steep drops in average global temperature in the years 1900-1910 and 1940-1950. However, the overall trend has generally been upwards since 1950 – but at different rates – so the picture is not a straightforward one.

As a result, there are some people who disagree with the theory that human activity is to blame for the rise in global temperature, and who instead put it down to natural processes – as illustrated by the second graph, which shows global temperature change from 900 to about 1900 (including a warm period and mini ice age). These people are known as climate change sceptics. Many of these sceptics don't produce their own data, but simply comment on the data that other people have produced.

your planet

There is enough water locked up in the Greenland ice sheet to raise global sea levels by 7 metres. But the ice will melt slowly, and it could take thousands of years for the sea level to rise to its maximum.

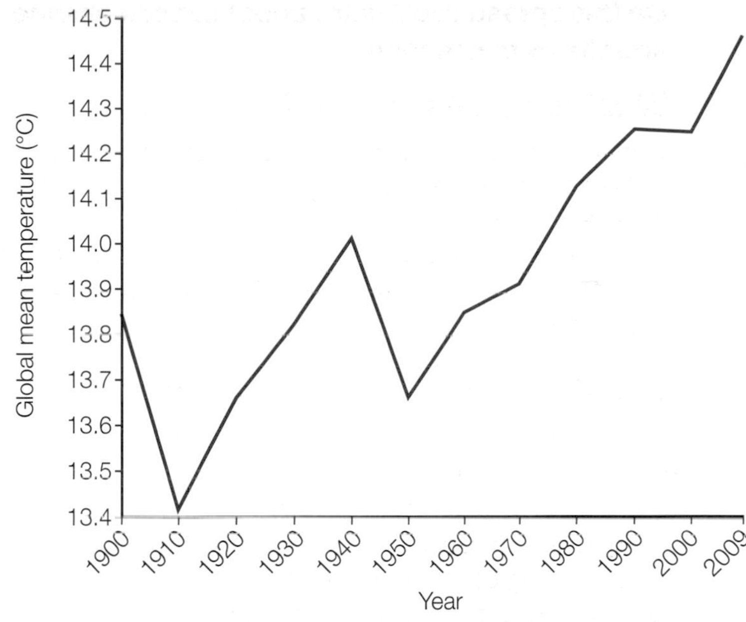

▲ *Changes in average global temperature, 1900-2009*

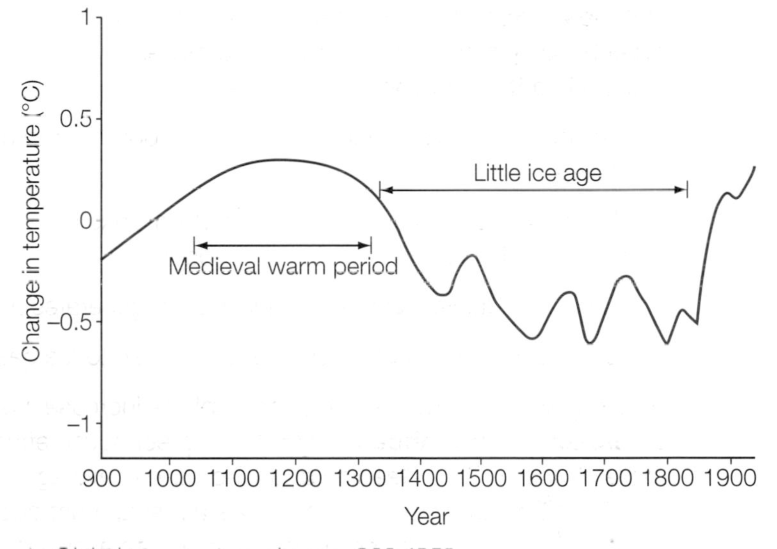

▲ *Global temperature change, 900-1950*

YOUR QUESTIONS

1 Explain the difference between climate change and global warming in your own words.

2 a Draw your own version of the diagram showing the greenhouse effect.

b Add labels to show how the greenhouse effect works.

c Add labels in a different colour to show how the enhanced greenhouse effect works.

On this spread you'll find out about some of the consequences of climate change for the UK and the rest of the world.

Met Office warns of catastrophic global warming in our lifetimes

Unchecked global warming could bring an average temperature rise of 4°C within many people's lifetimes, according to a report prepared by the Met Office. 'We've always talked about these severe impacts only affecting future generations, but people alive today could see a 4°C rise,' said Richard Betts. 'People will say it's an extreme situation, and it is, but it's also possible.'

Adapted from an article in *The Guardian* by David Adams, 28 September 2009.

Consequences for the UK

The headline in the newspaper article above is right – global warming could be catastrophic. But how might it affect the UK?

Temperature	Impacts
Average temperature would rise (see the first map on the right for estimated changes by 2050).	• There would be more heat-related deaths, but fewer cold-related deaths. • People could take more holidays at home – to enjoy the 'Mediterranean' weather.
Rainfall There would probably be drier summers and wetter winters, with rain falling in heavy downpours (and less snow). The two lower maps on the right show estimated changes to precipitation by 2050.	• Without irrigation, some areas might become too dry to grow existing crops. • Crops grown in warmer climates (such as grapes) could become common in southern England.
Rising sea levels Sea levels could rise by 30 cm by 2050.	• Low-lying coasts, particularly in eastern England, could flood. • Some coastlines, such as Holderness, could erode more rapidly. • Trying to hold back rising sea levels with defences is hugely expensive. Without them, even cities like London could be at risk, which would be a disaster for the UK economy as well as for people's homes.
Extreme weather events Would become more common, e.g. • heat waves, like in 2003, when temperatures reached 38°C • flooding, like in summer 2007 • storms like the great gales in 1987 and 1990.	• More heat-related deaths. • Flooding of homes and businesses. • Expensive inland flood defences might have to be built. • Insurance costs would rise.

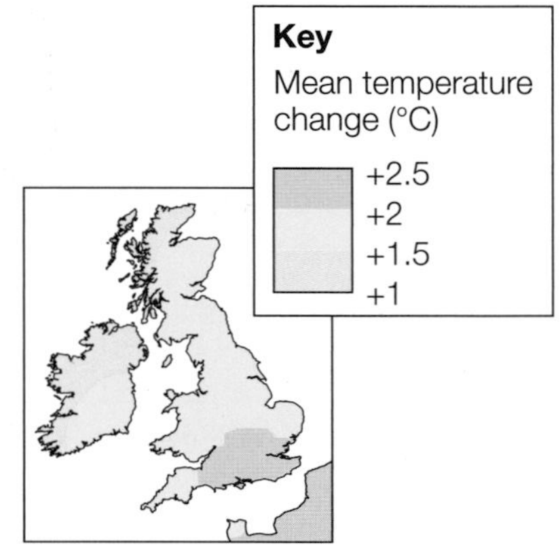

▲ *Estimated temperature changes by 2050*

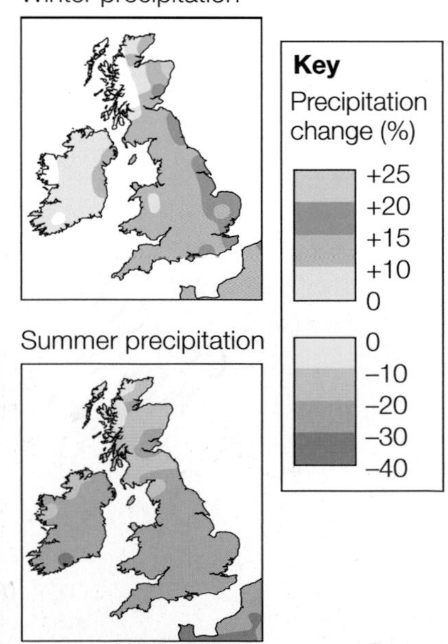

▲ *Estimated changes in rainfall by 2050*

Consequences for the rest of the world

No-one can escape the impacts of climate change. The text boxes below show some of the possible consequences around the world.

Because of melting sea ice, the North-West Passage between the Atlantic and Pacific Oceans could become open to commercial shipping.

Alpine ski resorts might be forced to close because of a lack of snow.

Mediterranean beaches would disappear because of rising sea levels.

Glaciers globally would continue to melt and retreat causing floods for some people and shortages of freshwater for others.

There would be increased drought and desertification, such as in the Sahel region of sub-Saharan Africa.

Malaria would increase as mosquitoes spread over wider areas.

Hurricanes, typhoons and cyclones would probably become more frequent and intense.

Competition for scarce water resources would increase – and could lead to conflicts, such as in the Middle East.

Entire low-lying countries, like the Maldives and many Pacific island nations, might disappear as sea levels rise.

People would be forced to depend on poorer-quality water sources, which would lead to an increase in water-borne diseases like cholera.

Expected changes to rainfall patterns mean that most of Africa would be unable to grow as much food, and there would be more-frequent and intense famines.

Between 20-50% of species in Africa could become extinct, because fragile habitats might not survive.

People displaced by flooding in low-lying countries such as Bangladesh would have to move elsewhere.

Increasing average temperatures are melting the Arctic ice. The amount of permanent sea ice decreased by 14% between 2004 and 2005. That's equivalent to an area three times the size of the UK. Until recently, 80% of the radiation from the sun was reflected back from the polar ice caps. As the ice melts, the area of ocean increases. Oceans are dark and, instead of reflecting radiation, they absorb it and convert it to heat. This speeds up the warming effect.

The melting Arctic sea ice means that polar bears would face extinction, because they would be unable to hunt.

Entire ecosystems, such as the Arctic tundra, could be lost as the planet warms and other plants take over.

YOUR QUESTIONS

1. Write a diary extract for one week twenty-five years from now. Describe how your life has changed as a result of global warming.
2. Draw up a table with two columns headed 'Costs of global warming to the UK' and 'Benefits of global warming to the UK'. Complete your table using the information opposite.
3. Classify the possible consequences of global warming around the world as economic, social, environmental, and political. Which list is the longest? Why do you think this is?

   Hint: 'Economic' means to do with money and economy. 'Social' consequences affect people. 'Environmental' will affect the environment. 'Political' will have impacts for a country and its government.

On this spread you'll explore responses to the threat of climate change at international, national, local and individual levels.

your planet

Carbon credits are measured in tonnes of carbon dioxide. 1 credit = 1 tonne of carbon dioxide.

International and national responses

Pages 62-63 warned us that global warming could be catastrophic, so we need to act before it's too late.

The Stern Review

In 2006, the British government published a review of global warming by Sir Nicholas Stern. It focused on the impacts of global warming and the actions needed to deal with them.

The government's response was to:

- set targets to reduce carbon emissions by 30% by 2020 and 60% by 2050
- invest in green technology, creating 100 000 new jobs
- create a $20 billion World Bank fund to help poorer countries to adapt to climate change.

Carbon credits

This is a system aimed at reducing greenhouse gas emissions. Companies pay for a number of 'credits', which allow them to emit a certain amount of carbon. The idea of this system is that it will encourage companies to produce fewer carbon emissions, or better still, none at all. Carbon credits can be bought or sold if companies produce more, or fewer, emissions than planned.

The Kyoto Protocol and Copenhagen Accord

The Kyoto Protocol was a global agreement made in 1997 to reduce greenhouse gas emissions. World leaders met in Copenhagen in late 2009 to try to agree a new deal on climate change – because the Kyoto Protocol's targets to reduce emissions were due to expire in 2012. The result was the Copenhagen Accord, which:

- recognised the need to limit global temperature rise to no more than 2°C above pre-industrial levels
- promised to give $30 billion of aid to developing countries over the following three years – with a goal of providing $100 billion a year by 2020 – to help them to deal with the impacts of climate change.

However, many people were disappointed that the deal was a not a legally binding agreement, and that it did not include targets to cut greenhouse gas emissions. US President Barack Obama said 'This progress is not enough'.

Tax

Many countries are trying to tackle vehicles' carbon dioxide emissions through tax changes. By 2010-11, around 9.4 million motorists in the UK had to pay more road tax. This was aimed at punishing 'gas-guzzling' and polluting vehicles like the Land Rover on the right. From 2010, the most-polluting cars paid up to £455 a year.

Local responses

One of the ways in which London has responded to the threat of climate change is through targeting traffic – to try to reduce congestion and pollution.

London's congestion charge

Since 2003, drivers have had to pay £8 a day to drive in the Central London Congestion Zone. This money is added to London's transport budget to help improve public transport. As a result, since 2003, every London bus has been renewed, and the old heavily polluting buses have been removed from service.

The scheme has been a success:

- Traffic levels have fallen by 15%, and there is 30% less congestion.
- Average traffic speeds have increased and accidents have fallen by 5%.
- Emissions of nitrous oxide and carbon dioxide have fallen by 12% within the zone.

To support the environmental effects of the congestion charge, the Greater London Low Emission Zone was brought in in 2008. The most-polluting diesel lorries either had to meet emissions standards or pay a daily charge of £100-£200 to enter the area.

Individual responses

How can you help to reduce greenhouse gas emissions and do your bit to help limit the threat of global warming? There are a range of options, from turning off the lights to signing up to initiatives like the 10:10 campaign, which aimed to get individuals, companies and institutions to reduce their carbon emissions by 10% during 2010. So, turning off your TV and DVD players, using low-energy light bulbs, buying less and recycling more, and turning the heating down will all help to make a difference.

▲ *The 10:10 symbol was a metal tag which could be worn on your wrist or around your neck. It was made from scrap metal salvaged from aeroplanes.*

YOUR QUESTIONS

1. Explain in your own words how carbon credits work.
2. List all the actions that you and your family could adopt to reduce your carbon emissions. Rank them from the most effective to the least effective.
3. Discuss in small groups whether US President Barack Obama was right when he said that the Copenhagen Accord did not represent enough progress on climate change.
4. 'Carbon credits just encourage companies to carry on producing greenhouse gas emissions.' How far do you agree with this statement? Explain your views.
5. Work in small groups. Imagine that your group works for Richard Branson and Al Gore. They have offered a £25 million prize for the best idea for dealing with climate change. You will need to review the information on pages 60-65 and decide whether your idea should deal with the consequences of climate change, or whether you will aim to reduce the emissions of greenhouse gases. Once you have decided, you should present your ideas to the rest of the class.

On this spread you'll find out what hurricanes are, how they form and the hazards they bring.

What are hurricanes?

Hurricanes, typhoons, cyclones (they're all names for the same thing) are particularly powerful **tropical revolving storms**. They're intense, destructive, low-pressure weather systems – and one of nature's most lethal weapons. They have very strong winds of over 120 km/h and torrential rain (250 mm can fall in one day). They bring chaos and misery.

Hurricane Katrina, shown in the satellite image, began life as a tropical depression near the Bahamas. As it became more intense, it was upgraded to a tropical storm – and then a hurricane.

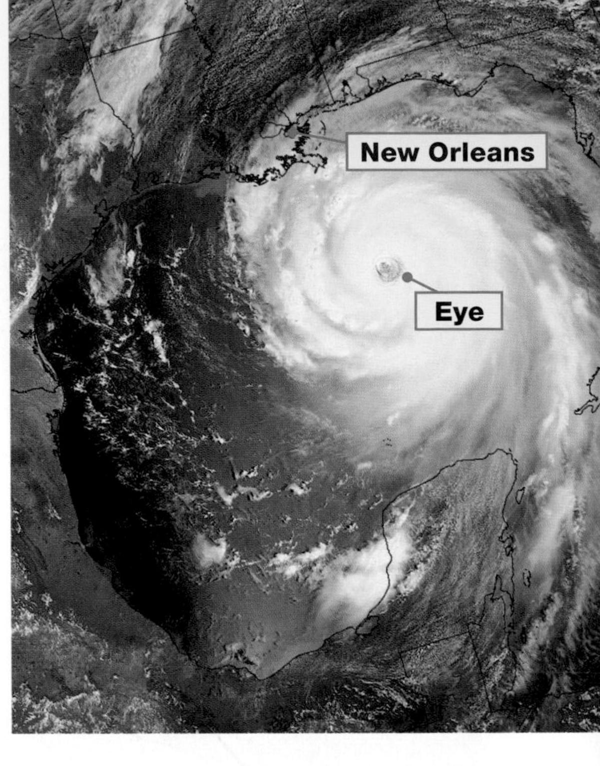

▲ *Hurricane Katrina heading for New Orleans in August 2005*

How do hurricanes develop?

Certain conditions are needed:

- Warm seas – it must be at least 27°C to a depth of 60 metres. This is where the storm gets its energy from.
- Latitude – hurricanes develop between 5° and 15° north and south of the Equator. Trade winds spiral into the storm because of the Earth's rotation. The spiralling is known as the **Coriolis force**.
- Low atmospheric shear – winds have to be constantly blowing from ground level to 12 km above ground level, otherwise the hurricane will shear – that means be pulled apart.

Geographers are not completely sure how hurricanes form, but it seems to be like this:

- A strong upward movement of air draws water vapour up from the ocean.
- As the air rises, it spirals, cools and condenses – releasing huge amounts of heat energy, which powers the storm.
- Colder air sinks down through the centre of the hurricane to form the eye.
- When the hurricane reaches land it rapidly decreases in strength, because the source of heat energy and moisture (the ocean) has disappeared.

your planet

The energy produced by a single hurricane could supply the whole of the USA with all of its electricity for six months.

▼ *A cross-section through a hurricane*

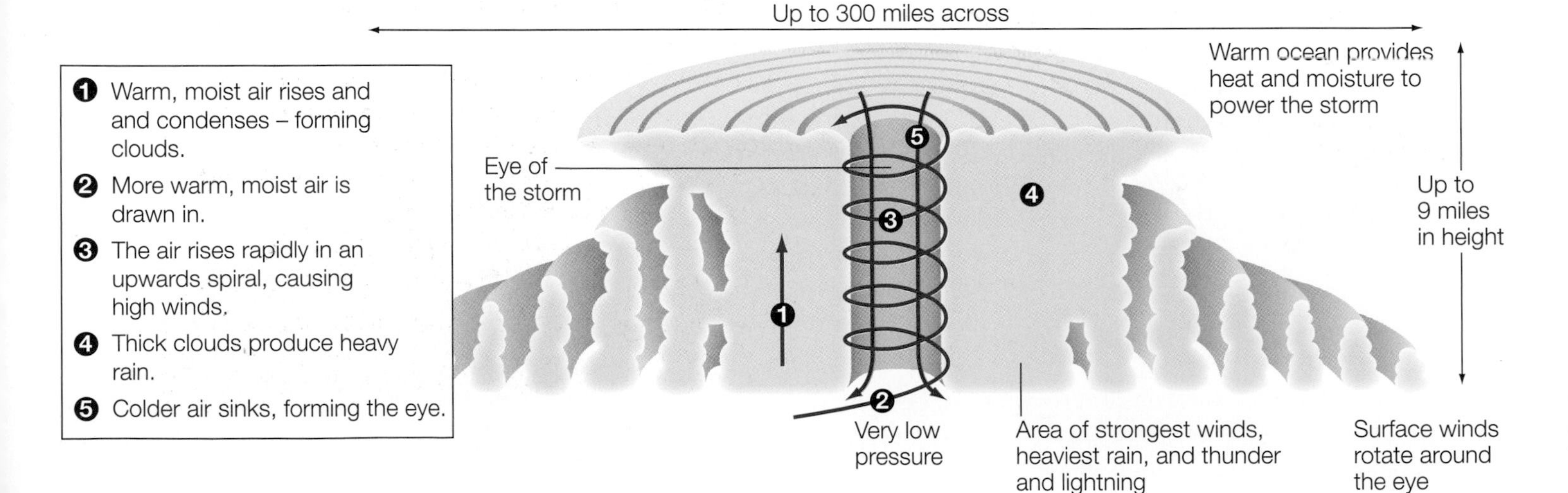

Where do hurricanes occur?

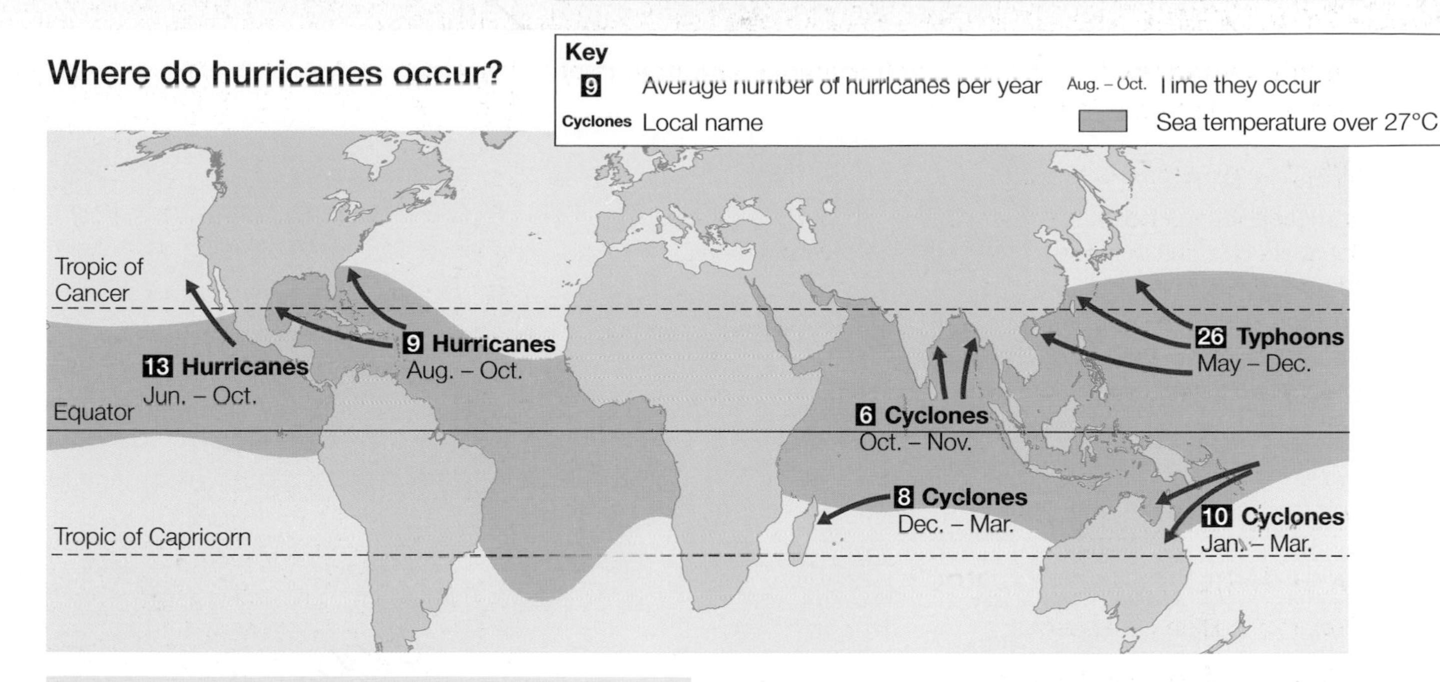

Hurricane facts and hazards

Hurricanes are the world's most violent and destructive storms, and a major climate hazard. They bring with them a range of problems:

- Strong winds (up to 250 km/h), which destroy homes and businesses, and disrupt transport and power.
- **Storm surges**, which are rapid rises in sea level – caused by the low pressure and strong winds. Hurricane Katrina hit New Orleans in 2005 with an 8.5-metre storm surge.
- Torrential rain.
- Flooding and landslides, caused by the torrential rain.

The average lifespan of a hurricane is 7-14 days.

Each hurricane is given a name. A list of names is produced by the National Hurricane Center in the USA – beginning with A and alternating between men and women's names. The list of names for 2009 began with Ana, Bill, Claudette and Danny, and ended with Victor and Wanda.

How are hurricanes measured?

Hurricanes are measured using the **Saffir-Simpson scale**, which is based on the hurricane's strength, or intensity.

Category	Wind speed	Storm surge	Damage
1	119-153 km/h	1-2 metres	minimal
2	154-177 km/h	2-3 metres	moderate
3	178-209 km/h	3-4 metres	extensive
4	210-249 km/h	4-6 metres	extreme
5	>249 km/h	>6 metres	catastrophic

YOUR QUESTIONS

1 Define or explain these key terms from the text: hurricanes, tropical revolving storms, Coriolis force, storm surge, Saffir-Simpson scale.

2 Use the map above to describe the global distribution of hurricanes.

3 Choose one part of the world from the map above (e.g. south east USA, south east Asia, etc.) and investigate a hurricane (or typhoon or cyclone) in that area. Do NOT investigate either Hurricane Katrina (2005) or Cyclone Nargis (2008), because they will both be discussed in detail on the following pages.

a Complete a table which gives the name of the hurricane, its category, rainfall and wind speeds, and the impacts that it had.

b Draw and annotate a diagram to show how the hurricane formed.

On this spread you'll investigate a hurricane in a rich part of the world.

Hurricane Katrina

29 August 2005.
Hurricane Katrina hit the coast of Louisiana in the southern USA at 6.10 in the morning. 1464 people died in Louisiana – but Kioka Williams survived.

'Oh my God, it was hell. We were screaming, hollering, lights were flashing. It was complete chaos' Kioka said. She had to hack through the ceiling of the shop where she worked as floodwaters rose in New Orleans.

Katrina's impacts

Hurricane Katrina's impacts on Louisiana were unbelievable. When it hit the coast it was a Category 3 hurricane. It caused huge storm surges 8.5 metres high, which smashed into the whole Mississippi Gulf coast. The storm surge overwhelmed floodwalls in New Orleans.

Social impacts

- Residents in New Orleans were ordered to evacuate the city, but 20% stayed put because they had no transport or money to enable them to leave.
- One million people were left homeless.
- There was a lack of clean water, food and toilet facilities in the city.
- Looting and disorder became serious problems.

Economic impacts

- Nearly everyone in New Orleans became unemployed.
- The cost of repair and reconstruction in Louisiana and Mississippi ran into billions of dollars.
- The total economic impact of Hurricane Katrina was estimated to be over $150 billion.
- Oil and natural gas production in the Gulf of Mexico was affected, as well as imports of oil and natural gas through the area's ports.

Other impacts

- Communications networks failed – many telephones and mobiles didn't work; Internet access and local TV stations were disrupted.
- 1.7 million people lost electricity.
- Most major roads into, and out of, New Orleans were damaged.
- The levees and floodwalls protecting New Orleans were breached – by 31 August, 80% of New Orleans was under water.
- Buildings suffered extensive damage.

Preparing for hurricanes

People in richer countries, such as the USA, often cope with disasters like hurricanes better than those in poorer countries. This is because they have the money and expertise to plan for, and deal with, these extreme events. The National Hurricane Center in Florida:

- provides hurricane predictions to US states and surrounding countries.
- aims to make people more aware of the risks they face, and actions to take if a hurricane hits. They run a Hurricane Preparedness Week every year. In 2010, it ran from 23-29 May (see below).
- offers advice to families, and provides Family Disaster Plans.

National Hurricane Preparedness Week

May 23-29, 2010

HISTORY	HURRICANE HAZARDS			FORECAST	PREPARE	ACT
S	M	T	W	Th	F	S
MAY 23	MAY 24	MAY 25	MAY 26	MAY 27	MAY 28	MAY 29
Hurricane Basics **Hurricane History**	**Storm Surge** **Marine Safety**	**High Winds** **Tornadoes**	**Inland Flooding**	**Forecast Process**	**Disaster Prevention**	**National Day of Family Preparedness**

The job of the Federal Emergency Management Agency (FEMA) in the USA is to reduce loss of life and property. It protects people and places by making sure that areas are prepared for disasters, as well as giving emergency help and aiding recovery.

Protecting New Orleans

The flood defences in New Orleans were left in tatters following Hurricane Katrina's onslaught. However, the levees and floodwalls have been repaired and strengthened, floodgates have been built and pumping stations improved. The rebuilding and strengthening work was expected to continue until 2013.

However, Louisiana has other problems as well as its built defences. Its wetlands are rapidly disappearing. Why does this matter? Well, wetlands provide barriers which absorb the energy of storm surges and protect inland areas. A plan called *Coast 2050* aims to recreate the mixture of swamp, marsh and barrier islands that used to exist in Louisiana, in order to help protect places like New Orleans in the future.

YOUR QUESTIONS

1. Draw a spider diagram to show how the USA prepares for hurricanes.
2. Write a radio broadcast about Hurricane Katrina. You need to tell your listeners when the hurricane hit Louisiana, what happened, and how New Orleans will be protected in future.
3. Look back at pages 66-67. Use the information on 'Hurricane facts and hazards' to explain Hurricane Katrina's impacts.

On this spread you'll find out about a cyclone (hurricane) in a poorer part of the world.

Cyclone Nargis

In early May 2008, Myanmar (also called Burma) was hit by a massive hurricane, or cyclone – Cyclone Nargis. It was the eighth deadliest cyclone of all time – bringing high winds, heavy rain and storm surges. And it killed 140 000 people.

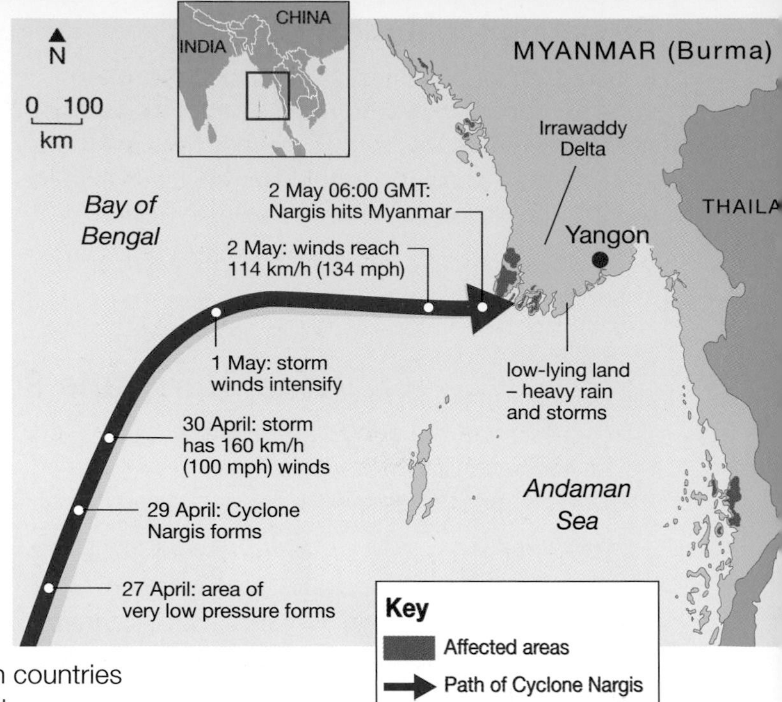

Why was Cyclone Nargis so deadly?

- Earlier environmental damage in Myanmar increased the effects of the cyclone. The country had previously destroyed over 80% of its coastal mangrove swamps. Like Louisiana's wetlands (see page 69), these mangroves would have provided a natural defence against storm surges. Destroying them left towns and villages unprotected.
- The government in Myanmar was suspicious of foreign countries offering aid, and didn't welcome or encourage their help.

Primary (immediate) impacts

- 140 000 people were killed, including 80 000 in the Irrawaddy Delta's Labbutta Township.
- 2.4 million people were affected, and over 2 million were left homeless.
- 95% of the buildings in the Irrawaddy Delta were destroyed.
- Transport links were swept away.
- Power lines were blown down.

Secondary (longer-term) impacts

- Families were left without clean water or electricity.
- Dirty water led to mosquitoes breeding, which helped to cause disease.
- The cost of rice increased by 50%, causing much hunger.
- Raw sewage leaked onto rice paddies, causing disease.
- There was difficulty travelling around the country.
- The damage to farmland meant that crops failed.
- Hundreds of thousands were forced to live on the streets without shelter (see the photo opposite).
- Many people were emotionally devastated at losing family members.
- US$10 billion was needed for rebuilding.

▼ *Cyclone Nargis destroyed this village*

Responses to Cyclone Nargis

In a country as poor as Myanmar, the impacts of a major natural disaster are always going to be severe. Preparing in advance for disasters is expensive, and – in a poor country – is bound to be limited. A country like Myanmar can't prepare in the same way as the USA (page 69). Rescuing people and rebuilding their lives is extremely difficult.

- Within hours of Cyclone Nargis hitting Myanmar, charities around the world launched appeals to help the people affected. But the government of Myanmar was very slow to accept the help offered. Ten days after the cyclone, aid had still only been given to a third of those in need.
- UN Secretary-General Ban Ki-moon called on Myanmar to let in more help. Journalists and TV crews were banned, but some smuggled cameras into the country to show what conditions were like after the cyclone.

▲ *Cold, wet and homeless after the cyclone hit*

your planet
The worst cyclone in history was the Bhola Cyclone in 1970, which killed 500 000 people.

Did the government of Myanmar make things worse?

Government action

- Journalists were banned from filming the destruction.
- Aid wasn't distributed effectively.
- Offers of help were refused at first.
- Mangrove swamps were destroyed.
- The army harassed volunteers.
- Leaflets suggesting many didn't need help were issued.
- Some aid workers weren't allowed into Myanmar.

YOUR QUESTIONS

1. Look at the photo on the opposite page. List as many words as you can to describe the situation in the village.
2. **a** List four primary impacts of Cyclone Nargis.
   b For each one, give a secondary (or longer-term) impact that would be likely to affect people later. Explain your thinking.
3. Write a letter to the government of Myanmar giving your opinions about its response to Cyclone Nargis.
4. **a** Draw up a table like the one on the right to compare Hurricane Katrina (on Spread 3.10) with Cyclone Nargis.
   b Write a paragraph which explains why so many people died in Myanmar, compared with the USA.

	Hurricane Katrina	Cyclone Nargis
Number killed		
Number made homeless		
Advance preparation for hurricanes/cyclones (in USA/Myanmar)		
Response to hurricane/cyclone (in USA/Myanmar)		

4 » Living world

What do you have to know?

This chapter is from **Unit 1 Physical geography Section A** of the AQA A GCSE specification. It is about different ecosystems, where you find them and what they're like, and how they are used and managed. The table shows how the pages in this chapter match the content in the specification.

Specification content	Pages in this chapter
What an ecosystem is, and how it works. How changing one part of an ecosystem affects the other parts.	p74-75
The global distribution of three ecosystems, and their characteristics. How vegetation adapts to climate and soils.	p76-79
How temperate deciduous woodlands are used and managed in a sustainable way.	Case study of Epping Forest p80-83
The causes of deforestation in tropical rainforests, and the impacts of deforestation.	Case study of the Atlantic Forest p84-85, p86
The sustainable management of tropical rainforests, and international cooperation.	p87 Case study of the Atlantic Forest p88-89
Using and managing hot deserts sustainably in different parts of the world.	Case study of the Australian Outback p90-91 Case study of the Sahara Desert p92-93

Your key words

- Ecosystem, biomes
- Environment
- Producers, consumers, decomposers
- Food chain, food web
- Continentality
- Leaching
- Pollarding
- Sustainable management
- Site of Special Scientific Interest (SSSI)
- Special Conservation Area (SCA)
- Endemic
- Carbon sink
- Slash and burn farming
- Conservation swaps (or debt-for-nature swaps)
- Ecotourism
- Hunting and gathering
- Retirement migration
- Irrigation
- Desertification

Exam help ...

Advice See pages 297-299 for information on how to be successful in your exams.

Practice See page 303 for exam questions on this chapter.

What if ...

- meerkats were introduced to the UK?
- all the world's forests were cut down?
- the world turned to desert?

4.1 » Introducing ecosystems

On this spread you'll learn what an ecosystem is, and how the different parts of an ecosystem depend on each other.

What is an ecosystem?

Look in any pond and you'll find a whole living world – a miniature **ecosystem**. Ecosystems are units made up of:

- living things (plants, animals, bacteria) and
- their non-living surroundings, or **environment** – the physical things that affect them, like the climate and the soil.

In any ecosystem, the living things interact with the environment and each other. For example, caterpillars in a wood breathe the air, feed on leaves, and get eaten by birds. If it gets too cold, they die.

An ecosystem can be big or small – ranging in size from the world in that pond, to a hedgerow, a wood, a tropical rainforest, or even the whole Earth.

A small woodland ecosystem

The photo below shows a small oak woodland ecosystem in the UK.

How do ecosystems work?

Every ecosystem works in the same way:

- The plants use sunlight, water, and nutrients from the soil to produce their own food (so they're called **producers**).
- The animals feed on the plants, or each other (so they're called **consumers**).
- Fungi and bacteria feed on dead and waste material, and make things break down or rot (so they're called **decomposers**) – they recycle nutrients for the plants to use again.
- Without plants, all other living things would die.

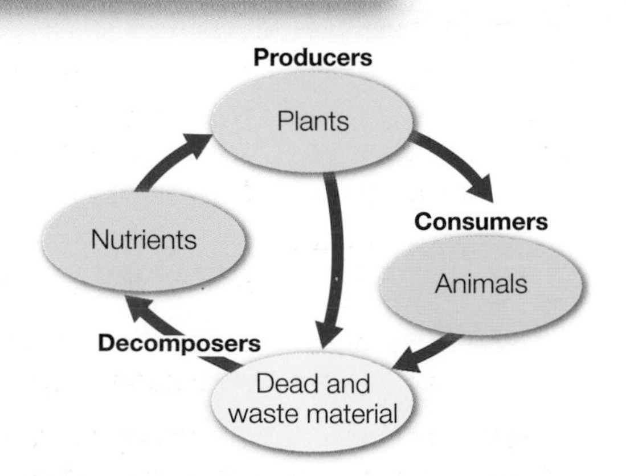

▲ *How living things are linked in an ecosystem*

Food chains and food webs

In any ecosystem, animals need to eat to survive – and whatever is eaten is part of the **food chain**. In the oak woodland opposite, it works like this:

oak leaf → caterpillar → wood mouse → fox

(The arrow means: 'it gets eaten by'.)

Often several consumers eat the same type of food. So, for example, caterpillars and aphids (a type of fly) both feed on oak leaves. Individual food chains then link up to form a food web. The diagram on the right shows part of the **food web** for an oak wood.

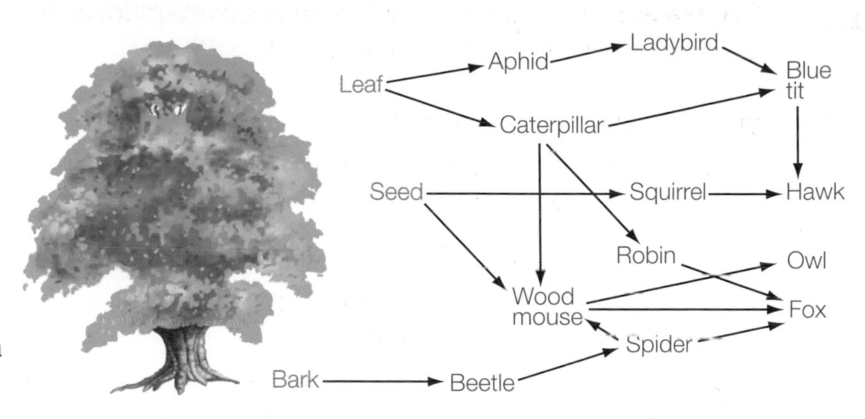

▲ *Part of a food web (complete food webs are large and complicated)*

Ecosystem processes

Ecosystems depend on two basic processes – recycling nutrients and energy flows.

Recycling nutrients

Nutrients continually circulate within ecosystems, as the diagram on the right shows.

Energy flows

Ecosystems work because there's a flow of energy through them. The main source of energy is sunlight, which is absorbed by plants and then converted by photosynthesis. Energy passes through the ecosystem in the food chain. Each link in the chain feeds on – and gets its energy from – the link before it.

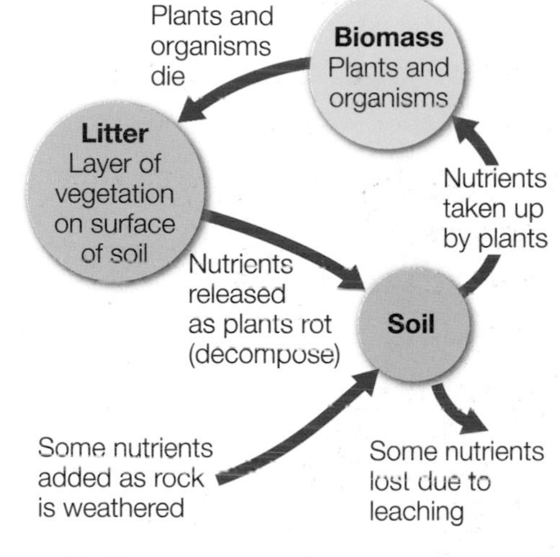

▲ *Nutrient recycling*

▼ *Energy flows*

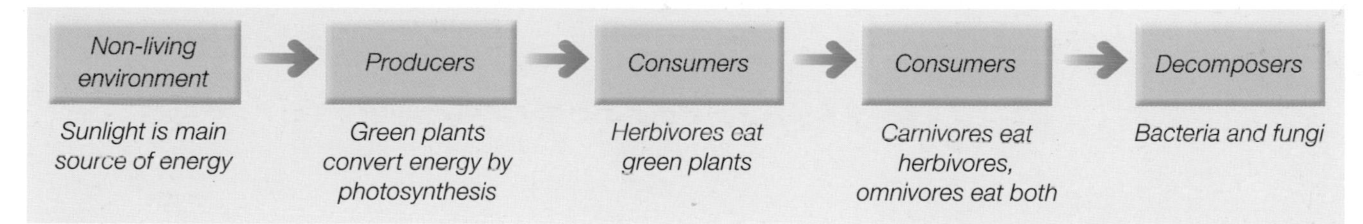

Changes to ecosystems

Different parts of an ecosystem depend on each other, and there's a balance between them. A change in one part of an ecosystem will affect other parts and upset the balance. Ecosystems around the world are facing changes, or threats, including:

- **climate change** – which can affect where species can live, when they reproduce, and the size of their populations
- **habitat change** – the conversion of land for farming can lead to a loss of habitat for huge numbers of species
- **pollution** – which, for example, can reduce oxygen levels in wetlands and rivers (killing fish), and also cause rapid plant and algal growth (called algal blooms).

YOUR QUESTIONS

1. Start a dictionary of key terms for this chapter. Begin with ecosystem, environment, producer, consumer, decomposer, food chain, food web.
2. Look at the food web above and identify two food chains. Write them out.
3. What are the two basic processes that ecosystems depend on? In pairs, describe one each.
4. Explain how changing land use can affect an ecosystem.

4.2 » The global distribution of ecosystems

On this spread you'll find out about the distribution of the world's main ecosystems, and learn why they're so different.

The big ones

The world can be divided up into eight big ecosystems, or **biomes**. Each one has its own type of vegetation. The locations and characteristics of each biome are mainly determined by climate. This is because climate affects the growth conditions for vegetation. It does this through:

- temperature – especially the seasonal pattern and the length of the growing season
- precipitation or rainfall – particularly the total amount and how it's distributed throughout the year
- the number of sunshine hours – which determines the amount of light available for photosynthesis
- rates of evaporation, transpiration, and humidity.

The map shows the distribution of the world's main ecosystems/biomes. The rest of this chapter will look at three in more detail: temperate deciduous forests, tropical rainforests and hot deserts.

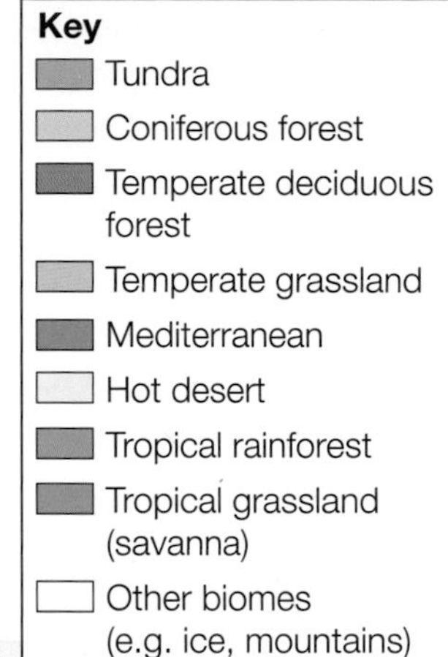

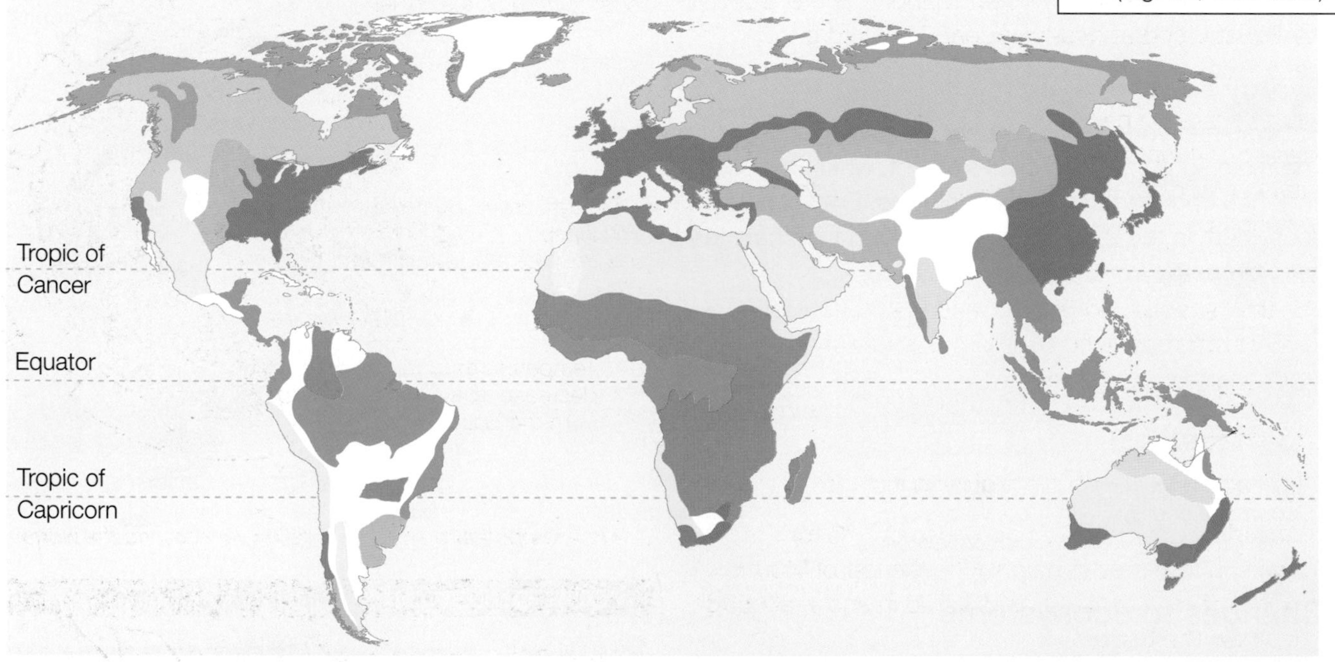

The map above makes it look as if there's a clear cut off between one ecosystem and another. But it's not quite like that in real life. Ecosystems change gradually as you move away from the tropical regions towards the North and South Poles, as the diagram on the right shows.

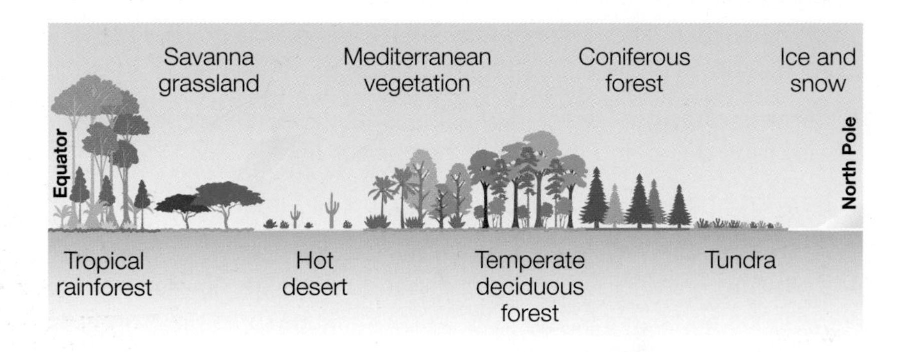

Explaining the global distribution of ecosystems

Temperature

Average temperature is the main factor affecting plant growth. Temperature gradually decreases as you move away from the Equator.

In the Tropics, the sun's rays are at a high angle in the sky for the whole year – and are concentrated over a smaller area than at the Poles. They provide a lot of heat and sunlight, so plants grow well here and vegetation is dense.

In Polar areas, the sun's rays are less concentrated. The lack of heat and light limits vegetation growth, so plants are stunted and low growing (see right).

Precipitation

Around the world, precipitation is more likely in some places than in others. Precipitation happens when air masses meet and the air rises in low **pressure belts**. These belts give rainfall all year round, and are found at the Equator and at mid-latitudes. Forests grow in both these areas.

In high-pressure belts, the air is descending and you get dry conditions – creating deserts. The North and South Poles are high-pressure areas with dry conditions.

The whole pattern of pressure belts changes with the seasons. So Mediterranean and tropical areas sometimes become low-pressure zones and experience rainy seasons.

Local factors

Other factors affecting plant growth include:

- altitude – temperatures decrease by 1°C for every 100 metres in height. So, the top of Mount Kilimanjaro in Tanzania is covered in snow, although it's near the Equator.
- **continentality** – this is the term for distance from the sea. Away from the sea, the land heats up in the summer and cools quickly in the winter. This increases the annual temperature range and reduces precipitation.
- nutrient-rich environments – environments rich in nutrients encourage plant growth. Nutrients are supplied by the soil or ocean currents.
- geology (rock type), soils, relief, and drainage of water.

▲ *As you head towards the Poles, the tundra vegetation becomes more low growing, and plants have a short lifecycle adapted to the limited growing season*

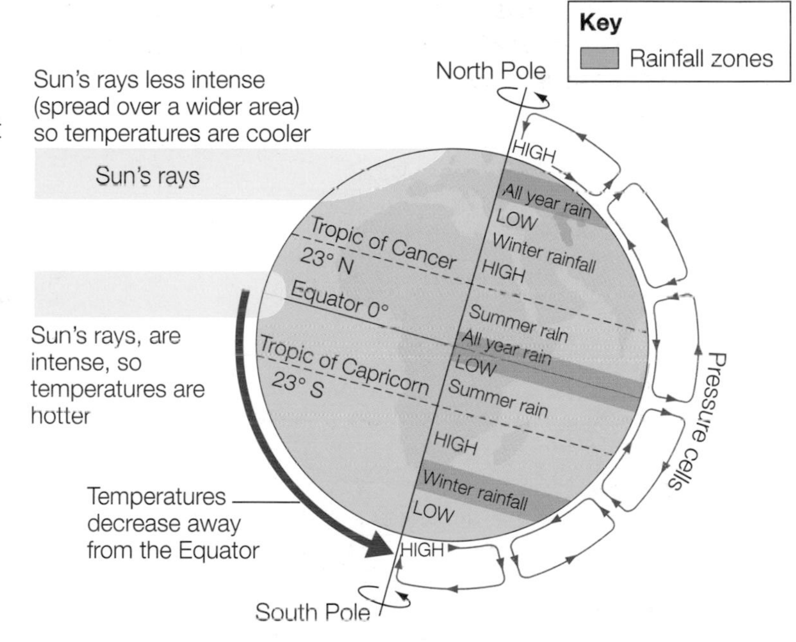

▲ *How temperature and precipitation vary around the world*

YOUR QUESTIONS

1. Add these terms along with a definition to your dictionary of key terms for this chapter: biomes, pressure belts and continentality.
2. List the ways in which climate affects plant growth.
3. Choose either temperate deciduous forests, tropical rainforests, or hot deserts. Use the map opposite to describe the global distribution of your chosen ecosystem.
4. Draw two large circles to represent the Earth.
   a. Annotate one to explain how temperature affects the distribution of ecosystems.
   b. Annotate the second to explain how precipitation affects the distribution of ecosystems.

On this spread you'll learn about the climate, soils and vegetation of three different ecosystems.

Temperate deciduous forests

Climate

Temperate deciduous forests are mainly found in Western Europe, including the UK, and the eastern parts of North America and Asia. The climate in these areas isn't extreme – summers are warm and winters are cool.

- The annual temperature range in these areas is low.
- Precipitation can occur throughout the year.
- There is a long growing season.

Soils

- The most common type of soil is known as brown earth, which is reddish-brown in colour.
- Deciduous trees lose their leaves every year. The fallen leaves then slowly rot – helping to keep the soil fertile.
- Minerals are slowly washed (or **leached**) through the soil.

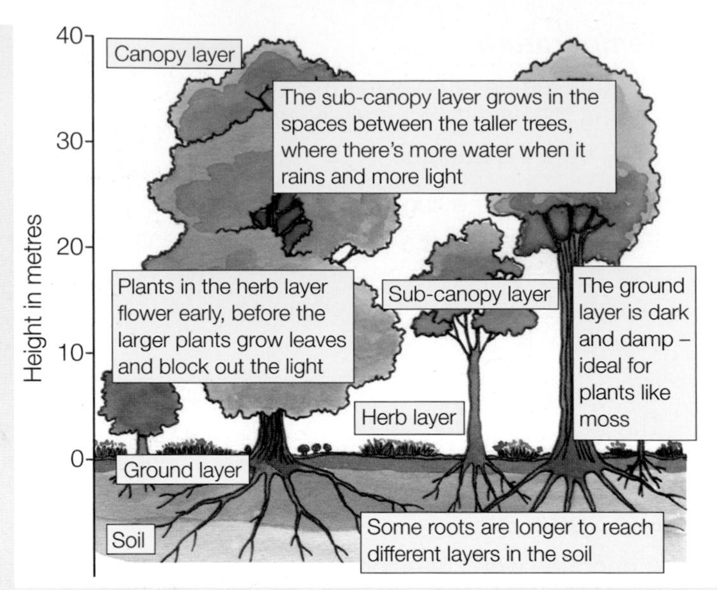

Vegetation

Deciduous trees lose their leaves in the winter, when the light level and temperature falls. The vegetation in temperate deciduous forests grows in layers (see the diagram). As you can see, the plants are adapted to the climate and soil in other ways too.

Tropical rainforests

Climate

- The average daily temperature is about 28°C. It never goes below 20°C, and rarely above 35°C.
- At least 2000 mm of rain falls a year.
- The atmosphere is sticky – it's hot and humid.
- There are no real seasons. Each day's weather is the same – starting off hot and dry, with thunderstorms and heavy rain in the early evening.

Soils

- Soils are red in colour and rich in iron.
- They have a thick layer of litter (dead leaves, etc.), but only a thin fertile layer – because the leaves rot quickly in the humid conditions.
- The soils in tropical rainforests are not very fertile, and not particularly good for plants to grow in, despite appearances.
- Nutrients are quickly washed out of the soil because of the heavy rainfall.

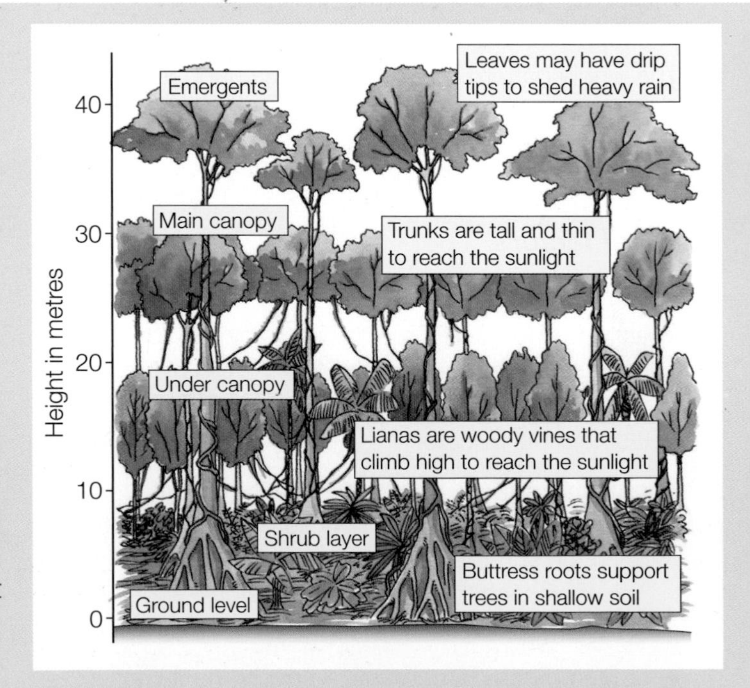

Vegetation

The vegetation in rainforests grows in distinct layers (see the diagram), and has adapted to the climate and poor soils.

Hot deserts

Climate

- Hot! Daytime temperatures can get up to over 40°C in the summer. But it can get very cold at night (below freezing), because there's no cloud to keep the heat in. So, there's a big daily temperature range.
- And dry! With less than 250 mm of rain a year. It might not rain for months or years – and when it does, it can come in one torrential downpour.
- There are two seasons – summer, when the sun is high in the sky and it's very hot, and winter when, although it's very warm compared to the UK, it's cooler than the desert summer.

Soils

- Desert soils are rocky, sandy and grey in colour.
- They are thin, and can have a crust caused by the impact of the infrequent heavy rainfall.
- Evaporation draws water up through the soil, leaving salts deposited near the surface.
- The soil is dry, but it can soak up water quickly when it rains.

Vegetation

The vegetation has adapted to survive in the harsh desert climate.

YOUR QUESTIONS

1 Explain the term leaching and add it to your dictionary of key terms.

2 Look at the climate graphs below.

- **a** Match the three ecosystems on this spread to graphs **A**, **B** and **C**.
- **b** Design a table to compare the climates, using the three graphs. Include maximum/minimum temperatures and total annual rainfall.

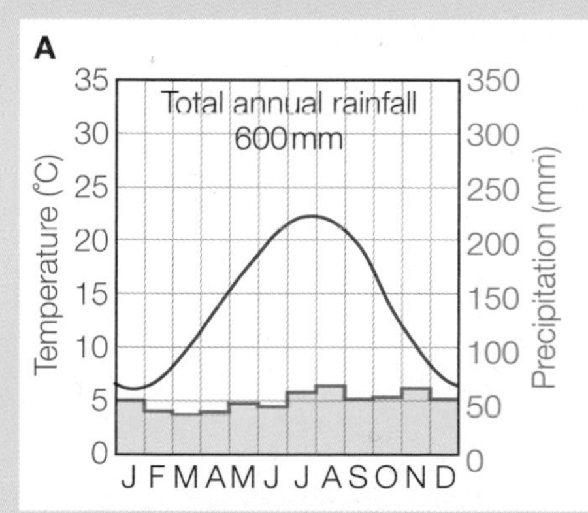

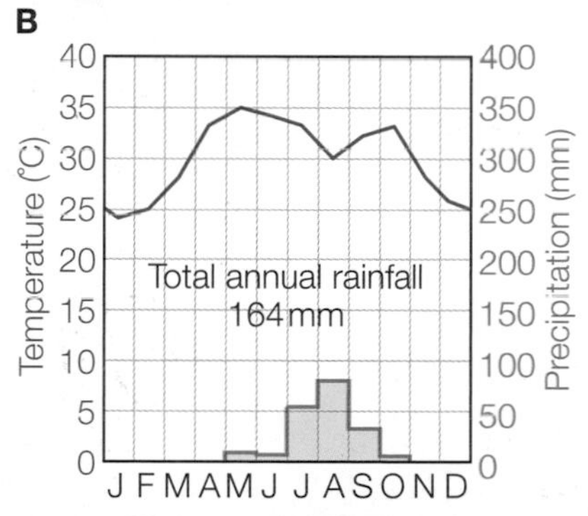

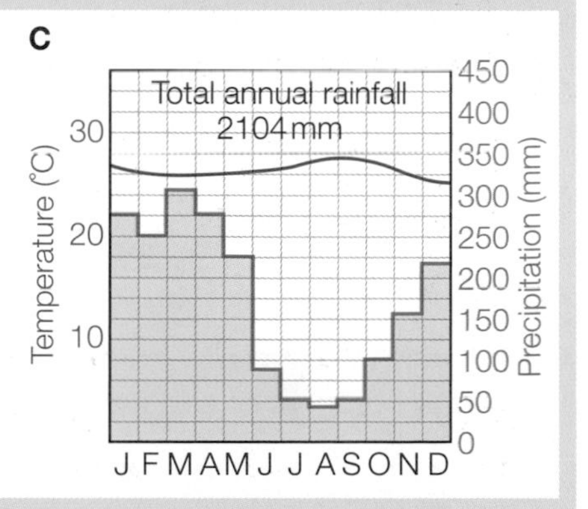

3 Work in small groups. Choose one of the ecosystems on this spread. Produce ***either*** a wall chart ***or*** a PowerPoint presentation which:

- summarises the climate (use the information in the text boxes and your answers to question 2 above)
- describes the soils
- describes how the vegetation has adapted to the climate and soils.

4.4 » Epping Forest

On this spread you'll explore Epping Forest, a temperate deciduous forest in Essex, and find out what it's like and how it's used.

A beautiful forest

In May 1882, a huge crowd flocked to see Queen Victoria visit Epping Forest. She declared: 'It gives me the greatest satisfaction to dedicate this beautiful forest to the use and enjoyment of my people for all time'.

Queen Victoria was right – Epping Forest ***is*** beautiful. And it's Greater London's largest public open space. It stretches for over 12 miles from East London to just north of Epping in Essex. And it's almost 6000 acres in size – that's big.

Saving Epping Forest

In the second half of the nineteenth century, Londoners started to enclose parts of Epping Forest for development. The rights of local people to graze animals and use the trees there were being ignored. In 1878, an Act of Parliament was passed, called the Epping Forest Act. This gave the City of London ownership and care of Epping Forest – and it's still responsible for it today. Epping Forest now combines the roles of scenic open space, important wildlife habitat, and recreational resource.

▲ *Woodland in Epping Forest*

A temperate deciduous woodland

Epping Forest is mostly ancient temperate deciduous woodland, but it also includes areas of grassland and wetlands. The woodland mostly consists of beech trees, but it also includes oak and hornbeam. The soils in the forest vary, depending on the rock below them. They include clay, gravel soils and loam soils. The vegetation that grows on the surface depends on the soil below.

Epping Forest supports a variety of wildlife:

- The trees provide nesting sites for birds like tree creepers and nuthatches.
- Dead wood left to rot on the ground provides food for wood-boring beetles (including some rare and endangered species).
- The many bogs, pools and ponds provide homes for large numbers of wild fowl, aquatic plants and dragonflies.
- Grey squirrels, rabbits, muntjac deer and fallow deer also live there.

▲ *At least 48 species of bird are found in Epping Forest, including nightingales, woodpeckers and sparrowhawks, and the nuthatch shown here*

your planet

Why do forests matter? They're crucial for maintaining biodiversity. There are around 500 rare and endangered insect species in Epping Forest alone. In the last century, 46 broadleaved woodland species have become extinct in the UK.

Recreation in Epping Forest

Millions of people visit Epping Forest every year. With its woods, grassy plains and attractive ponds and lakes, there's something for everyone. There are many footpaths for walkers (including easy access paths for people with limited mobility), 50 km of rides for horse riders, and plenty of space for cyclists. And that's not all ...

- There are over 60 football pitches for hire on Wanstead Flats.
- There's also an 18-hole golf course at Chingford.
- Refreshments are available throughout the forest, ranging from tea stalls and ice-cream stands ... to pubs and cafes.

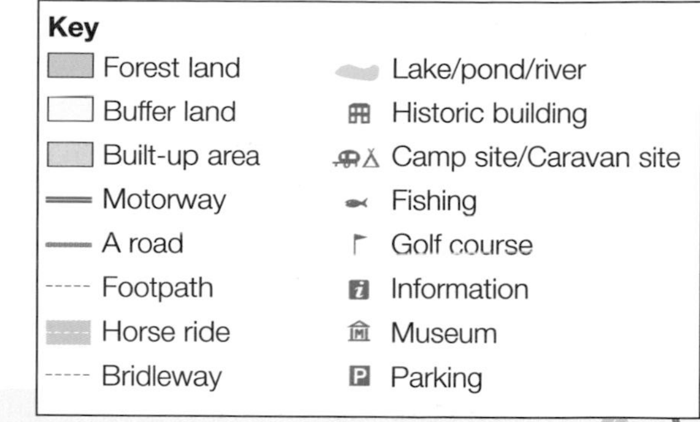

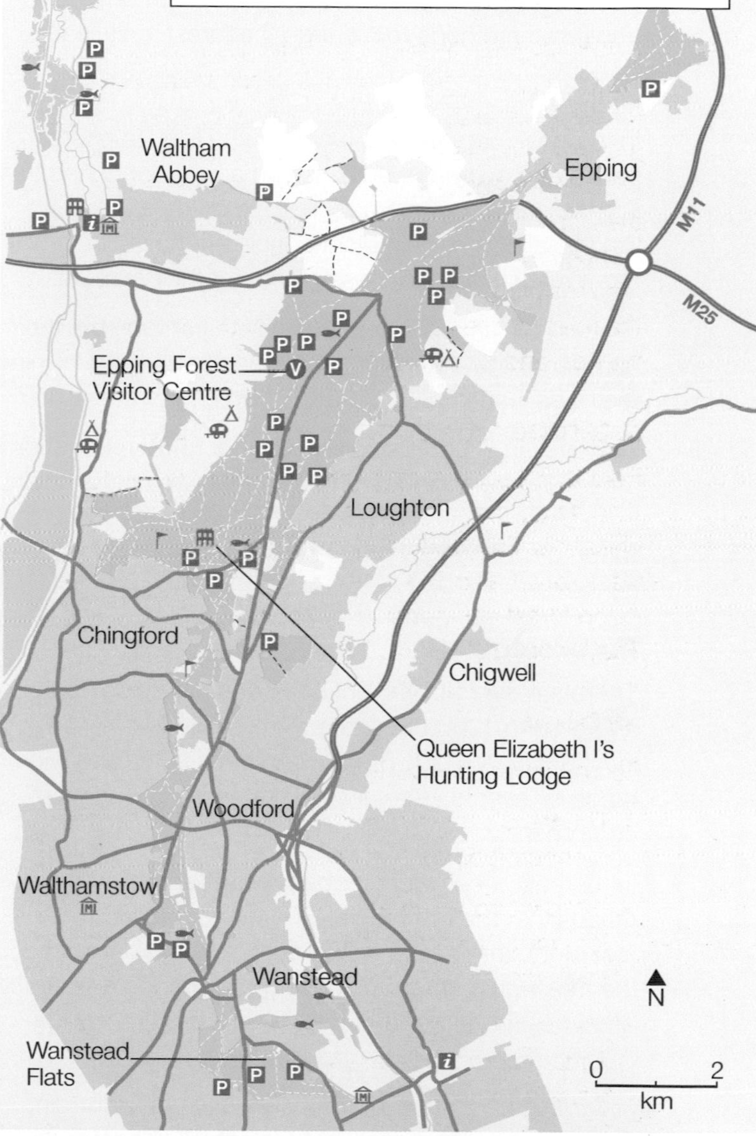

▲ *A day out in Epping Forest*

▲ *Epping Forest was once a royal hunting ground, and you can still visit Queen Elizabeth I's Hunting Lodge. It was originally built by Henry VIII in 1543 – to allow his guests to see him hunting, and to shoot their crossbows at the deer from the upper floors. It also helped Henry to show off his power and wealth.*

YOUR QUESTIONS

1. Why was the Epping Forest Act of Parliament passed?
2. Describe the different roles, or uses, of Epping Forest today.
3. Design a leaflet to attract people to Epping Forest. Tell them where it is, what it's like, and what they can see and do there.

4.5 » Managing temperate deciduous forests

On this spread you'll find out how Epping Forest is managed, and about The National Forest.

Conservation

Since 1878, the City of London has been in charge of managing Epping Forest, to 'conserve and protect it as an open space for the recreation and enjoyment of the public'.

Epping Forest's natural habitat has developed over a thousand years of use by people and their animals, and the City of London has reintroduced many traditional methods to manage the forest sustainably.

Grazing

In the past, commoners (people who lived in a forest parish and owned at least half an acre of land) had the right to graze their animals in Epping Forest. Grazing like this allows certain plant species to thrive. In turn, many insects rely on those plants for their survival.

Grazing continued into the twentieth century, but the number of cows declined because of changing farming methods – and BSE in 1996 put an end to the tradition.

However, in 2002, a small number of English Longhorn cattle were reintroduced onto a small area of heathland in the forest. The herd has now grown to 50 cows, which graze in small groups looked after by a herdsman.

Pollarding

There are over 50 000 ancient trees in Epping Forest, and around 1200 of them are 'keystone trees' – all at least 300 years old. The ancient trees support a wide range of insects, other invertebrates, fungi, mosses and lichens.

In the past, trees were lopped or **pollarded** – this means that the top branches were cut off, leaving the rest of the tree intact. New shoots would appear and the tree would carry on growing. Commoners used the branches for timber and fuel.

Now, In order to preserve the trees, traditional pollarding is being adopted again. Without pollarding, the crowns of the trees would eventually become too heavy and they would topple over or split – and die.

Sustainable management means managing in a way that meets people's needs – both now and in the future – and also limits harm to the environment.

▲ *English Longhorn cattle are a rare breed. They are known for being docile (despite the big horns!) and for being able to thrive on rough grazing.*

▲ *Pollarding helps to extend the life of the trees and provides nesting sites for birds*

Protection

More than two-thirds of Epping Forest has been designated a **Site of Special Scientific Interest** (SSSI) and a **Special Conservation Area** (SCA).

- SSSIs are places that are important because of their plants and animals, or the geology and geography of the area. So they need to be protected.
- SCAs are areas that have been given special protection by the EU. They provide increased protection for wild animals, plants and habitats – and are an important part of global efforts to conserve the world's species.

The National Forest

The National Forest was created in 1990 and covers parts of Leicestershire, Derbyshire and Staffordshire. It is a new forest and over 7 million trees have been planted there.

One of the main purposes of The National Forest is to supply wood for the UK's needs. Around 85% of the UK's wood is currently imported, which costs about £8 billion a year. But the amount of wood produced in British forests is now increasing – from 4 million m^3 in the 1970s to nearly 9 million m^3 now.

Using The National Forest to supply wood products helps to develop a forest economy in the area. Woodland 'thinnings' are used to produce woodchips, and more mature trees are cut down for timber.

Forest businesses plant and maintain the forest – coppicing, extracting and processing the timber to make and sell wood products. By adding value to the wood and the waste, they ensure that the forest is managed sustainably for the future.

▲ *Logging in the National Forest*

YOUR QUESTIONS

1 Add these terms, along with a definition, to your dictionary of key terms for this chapter: sustainable management, pollarding, Site of Special Scientific Interest, Special Conservation Area.

2 Draw a simple pollarded tree. Annotate your drawing to show how pollarding is done and the benefits it has.

3 Choose ***either*** Epping Forest ***or*** The National Forest.

a Describe how the forest is managed.

b Do you think that the Forest is being managed sustainably? Explain your answer.

Hint: This is an either/or question. You don't need to write about both forests. Just choose one!

4.6 » The Atlantic Forest

On this spread you'll learn about deforestation in the Atlantic Forest, a tropical rainforest.

The Atlantic Forest

The golden-headed lion tamarin lives in one place and one place only – the Atlantic Forest of South America. It lives nowhere else on Earth. Not only that, but it's an endangered species, because its rainforest habitat has declined dramatically over recent decades.

The Atlantic Forest is an area with a mind-boggling number of species, and many – like the golden-headed lion tamarin – are **endemic** (they only live there). The table below shows the number of species found in the Atlantic Forest – and the percentage of each type that are endemic.

▲ *The endangered golden-headed lion tamarin*

	Number of species	Percentage of endemic species
Plants	20 000	40%
Mammals	264	27%
Birds	934	15%
Reptiles	311	30%
Amphibians	456	62%
Freshwater fish	350	38%

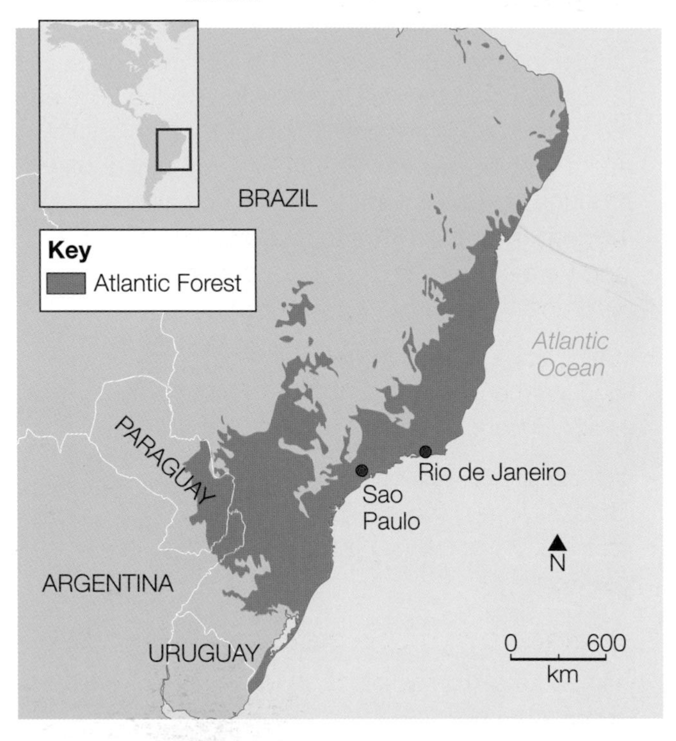

▼ *The Atlantic Forest. The vegetation looks dense and luxuriant, but appearances can be deceptive. This is a fragile and threatened ecosystem.*

Under threat

You've probably read about deforestation in the Amazon Rainforest – the Atlantic Forest's famous neighbour. It's claimed that one hectare of Amazon Rainforest is cleared every second – amounting to 20% of the rainforest so far. But the Atlantic Forest is really staring destruction in the face – over 90% of it has already gone. What's left is fragmented, and many species there are clinging on to survive.

Deforestation in the Atlantic Forest

Two groups of Amerindians live in the Atlantic Forest – the Tupi and the Guarani. About 134 000 Amerindians live there today, and have done for hundreds of years – watching their rainforest home disappear.

Deforestation here began as far back as the sixteenth century, when Europeans first settled on the coast and began clearing the rainforest for timber. The cleared land was then turned into cattle ranches and sugar plantations.

Population pressure – caused by rapid population growth in coastal cities like Rio de Janeiro and Sao Paulo – has led to more rainforest clearance and the Atlantic Forest has shrunk even further. About 70% of Brazil's 200 million people live in the Atlantic Forest area in southeast Brazil.

Other recent causes of deforestation in the Atlantic Forest have been:

- the need to make **debt repayments**. In the late 1980s, Brazil's government had to repay large loans taken out between 1965 and 1985. Brazil's farmers were put under pressure to produce more and earn more to help repay the **debt** (see pages 242-243 for more about debt and debt repayments). As a result, more rainforest was cleared.
- the pressure of **large-scale farming** – initially to grow cash crops like coffee, tobacco and eucalyptus. Once the soil's fertility drops (see right), the farmland is often converted into huge cattle ranches.
- **logging** – for expensive timber, such as mahogany, and for basic wood to make timber products like pulp and paper.
- an increase in the number of **small farms**, as landless people move into the area from cities in Brazil, Paraguay and Argentina.
- the **expansion of heavy industry** as a result of Brazil's development. A lot of land has been cleared for industry, and this area is now Brazil's industrial centre.

The impacts of deforestation

The impacts of the rainforest clearance have been enormous:

- The area's Amerindians have watched their land and way of life disappear before their eyes.

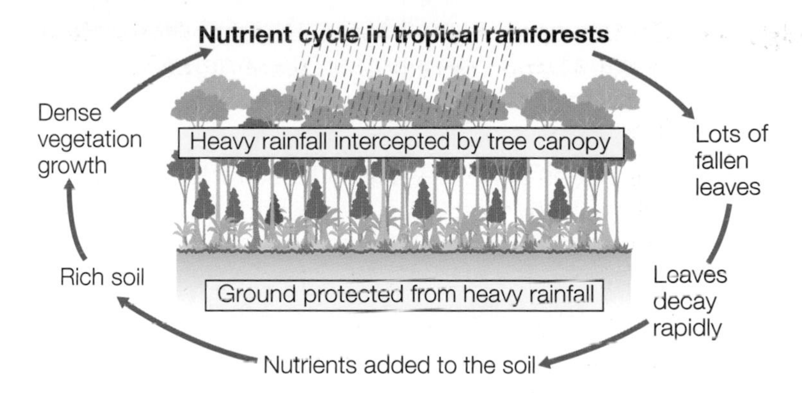

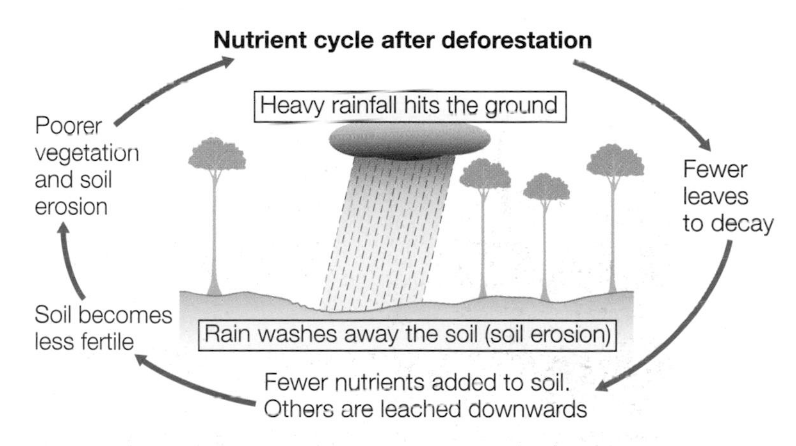

▲ *The impact of deforestation on the nutrient cycle*

- With no rainforest vegetation to protect the soil, heavy rainfall washes it away (see the diagram). The farmers then need to clear more land.
- When rainforest is cleared and the land is intensively farmed, it loses its fertility within 20 years.
- Deforestation breaks the nutrient cycle, which the soil depends on for its fertility (see the diagram).
- The loss of so much forest has helped to contribute to global warming. The rainforest absorbs carbon dioxide from the atmosphere. But clearing the trees and burning them just adds to the problem.

YOUR QUESTIONS

1 Explain this term to your partner and add it to your dictionary of key terms for this chapter: endemic.

2 a Draw a spider diagram to show the causes of deforestation in the Atlantic Forest.

b Draw another spider diagram to show the impacts of deforestation.

3 Work with a partner to write a two-minute news item about the Atlantic Forest. You need to tell people where it is, why it's important and how it has been destroyed.

4 At least one species of bird from the Atlantic Forest is now extinct in the wild (a few exist in captivity in Rio de Janeiro). Is that important? Explain your answer.

Hint: You won't find the answer on this spread. This question is meant to make you think – but you could begin by reminding yourself about food chains and food webs (see page 75).

On this spread you'll find out about other causes of deforestation – this time in the Amazon Rainforest – and begin to learn how rainforests can be managed.

Deforestation in the Amazon Rainforest

What's been happening to the Atlantic Forest (see pages 84-85) has also been happening to the Amazon Rainforest – and many other rainforests around the world. Other causes of deforestation include mining, road building and slash and burn.

Mining

The Carajas Mine in Brazil is huge – it's the world's largest iron ore mine. Where there was once rich tropical rainforest, a vast area has now been scarred by opencast mining.

The Amazon Rainforest developed above a rich base of minerals. Carajas isn't just rich in iron ore – manganese, copper, tin, aluminium and gold are also found there. Huge areas of rainforest have been cleared to get to the minerals – causing total devastation.

▲ *Carajas Mine, Brazil*

Road building

Big projects like mines need roads to reach them, and to transport the mined ore to processing plants and the coast for export. Elsewhere in the Amazon Rainforest, other roads have been built for basic communications, or for loggers and farmers. The new roads built across Brazil have led to uncontrolled development and deforestation up to 50 km on either side.

Slash and burn

Slash and burn is the traditional method of farming used by Amerindians in the Amazon Rainforest. A small area of land is cleared and farmed for a few years, before the people leave it and move on. Many see this type of farming as sustainable, because only a small area is cleared – and, when the people move on, the forest is left to regenerate and recover.

▲ *New roads may bring benefits for a country's development, but they also lead to deforestation*

The impacts of deforestation

Cause of deforestation	Economic impacts	Social and political impacts	Environmental impacts
Mining	It allows a country to earn foreign money to pay off its debts and support its development.	It provides jobs. It destroys the traditional Amerindian way of life.	A massive amount of rainforest is cleared, which can lead to flooding.
Road building	It helps a country to develop economically.	It helps to improve communications within the country.	It opens up more rainforest and leads to further deforestation.
Slash and burn	This is subsistence farming – growing crops to live off, rather than to trade.	It allows Amerindians to continue their traditional farming lifestyle and maintain their culture.	Only a small area of rainforest is cleared at a time. The forest is allowed to regenerate and recover.

Managing the rainforests

Large-scale deforestation is bad news. But there are ways of managing rainforests in order to halt or reduce the destruction, and in some cases repair the damage.

Selective logging

This is a technique where individual trees are felled only when they're mature. The idea is that the rainforest canopy is then preserved, which protects the ground below and also helps slower-growing hardwoods, like mahogany. But, worryingly, a team of scientists found that in parts of northern Brazil, nearly a third of the selectively logged forest had been completely cleared within four years. The roads left behind by the selective loggers allowed other people to follow them in and open up the forest further. Also, for every tree that's selectively logged, up to 30 other trees can be damaged or destroyed getting the logged tree out of the forest.

▲ *Mahogany being selectively logged*

Reducing debt

Conservation swaps, or **debt-for-nature swaps**, are a way of reducing a country's debt and benefiting nature and conservation at the same time. The most common type of debt-for-nature swaps work like this. A country (e.g. the USA) that is owed money by another country (e.g. Peru), cancels part of the debt in exchange for an agreement by the debtor country to pay for conservation activities there. Non-Governmental Organisations (NGOs), like the WWF, often help to arrange the swaps.

Promoting responsible management and use

The Forest Stewardship Council is an NGO that promotes the responsible management of the world's forests. Approved companies can use its logo to show that their wood products have been produced responsibly. Consumers – that's people like you and me – can then make a choice between buying approved products, with the logo, or products produced in a less responsible way (hopefully reducing demand for them).

▲ *A jaguar, one of the rare species whose habitat in the Peruvian rainforest has been preserved after debt-for-nature swaps between Peru and the USA*

YOUR QUESTIONS

1 Add the terms slash and burn and debt-for-nature swaps, along with their definitions, to your dictionary of key terms.

2 Download a rainforest photo. Add labels to it to show how it could be managed.

3 Explain why selective logging might not always help the rainforest.

4 Complete a table for all of the causes of deforestation given on page 85 and on this spread. It should show their economic, social, political, and environmental impacts.

Hint: Use the table opposite as a start. Copy it out and add any extra information that you can think of. Then add extra rows for the causes and impacts given on page 85. Use the spider diagrams in question 2 there to help you complete the table.

4.8 » Managing the Atlantic Forest

On this spread you'll explore the different ways in which the Atlantic Forest is being managed sustainably.

Conservation and protection

Conservation International is an organisation that helps communities, societies and countries to protect and value a range of different ecosystems. Their view of the Atlantic Forest is that – although the past has been grim – the future is looking brighter.

- Nearly a quarter of the remaining Atlantic Forest is under some form of protection – as National and State Parks, biological reserves, and ecological stations. The German government has invested in the protection of a number of areas.
- Conservation corridors are being established to link up fragmented areas of rainforest. The World Bank is helping to fund these corridors, which enable species to move, feed, and breed between different areas.
- The Critical Ecosystem Partnership Fund provides grants to help conserve threatened species, helps private landowners to manage their land sustainably, and provides support for other conservation projects.

▲ *REGUA replants the forest using native plants grown in their own nursery*

Restoration and education

REGUA (Reserva Ecológica de Guapiaçu) is an NGO – made up of local landowners and members of the community – that aims to protect the Atlantic Forest. It's trying to restore cleared areas of rainforest that have been slow to regenerate naturally. By 2008, REGUA had planted over 38 000 trees (of more than 50 different species). This work is having real success.

Another one of REGUA's main objectives is to help local people to understand and value the unique and threatened environment that they live in:

- REGUA provides an environmental programme for local schools. The students visit the forest to learn about REGUA's conservation work.
- Older local people are also kept informed of REGUA's work through local meetings and regular communications.
- Finally, REGUA has plans to create a research and education centre to help them do more work.

▼ *Education taking place at REGUA*

Ecotourism

The Brazil Travel Information website uses the following words to start its section about ecotourism in Brazil: '... diverse ecosystems ... heart-stopping landscapes ... waterfalls ...' No wonder, then, that Conservation International uses a photo of the waterfall on the right to introduce its website section about the Una Ecopark.

Ecotourism is when people visit a place because of its natural environment – and cause as little harm as possible. It aims to put back into the environment as much as it takes out, by conservation and education.

The Una Ecopark in the Atlantic Forest is located in one of those 'diverse ecosystems' with a 'heart-stopping landscape'. It:

- is a private reserve – part of Conservation International's work (in partnership with other organisations) to conserve the Atlantic Forest
- has a visitor centre to educate tourists and local people about the Atlantic Forest
- acts as a research and study centre for the Atlantic Forest, alongside the visitor centre
- has a canopy walkway, so that visitors can experience nature at first hand and appreciate why the forest needs conserving
- provides economic opportunities for the local community through nature-based tourism
- uses part of its entrance fee to support conservation projects in the region
- aims to show that income from ecotourism depends on conserving the Atlantic Forest as it is, and that – instead of destroying the rainforest to make money – local people can make a good living from ecotourism by conserving it instead.

▲ *Deep in the forest, visitors experience walking in the tree canopy 20 metres above the ground – on a bridge constructed using the living trees and rope!*

Making farming more sustainable

Conservation International is also helping to save key fragments of the Atlantic Forest by introducing local cocoa farmers to more-efficient farming methods. Elsewhere, other organisations are working with farmers to find more-sustainable ways of making money, instead of just clearing more land. This includes ecotourism.

YOUR QUESTIONS

1 Write your own definition of ecotourism and add it to your dictionary of key terms for this chapter.

2 Work with a partner to write a two-minute news item following up on the one you wrote on page 85. This one should tell people how the Atlantic Forest is being managed in a sustainable way.

3 Discuss as a class: 'The rainforest is more valuable when left intact than when destroyed'.

4.9 » The Australian outback

On this spread you'll find out how hot deserts provide opportunities for economic development in a richer part of the world.

Tourism

Watching the sunset at Uluru in Australia's outback is on the BBC's list of '50 things to do before you die'. It's the most visited spot in Australia, and a sacred site for the aboriginal Anangu people. The surrounding environment is very sensitive, but – despite this – the number of visitors has risen dramatically from 5000 in 1961 to 400 000 in 2005.

Tourists do bring economic benefits to the local people, e.g. when they buy aboriginal arts and crafts, but there are problems too:

- Aboriginal culture is often exploited and adapted to provide entertainment.
- People come for the 'experience' of the sunset at the sacred rock, but may learn nothing about aboriginal culture or beliefs while they're there.
- The Anangu have no role in the management or development of the tourist resort where most visitors stay.

▼ *Several kilometres from Uluru, the view is spectacular at sunrise and sunset*

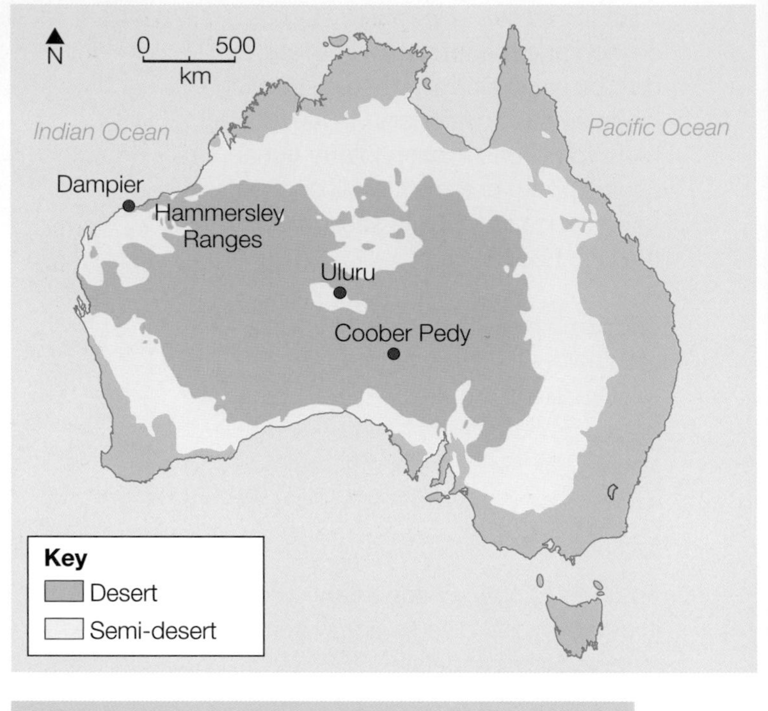

Farming

It can be difficult to make a living from farming in the outback. The soils are poor, with little organic matter to retain moisture, and plants are low in nutrients. If water is available, there's just about enough grass to feed cattle or sheep – but only in quite low numbers. So, in order to make any money, farms are huge – some the size of Wales!

Hunting and gathering

Australia's aboriginal peoples have traditionally survived by **hunting and gathering** – finding edible plants and animals in the outback.

- They created conditions in which grubs could live and breed.
- They built dams across rivers to catch fish, and to make pools where birds would gather.
- They used fire to drive out animals for hunting, to clear wood, and to allow grass to grow. As a result, fire-tolerant plants (eucalyptus trees) came to dominate the landscape.

Australia now has a growing 'native food' industry, based on traditional aboriginal knowledge of what's edible in the outback.

▲ *For aboriginal groups who live in the desert, there's a huge variety of food available, including seeds, grubs, and meat such as kangaroo and crocodile. Witchetty grubs are the larvae of moths and beetles, which can be eaten raw or cooked. They taste like scrambled eggs and peanut butter with a crispy 'chicken skin' coating.*

Mining

Most people in the outback work in mines. Australia has some of the world's largest reserves of quality iron ore, silver and gems (such as opal). Coober Pedy in South Australia is the opal capital of the world – an estimated 70% of the world's opal comes from there.

Most of Australia's iron ore is sold to China. Every day, several trains (each one 2.2 km long) run from the Hammersley Ranges in Western Australia to the port of Dampier. Huge ships are then loaded up with 250 000 tonnes of iron ore each before they head for China.

The world's three largest mining companies control 80% of the iron ore market. In 2008 – because China's demand was so great – they raised the price of iron ore by 70%.

▲ *Australian wagons filled with iron ore heading for port and then China. In 2007, China took as much iron ore as Australia could provide.*

your planet

Coober Pedy is an English form of an aboriginal name that means white man in hole or burrow. It's so hot there that most people live below ground to keep cool.

Retirement

The Sonora Desert in the USA attracts retired people moving to places like Phoenix for the sunny climate and open spaces. This movement is called **retirement migration**. In Australia, most retired people who move go to the coast, but there are some retirement villages in the outback, at Whyalla and Mount Gambier in South Australia.

YOUR QUESTIONS

1. Add these terms to your dictionary of key terms for this chapter, along with a definition: retirement migration, hunting and gathering.
2. Draw a mind map to show how the activities on this spread (mining, etc.) provide opportunities for economic development.
3. Describe how tourism in the outback is changing, and the benefits this could bring.
4. Is hunting and gathering sustainable? Make a table to show ways in which it is and isn't sustainable. Do the same for farming and compare your tables.

Managing the challenges

Farming

Farming in the outback is very challenging, because of the lack of water. Farmers there have two main water sources:

- Most farms have dams and reservoirs to store water for sheep and cattle.
- The farmers also use boreholes to tap into underground water supplies.

Although farmers can currently meet the challenges of the harsh desert environment, recent droughts in Australia have put pressure on the land and water supplies. This has led people to question whether both water and land are being used sustainably.

Tourism

The new Uluru Aboriginal Cultural Centre educates visitors about aboriginal culture and history. Its displays include photos, spoken histories, aboriginal language learning, videos and artefacts. Aboriginal guides also lead outback walks to inform visitors about bush food, as well as the significance of Uluru as a sacred site and other cultural subjects.

The new Cultural Centre provides economic as well as cultural benefits. The income from the admission fees goes to the Anangu community. Today over 30 aboriginal people work in the park, and the park's management is dominated by aboriginal owners.

4.10 » The Sahara Desert

On this spread you'll find out how hot deserts provide opportunities for economic development in a poorer part of the world.

Energy

There are believed to be large oil and gas reserves underneath the Sahara Desert. But they're difficult to find – and extracting and transporting the oil and gas is even harder. For Algeria, though, oil and gas are big business – half of the money the country earns comes from them. However, drilling for oil and gas at the Hassi Messaoud oilfield isn't easy:

- It's difficult to get there for a start, because the oilfield is deep in the desert, so the workers have to travel in and out by plane.
- 40 000 people work at Hassi Messaoud. They have to pump their water supplies from underground aquifers and fly in their food.
- Then they have to drill hundreds of metres down to reach the oil and gas supplies.
- Finally, pipelines carry the oil for hundreds of kilometres across the Sahara to ports on the North African coast.

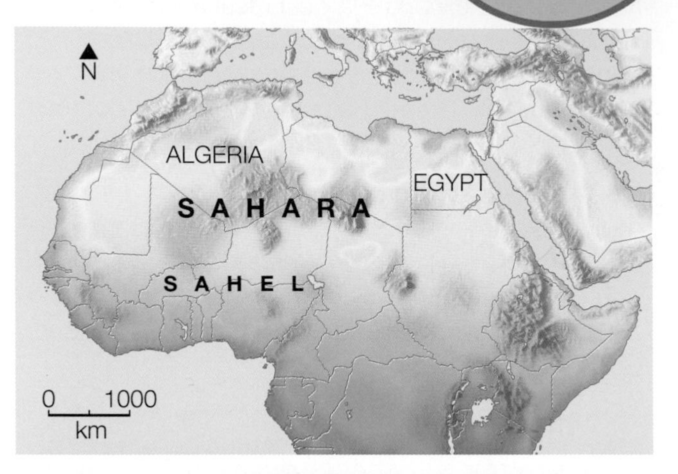

Preparing for a sustainable future

Most people agree that our use of fossil fuels like oil and gas is not sustainable. So Algeria is beginning to prepare for a future without them. Work has begun on constructing the country's first solar power plant in the Sahara Desert. The plan is to cover large areas of desert with solar panels to turn the sun's energy into electricity. The aim is to export solar power to Europe through cables below the Mediterranean Sea.

▲ *Solar panels in the desert*

Farming and irrigation

Egypt is hot and dry. It also has a soaring population, which has grown from 20 million to 79 million in the last 25 years – and is expected to keep on growing rapidly. Most Egyptians live in the heavily **irrigated** Nile Valley. The irrigated land means that farmers can grow more food, both to feed Egypt's growing population and for export. 13% of Egypt's GDP comes from farming, and it employs 32% of the labour force.

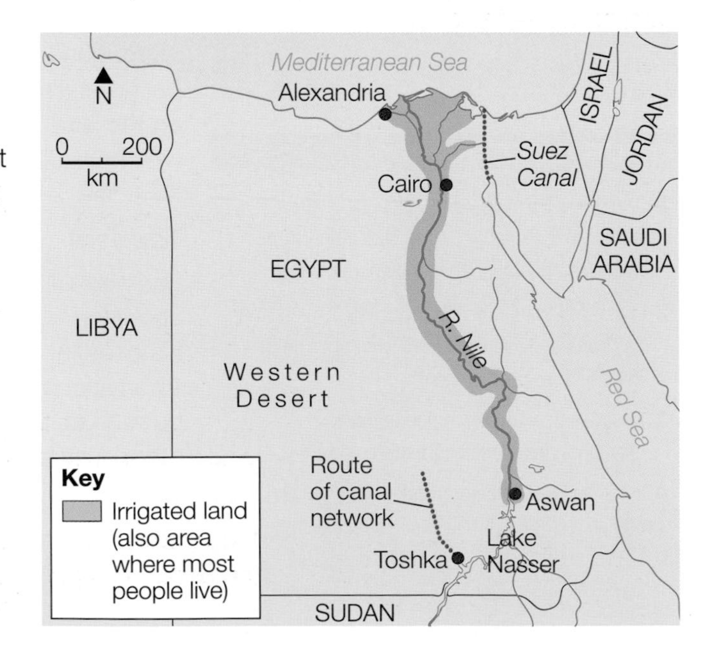

Salinity

However, Egypt's irrigated land is increasingly suffering from salinity. Irrigation water contains mineral salts. When the water evaporates from the surface of the soil, the salt crystals are left behind. Most plants then die, and the land that irrigation was meant to improve is destroyed (see the diagram on the right).

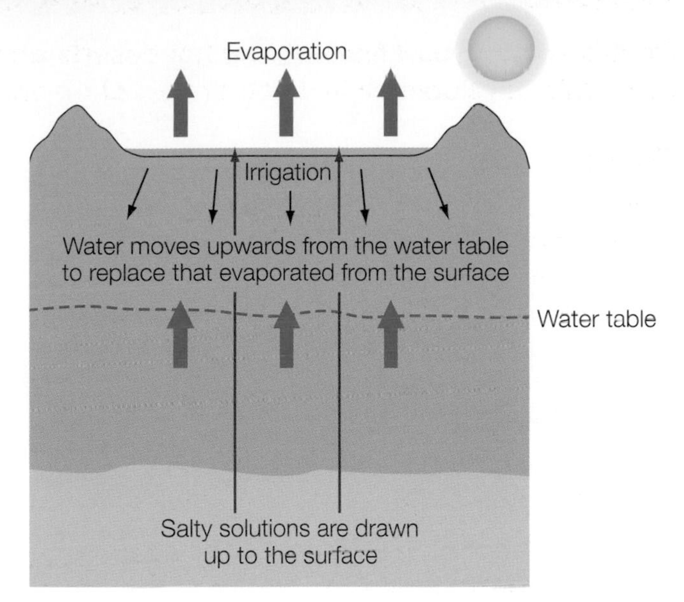

▲ *How salinity destroys irrigated land*

Preparing for the future

Because Egypt's farmland is increasingly being lost to urbanization, wind-blown sand and salinity, the Egyptian government has begun a scheme to irrigate more land away from the Nile Valley. The Toshka Project will cost $70 billion, and will use pumps and canals to transfer water from Lake Nasser into the Western Desert. It will:

- increase Egypt's irrigated land area by 30%
- enable high-value crops, such as olives, citrus fruits and vegetables to be grown
- provide food, electricity and jobs for 16 million Egyptians in new towns in the desert
- improve roads, railways, and telecommunications
- promote tourism.

Farming and desertification

The Sahel is a belt of land south of the Sahara. It is under intense pressure, and in some places the quality of the land has declined so much that it's turned into desert. This process is called **desertification**, and the main causes are shown on the right.

The ***climate*** in the Sahel is getting drier – on average it now rains less there than it did 50 years ago.

Cattle, sheep and goats are ***overgrazing*** the vegetation – leaving the soil exposed to wind erosion.

Population growth in the countries of the Sahel is increasing the demand for food. The resulting ***overcultivation*** of crops is using up the nutrients in the soil, so eventually nothing will grow.

Population growth is also increasing the ***demand for fuel wood***, so the land is being cleared of trees that bind the soil together and prevent soil erosion.

Sustainable farming

There are ways in which desertification can be stopped – and even reversed. The following examples have been tried successfully in the Sahel:

- Reducing the number of farm animals. This stops overgrazing and allows the protective vegetation to grow back.
- Growing crops as well as keeping animals. The animal manure is used to fertilise the soil and help the crops to grow.
- Planting more trees to protect the soil from wind and rain. The tree roots help to hold the soil together and prevent erosion.
- Building earth dams to collect and store water in the wet season. The stored water is then used to irrigate crops in the dry season.

YOUR QUESTIONS

1 Write a definition for irrigation and desertification. Then add them to your dictionary of key terms.

2 Draw two spider diagrams – one to show the causes of desertification and the other to show how it can be stopped.

3 a Describe the different economic activities in the Sahara Desert.

b Choose one and describe the challenges and problems that it faces.

4 a Describe how ***either*** Egypt ***or*** Algeria are preparing for the future.

b Which approach is more sustainable? Explain your answer.

5 » Water on the land

What do you have to know?

This chapter is from **Unit 1 Physical Geography Section B** of the AQA A GCSE specification. It is about rivers and landforms, flooding, and how rivers are managed. The table shows how the pages in this chapter match the content in the specification.

Specification content	Pages in this chapter
The shape of river valleys change as rivers flow downstream, due to the processes of erosion, transportation and deposition.	p96-101
Different landforms are caused by different processes as rivers flow downstream.	p96-101
The amount of water in a river changes because of different factors.	p102-103
Rivers flood because of physical and human causes, and flooding seems to be happening more often.	p104-105, 107
The effects of, and responses to, flooding vary in different parts of the world.	Case study of flooding in Cumbria p104-107 Case study of flooding in Pakistan p108-109
Managing floods – hard and soft engineering.	p110-111
Managing rivers to provide a water supply, and the need for sustainable supplies of water.	p112-113 Case study of Kielder Water p114-115

Your key words

- Tributary
- V-shaped valley
- Interlocking spurs
- River channel
- Erosion (abrasion, attrition, hydraulic action, solution)
- Transport (saltation, solution, traction, suspension)
- Load, bedload
- Waterfall, gorge, plunge pool
- Flood plain
- Meander (helical flow, thalweg, point bar, oxbow lake)
- Sediment
- Levées
- Bankfull
- Mudflats, saltmarshes
- Long profile, cross profile
- Discharge
- Antecedent rainfall
- Hydrograph, rising limb, lag time
- Confluence
- Hard engineering, soft engineering
- Water stress
- Water transfer-scheme

Exam help ...

Advice See pages 297-299 for information on how to be successful in your exams.

Practice See page 304 for exam questions on this chapter.

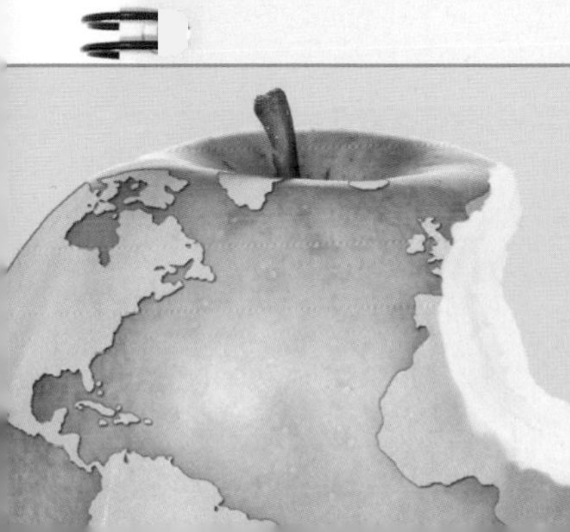

What if ...

- we ran out of water?
- the whole of London flooded?
- the UK had monsoon rains every year?
- salmon couldn't leap?

5.1 » River valleys in their upper course

On this spread you'll learn about the importance of erosion in a river valley's upper course.

Buckden Beck

The photo on the right is of Buckden Beck – a small stream, or **tributary**, which flows into the River Wharfe in northern England. It has just begun its journey to the sea. In the photo, the stream looks as if it's flowing quickly over the small rapids and waterfalls. However it's actually flowing quite slowly, because a lot of its energy is lost through **friction** with the stream bed.

▲ *Small rapids and waterfalls are typical of a river's upper course, which tends to have a steep **gradient**, or slope*

V-shaped valleys

The second photo shows the valley of Buckden Beck. Notice that:

- the valley has steep sides and a narrow bottom. It is **V-shaped**.
- the stream winds around ridges, or spurs of land, which jut into the river valley. These are called **interlocking spurs**.

▲ *Buckden Beck forms a typical river valley in its upper course*

River processes

A river's energy allows it to **transport**, or carry, boulders and stones downstream (especially when heavy rainfall increases the energy). The stones then wear away, or **erode**, the river's **channel**. The river also transports smaller material in its current.

How a river erodes its channel

By **abrasion** – where sand and pebbles are dragged along the river bed, wearing it away.

By **hydraulic action** – where fast-flowing water is forced into cracks, breaking up the bank over time.

By **attrition** – where rocks and stones wear each other away as they knock together.

By **solution** – where rocks such as limestone are dissolved in acid rainwater.

How a river carries its load

The material that a river carries is called its **load**. The size of the load that can be carried by a river depends on the amount of energy it has.

Waterfalls and gorges

In their upper course, rivers mainly erode downwards – or **vertically**. If the river flows from harder (**resistant**) rock onto softer, less-resistant rock, a step will form. Eventually, this might become a waterfall. The river increases in energy when it flows over the waterfall, which means that it can erode rapidly.

1 Waterfalls occur when a river crosses a bed of more-resistant rock.

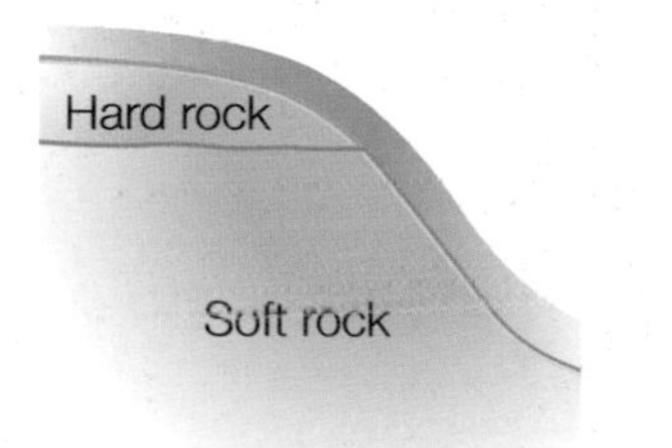

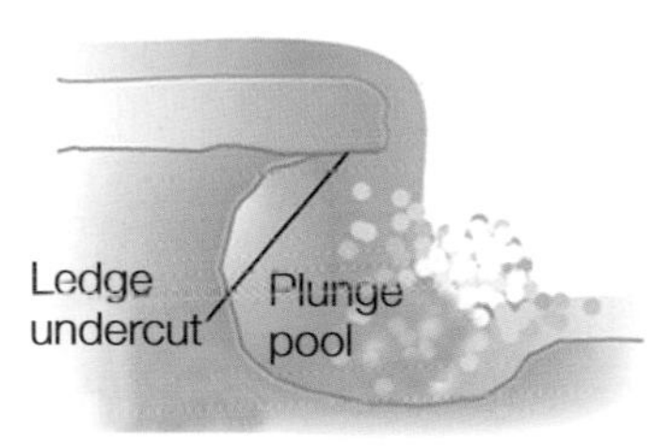

2 Erosion of the less-resistant rock underneath undercuts the hard rock above it. The river's energy creates a hollow at the foot of the waterfall, known as a **plunge pool**.

3 The less-resistant rock beneath is eroded more rapidly by abrasion and hydraulic action. This creates a ledge, which overhangs and eventually collapses.

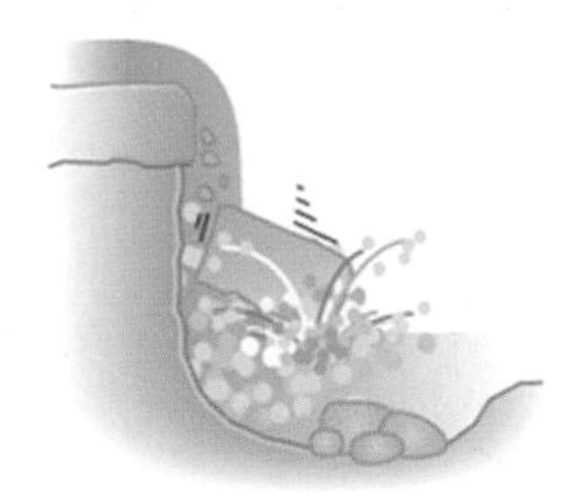

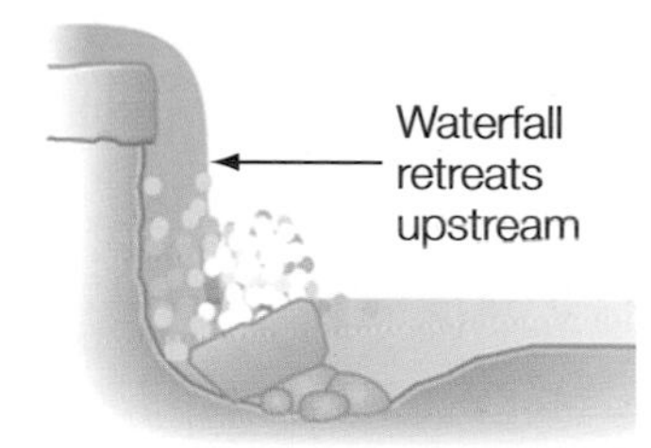

4 The waterfall takes up a new position, leaving a steep valley or **gorge**.

YOUR QUESTIONS

1 Work in pairs and test each other.

- **a** One partner should try to describe the four types of erosion.
- **b** The other partner should try to describe the four ways in which a river carries its load.

2 For each set of words below: **a** say which is the odd one out, **b** explain your choice.

- stream, tributary, river, waterfall
- solution, attrition, plunge pool, abrasion
- suspension, gorge, traction, load
- hydraulic action, saltation, suspension, traction

3 Draw a large simple sketch of the photo of Buckden Beck's valley. Add labels to show the features of the valley.

Hint: Start by drawing a frame the same shape as the photo. Draw lines to show the shape of the land and valley, but don't add lots of detail.

5.2 » River valleys in their middle course

On this spread you'll find out how rivers and their valleys change in the middle course.

The River Wharfe in its middle course

The photo shows the River Wharfe between Kettlewell and Starbotton, downstream from Buckden Beck. By now, the river is in its **middle course**. Several streams like Buckden Beck have joined the Wharfe, making it wider and deeper. In wet weather, the volume of water in the river can be so great that sometimes it floods over a broad flat area, known as the **flood plain**.

The valley shape has also changed. From the V-shaped valley of Buckden Beck, the flood plain has now broadened out the valley of the River Wharfe into a U-shape. The valley side rises steeply in the distance, where the flood plain ends. The OS map extract shows the wide, flat flood plain of the River Wharfe.

▲ *The valley of the Wharfe in its middle course. This is a* ***meander*** *and flood plain near Starbotton in Wharfedale.*

▼ *A 1:50 000 OS map extract of the River Wharfe in its middle course. Notice Buckden Beck in grid square 9477.*

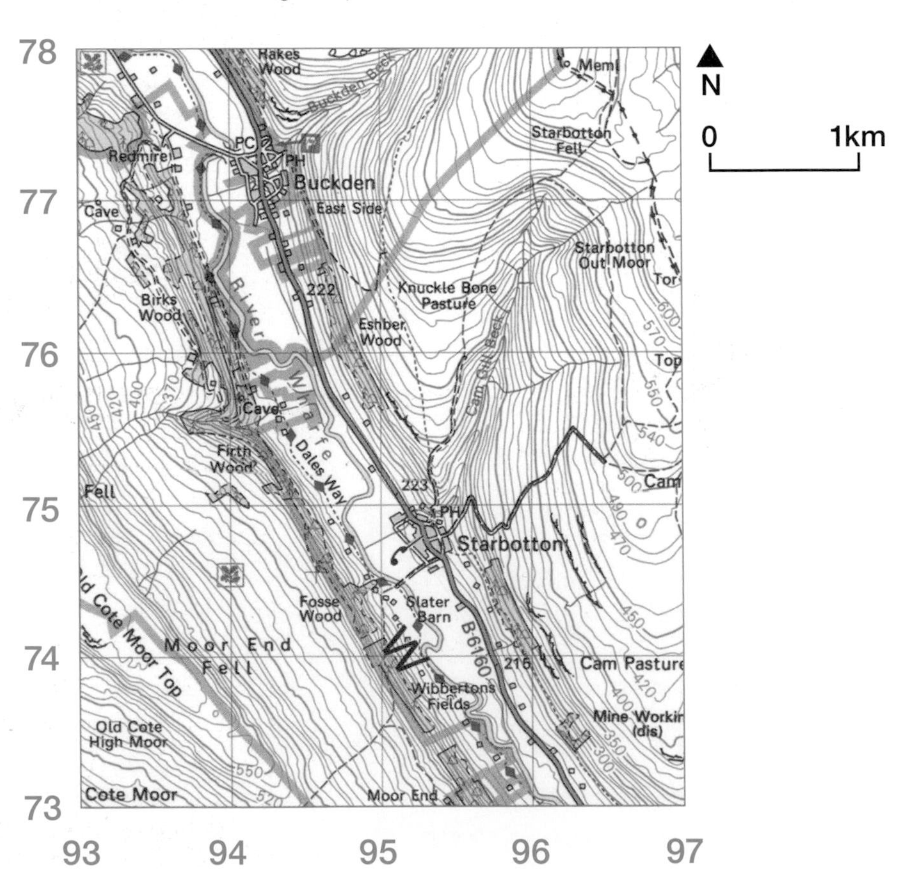

The river's energy

The River Wharfe has a gentle gradient by the time it reaches its middle course – but it actually flows faster than the upper course. This is because the greater volume of water in the river has more energy (once it's overcome friction). Now instead of eroding downwards, the river erodes sideways – or **laterally**.

Deposition

When a river no longer has enough energy to transport its load (such as in drier weather), it drops or deposits whatever it's carrying. It deposits the larger, heavier stones first, and then the smaller, lighter ones. The deposited material is called **sediment**. When the river floods over the flood plain, it deposits the lightest sediment – fine sand and clay – to form layers of alluvium, which is fertile and good for farming.

How meanders change the valley

Meanders are bends found in the river's middle course. They're natural features, which form as a result of both erosion and deposition.

1

The water in a river flows naturally in a corkscrew pattern. This is called **helical flow**.

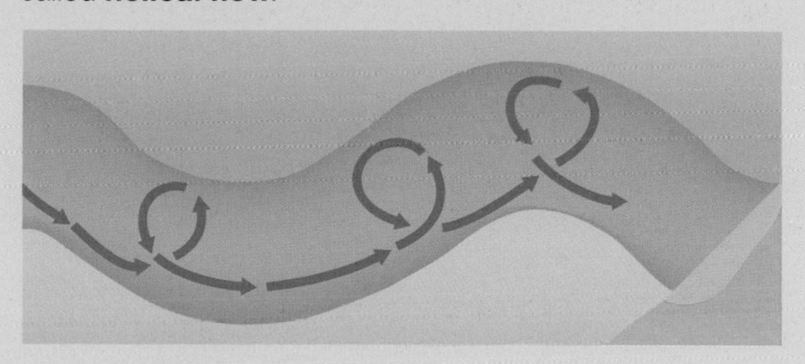

2

Helical flow sends the river's energy to the sides (laterally). The fastest current (called the **thalweg**) is forced to the outer bend (A), where it undercuts and erodes the bank to form a river cliff.

The helical flow then transports sediment from (A) across the channel to the inner bank (B), where the slower-moving water deposits it to form a **point bar**.

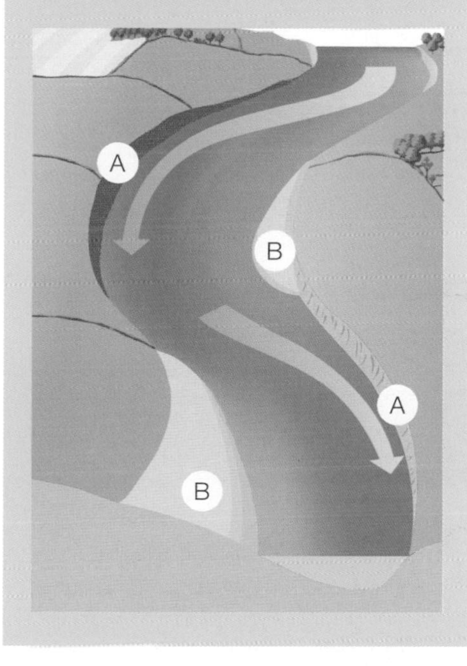

3

Continued erosion sometimes creates a narrow neck between two meanders (X). Eventually, the neck is cut through at Y, and the river creates a new channel for itself across the neck of the meander (an easier route for the water). The old meander then becomes an **oxbow lake** (Z), when deposition seals the ends – completely separating it from the river.

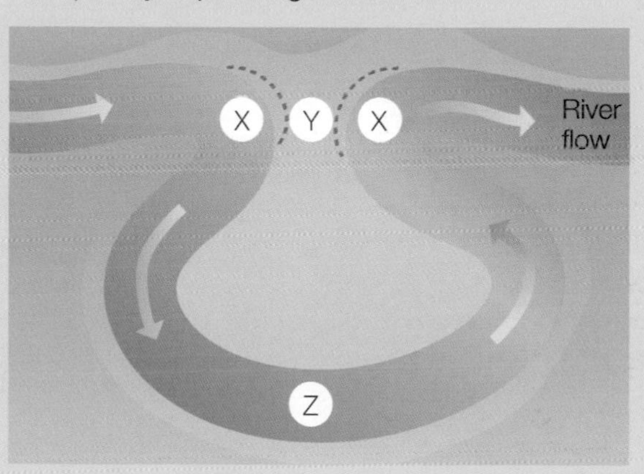

YOUR QUESTIONS

1 Draw a sketch of the photo opposite and add the following labels: flood plain, alluvium, lateral erosion, meander, thalweg, undercutting, point bar.

2 Using the information above to help, draw 3-4 labelled diagrams to show how an oxbow lake is formed.

3 The photo on the opposite page was taken at grid reference 957737 on the OS map extract, looking west towards Wibberton's Fields.

a Find the grid reference on the map.

b Describe the shape of the valley of the River Wharfe at this point.

c Explain why the valley (and the River Wharfe in its middle course) is different from the valley of Buckden Beck on page 96 and in grid square 9477 on the map extract.

5.3 » River valleys in their lower course

On this spread you'll find out how rivers and their valleys change in the lower course.

The lower course of the Wharfe

By the time the River Wharfe has reached its lower course, the differences with its upper course are really obvious! The river – once narrow with waterfalls – is now wide and deep, flowing over a gentle gradient. It has large meanders, it carries a large volume of water – and it has a wide, flat flood plain. This is low-lying and floods easily.

There are embankments beside the river. These can be either natural or artificial, and are known as **levées**.

- Natural levées form beside the river's bank where it first floods. As a river reaches **bankfull** i.e. before it spills onto the flood plain – it deposits sand and clay particles where the flow is slower. These build up beside the river and form a levée.
- Artificial levées are built by engineers to protect places from flooding. They are common in the UK along rivers like the Severn and the Ouse, and along some of the world's giant rivers, e.g. the Mississippi in the USA.

▲ *The lower part of the River Wharfe at Tadcaster*

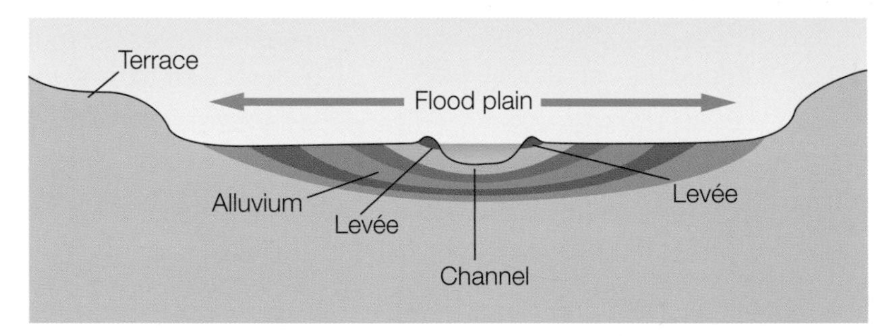

▲ *Levées and flood plains*

Towards the sea

The River Wharfe joins the Ouse and eventually the Humber and forms an estuary – where the river meets the sea. Water flows in two directions here. The river flows **outwards**, taking water out to sea, and the sea flows **inwards** at high tide. Twice a day, incoming tides meet the outgoing river and the water flow stops, forcing the river to deposit its sediment, or mud, creating a broad wide area of **mudflats**.

▲ *Mudflats and salt marshes on the river Humber, near its estuary. Notice the artificial stone levée in the top right of the photo.*

Because the estuary is tidal, it is submerged by seawater twice a day. Plants which grow there have to be able to withstand salt water as well as fresh. **Salt marshes** are very important for wildlife. Migrating birds use them as shelter during stormy weather, and the mud is rich in shellfish and worms. But salt marshes are under threat from ports and industries – as the photo on the right shows.

▲ *An aerial view of the Humber estuary at Immingham. Estuaries have enormous economic value – providing both shelter for shipping and flat land for industrial development. But this can conflict with their value as a haven for plants and wildlife.*

Long and cross profiles

As a river flows from its upper to its lower course, two kinds of changes happen.

There are changes in its **long profile**. As the diagram below shows, the long profile is the way in which the gradient of the river changes. It is steep in upland areas and gentle in the lower course.

There are also changes in the **cross profile**, or valley shape. The valley is V-shaped in the river's upper course and almost flat by the lower course (see the diagram on the right).

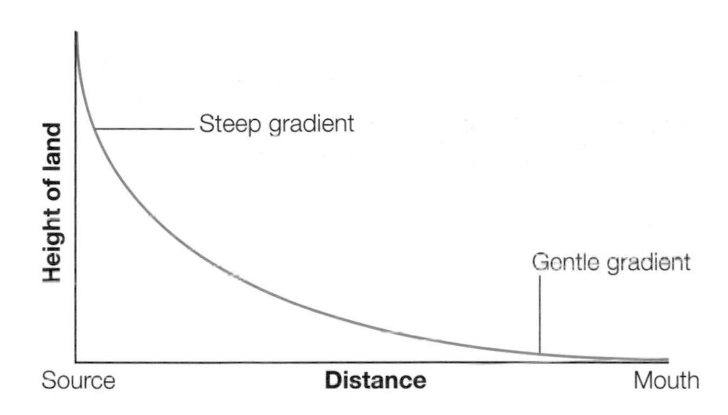

▲ *The long profile of a river – showing how river gradient changes as the river moves from its upper to its lower course*

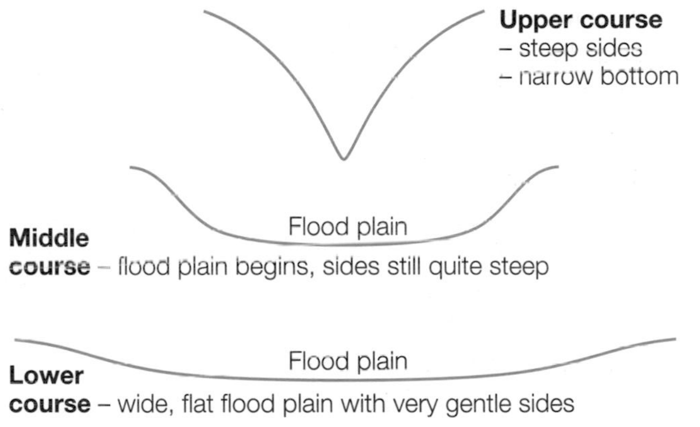

▲ *The cross profile of a river – showing how valley shape changes as the river moves from its upper to its lower course*

YOUR QUESTIONS

1 Explain what these terms mean: levées, bankfull, mudflats, salt marshes, long profile, cross profile.

2 Put the following phrases in the correct order to show how levées are formed:
- sand and clay particles are deposited where river flow is slower
- natural levées form next to the river bank where it first floods
- artificial levées are built to protect places from flooding
- these build up beside the river to form a levée
- as a river reaches bankful

3 Make a large copy of the table below and compare the river in its upper, middle and lower courses.

	Upper course	Middle course	Lower course
River features			
River processes			
Main landforms			
Outline sketch of valley shape			
Words to describe the long profile			

5.4 » What affects a river's discharge?

On this spread you'll find out why the amount of water in a river changes.

Too much discharge?

Cockermouth and Workington in Cumbria were hit by massive flooding in November 2009. The photo shows the remains of a bridge in Workington which collapsed. Kevin Bell worked at a nearby hotel: 'It's terrible, it shows the volume of water that went through to knock a bridge over' he said.

Kevin was right – the river was carrying a colossal amount of water. The correct term for this is **discharge**. Discharge is the volume of water flowing past a point in a river. It is measured in cubic metres per second.

▲ *The discharge of the river in Workington was so high that it destroyed bridges*

From rain to river

How does water get into a river? When it rains, very little falls directly into the river itself – most of it falls elsewhere. The diagram shows what happens next.

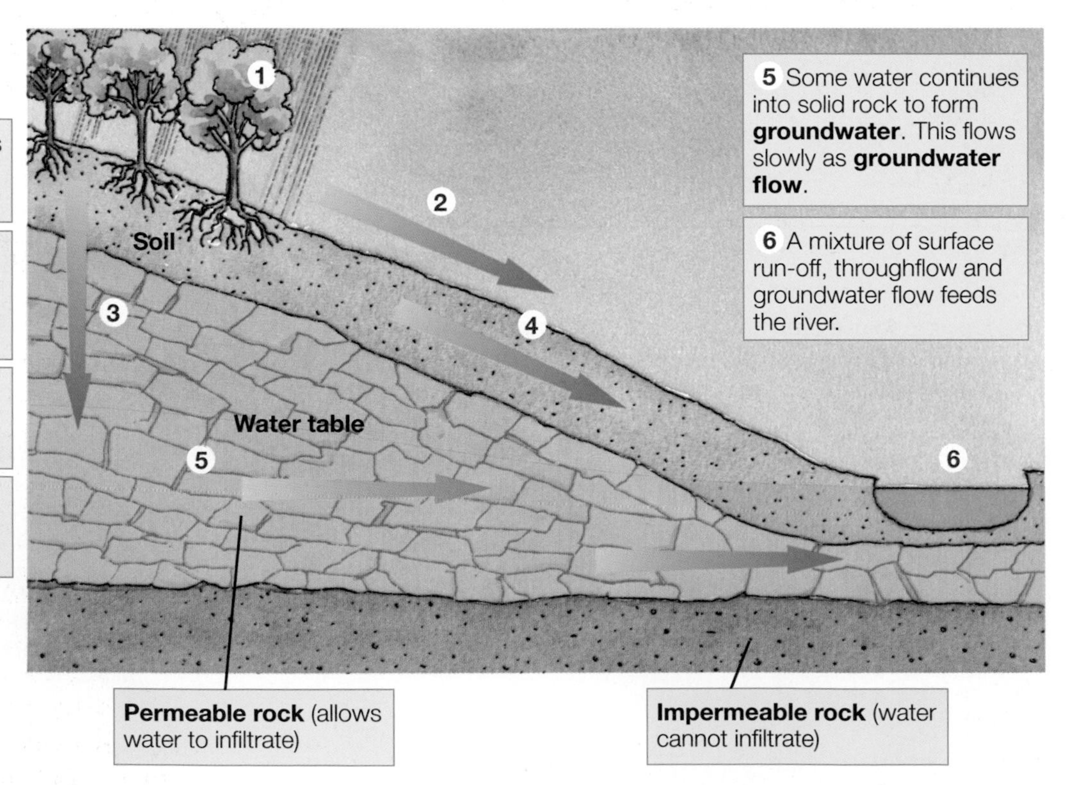

Factors affecting discharge

The diagram opposite shows how water gets into a river. But the volume of water in a river can change because of a range of factors.

Rock and soil type

- Permeable rocks and soils (such as sandy soils) absorb water easily, so surface run-off is rare.
- Impermeable rocks and soils (such as clay soils) are more closely packed. Rainwater can't infiltrate, so water reaches the river more quickly.
- **Pervious** rocks (like limestone) allow water to pass through joints in the rock, and **porous** rocks (like chalk) have spaces between the rock particles.

Land use

- In urban areas, surfaces like roads are impermeable – water can't soak into the ground. Instead, it runs into drains, gathers speed and joins rainwater from other drains – eventually spilling into the river.
- In rural areas, ploughing up and down (instead of across) hillsides creates channels which allow rainwater to reach rivers faster – increasing discharge.
- Deforestation means less interception, so rain reaches the ground faster. The ground is likely to become saturated and surface run-off will increase.

Rainfall

The amount and type of rainfall will affect a river's discharge:

- **Antecedent rainfall** is rain that has already happened. It can mean that the ground has become saturated (see the photo). Further rain will then flow as surface run-off towards the river.
- Heavy continual rain, or melting snow, means more water flowing into the river.

Relief

Steep slopes mean that rainwater is likely to run straight over the surface before it can infiltrate. On more gentle slopes, infiltration is more likely.

Hydrographs

A flood or storm **hydrograph** is a graph showing how a river responds to a particular storm. It shows rainfall and discharge. Once water enters the river, discharge increases (shown by the **rising limb** on the graph). The gap between the peak (maximum) rainfall and peak discharge (highest river level) is called the **lag time**.

Changes to land use can mean that water gets into the river faster. This makes the lag time shorter, the rising limb steeper, and the peak discharge higher.

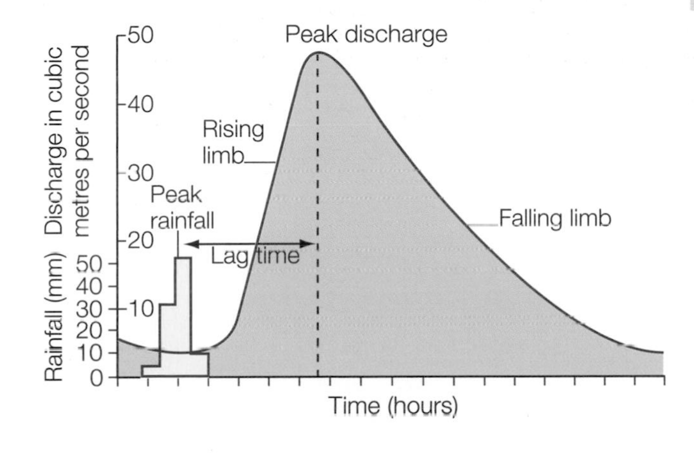

Weather conditions

- Hot dry weather can bake the soil, so that when it rains the water can't soak in. Instead, it will run off the surface, straight into the river.
- High temperatures increase evaporation rates from water surfaces, and transpiration from plants – reducing discharge.
- Long periods of extreme cold weather can lead to frozen ground, so that water can't soak in.

YOUR QUESTIONS

1. Define these terms: discharge, permeable, impermeable, antecedent rainfall, hydrograph, rising limb, lag time.
2. Explain how pervious rocks and porous rocks affect a river's discharge.
3. Explain how water gets into a river. Use the words in bold on the diagram on the opposite page in your explanation.

On this spread you'll find out what caused the Cumbrian floods in November 2009.

Cumbrian floods – policeman died saving lives

PC Bill Barker died a hero, doing the job he loved. Ignoring his own safety, he stood in the dark – at 4.40 in the morning – desperately directing traffic away from a bridge above a dangerously swollen river. Then the bridge collapsed and the torrent surged through, sweeping him away.

Adapted from the *Mail Online* website, 21 November 2009

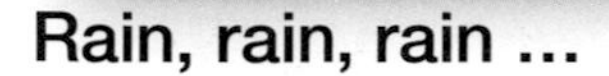

▲ *The water was up to 2.5 metres deep in Cockermouth town centre*

Rain, rain, rain ...

A massive downpour of rain (31.4 cm), over a 24-hour period, triggered the floods that hit Cockermouth and Workington in Cumbria in November 2009. The table shows actual rainfall in Cumbria from 18-20 November, compared with the normal 3-day average for that time of year.

Actual rainfall in Cumbria (18-20 November 2009)	
Seathwaite	37.2 cm
Shap	19.7 cm
Keswick	18.9 cm
St Bees Head	7.8 cm
Walney Island	7.3 cm
Carlisle	4.2 cm
Normal 3-day average for Cumbria in November	**3.0 cm**

What caused all the rain?

The long downpour was caused by a lengthy flow of warm, moist air that came from the Azores in the mid-Atlantic. This kind of airflow is common in the UK in autumn and winter, and is known as a 'warm conveyor'. The warmer the air is, the more moisture it can hold.

▼ *How it all happened*

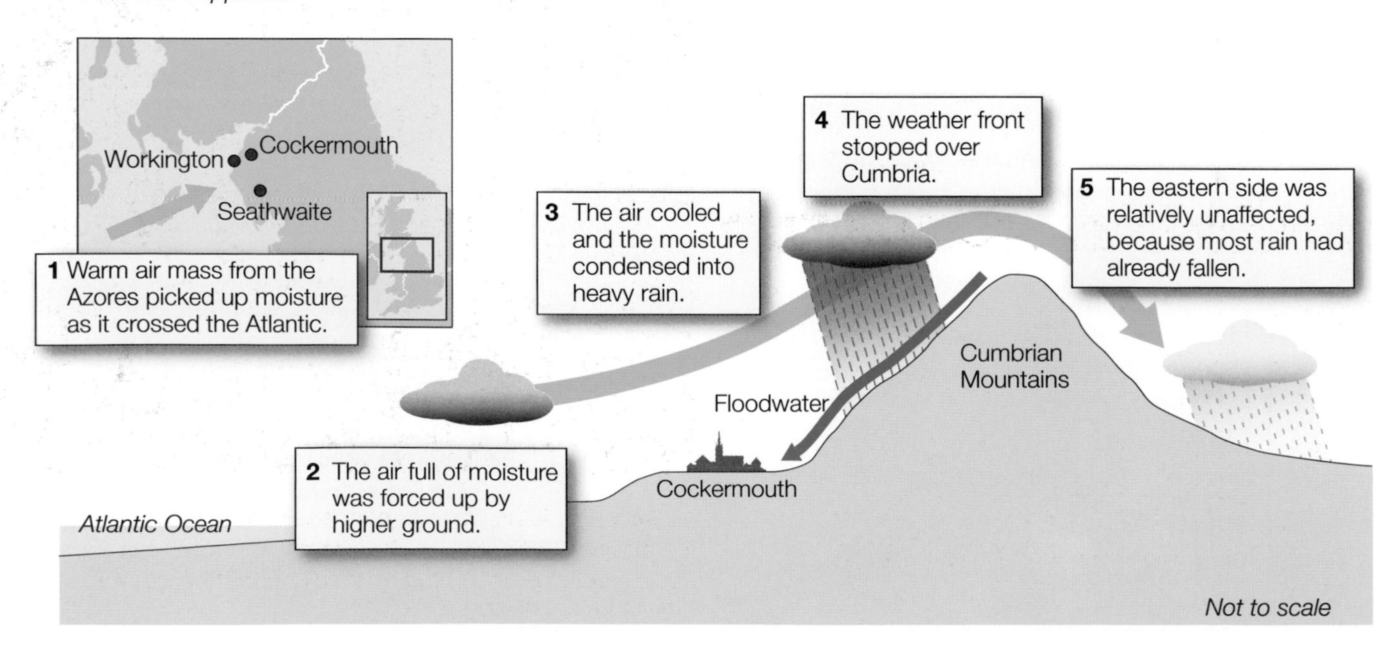

What else helped to cause the Cumbrian floods?

- The ground was already saturated, so the additional rain flowed as surface run-off straight into the rivers.
- The steep slopes of the Cumbrian Mountains helped the water to run very rapidly into the rivers.
- The rivers Derwent and Cocker were already swollen with previous rainfall.
- Cockermouth is at the **confluence** of the Derwent and Cocker (i.e. they meet there).

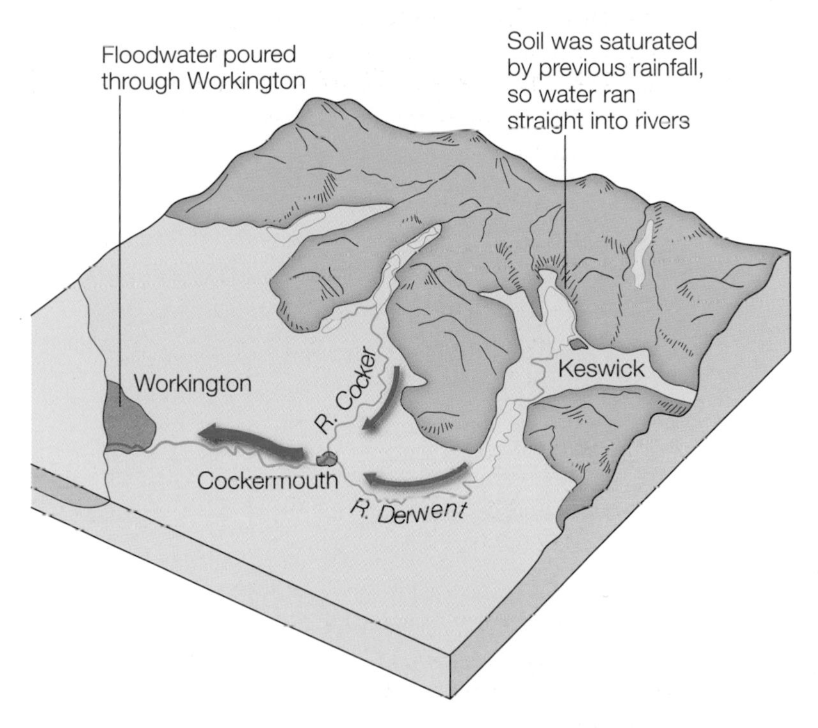

▲ *The heavy rainfall flowed rapidly off the mountains, into the rivers, through Cockermouth and on to Workington*

What factors cause flooding?

Rivers flood for a number of factors, or reasons:

- It was mainly physical factors that caused the 2009 Cumbrian floods.
- Heavy rain, sometimes coming in just short bursts, has been the cause of many floods in the past (e.g. the flash floods in Boscastle in 2004, see page 54).
- Rapidly melting heavy snowfalls can also lead to flooding (e.g. Boscastle again in 1963).

Human factors also play a part in many cases of flooding. Anything that increases the discharge of a river (see pages 102-103), can increase the likelihood of flooding. For example:

- building, particularly on flood plains, increases the area of impermeable surfaces – reducing infiltration and increasing surface run-off. This helped to cause the floods in June 2007 in Yorkshire and Humberside.
- the destruction of natural environments, e.g. by deforestation, increases the rate and amount of run-off. The destruction of grasslands and wetlands in South Africa (which absorbed extra water), increased the flooding in Mozambique in 2000.

▲ *Deforestation in Nepal increases the risk of flooding in locations downstream, like Bangladesh. The floods there in 2004 covered half the country, killed over 750 people, and affected over 30 million.*

YOUR QUESTIONS

1. Explain what the term confluence means.
2. Annotate a blank outline map of Cumbria to show what caused the floods there in 2009.
3. Look at the table of rainfall in Cumbria.
   - **a** Locate the places in the table on a map of Cumbria.
   - **b** Use the diagram opposite to explain the differences in rainfall between St Bees Head and Seathwaite.
4. How do humans help to cause flooding? Give two examples in your answer.

5.6 » The 2009 Cumbrian floods – 2

On this spread you'll explore the effects of, and responses to, the flooding in Cumbria in 2009.

The flooding in Cumbria was sudden and severe. One of the major effects was the destruction of key bridges.

Effects of the flood

- Over 1300 homes were flooded and contaminated with sewage (including the house in Cockermouth where the poet William Wordsworth was born).
- A number of people had to be evacuated, including 50 by helicopter, when the flooding cut off Cockermouth town centre.
- Many businesses were flooded, causing long-term difficulties for the local economy.
- People were told that they were unlikely to be able to move back into their flood-damaged homes for at least a year. The cost of putting right the damage was an average of £28 000 per house.
- Insurance companies estimated that the final cost of the flood could reach £100 million.
- Four bridges collapsed and 12 were closed because of flood damage. In Workington, all the bridges were destroyed or so badly damaged that they were declared unsafe – cutting the town in two. People faced a huge round trip to get from one side of the town to the other, using safe bridges.
- One man – PC Bill Barker – died.

Responses to the flood

- The government provided £1 million to help with the clean up and repairs, and agreed to pay for road and bridge repairs in Cumbria.
- The Cumbria Flood Recovery Fund was set up to help victims of the flood. It reached £1 million after just 10 days.
- Network Rail opened a temporary railway station in Workington.
- The 'Visit Cumbria' website provided lists of recovery services and trades, and people who could provide emergency accommodation.

▲ *The army built a temporary footbridge across the River Derwent in Workington, which re-united the two halves of the town. It was named 'Barker Crossing', in honour of PC Barker, and took just a week to build.*

Flood warning

The Environment Agency works out which places are at risk from flooding and issues warnings, when necessary. Three codes are used:

- Flood Alert
- Flood Warning
- Severe Flood Warning.

The codes tell people whether flooding is expected, and what they should do. The Environment Agency also lets people know when the risk of any further flooding has passed.

Is flooding happening more often?

In a word, yes. From the 1960s to the 1990s, big floods were pretty rare in the UK. However, serious floods now seem to be happening more often, as the table below shows.

Date	Places affected by flooding
1998	The Midlands
1999	North Yorkshire
2000	York, Shrewsbury, Lewes, Maidstone
2002	Glasgow
2004	Boscastle
2005	Carlisle
2007	Yorkshire, Lincolnshire, the Midlands, Gloucestershire and Boscastle
2009	Cockermouth/Workington

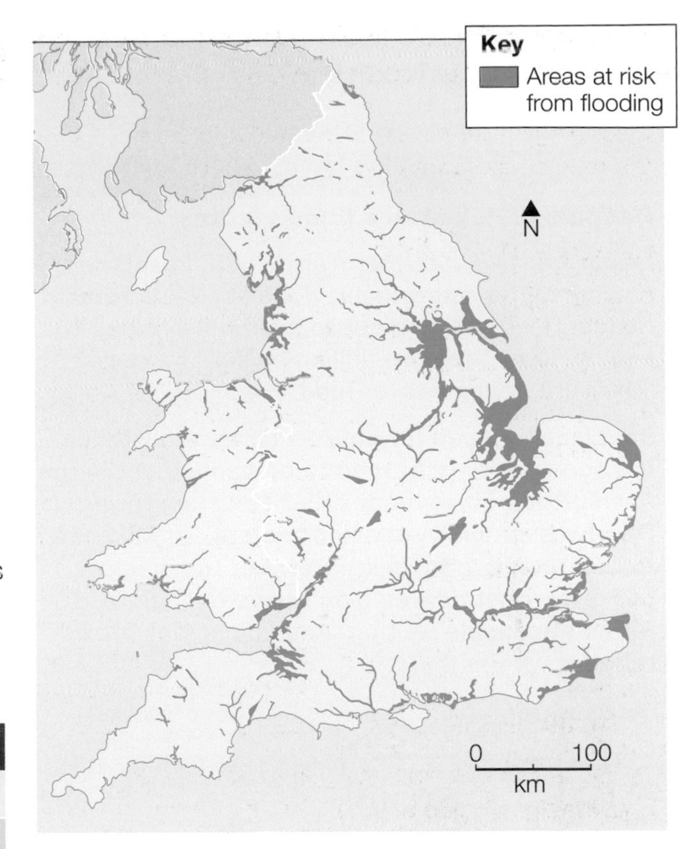

▲ *The areas of England and Wales at risk of flooding from rivers and the sea. Around 5 million people live in these areas.*

YOUR QUESTIONS

1 Go to the Environment Agency website (www.environment-agency.gov.uk) and search for flood warning codes. Create a leaflet to tell people what the codes are, what they mean, when they are used, and what they should do. Use your own words!

2 Draw a spider diagram on a sheet of A4 paper, like the diagram shown on the right. Complete the diagram, using information from this and the previous spread, with notes saying: what happened, where it happened, why it happened, when it happened, who was affected and who responded.

Hint: Drawing a spider diagram of the 5 Ws (Who? What? Why? Where? When?) will mean that you have some notes to help you remember your flooding case study.

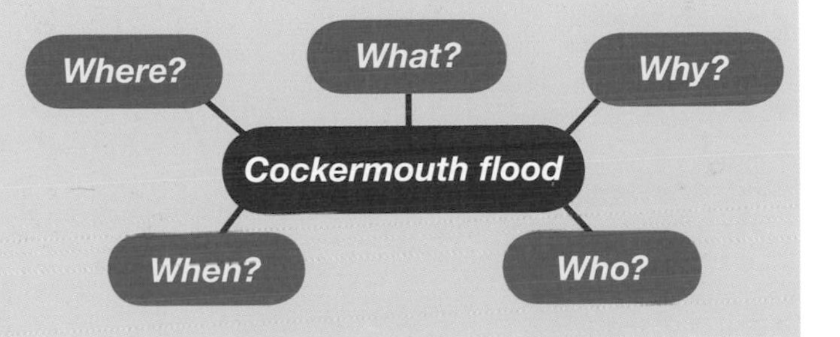

5.7 » The 2010 Pakistan floods

On this spread you'll learn about the effects of, and responses to, the floods in Pakistan in 2010.

Despair of Pakistan's flood victims

Liaqat Babar is a farmer from Sindh, in southern Pakistan. The catastrophic floods of 2010 brought hunger, loss and torment to his family. 'When I see my kids, I feel like killing myself' he says. 'They are crying out for food.'

So, Liaqat queued for hours – under a blistering hot sun – waiting for food to be handed out in the town of Daur. Like many other Pakistani towns, Daur was cut off by the floodwaters, so food had to be brought in by helicopter. But the single helicopter that arrived didn't have enough food for all the people waiting. And Liaqat was forced to leave with nothing.

(Adapted from an article on the BBC News website)

The Pakistan floods

At the end of July 2010, unusually heavy monsoon rains in northwest Pakistan caused rivers to flood and burst their banks. The map shows the huge area of Pakistan affected by flooding, and the graphs show how the floodwaters gradually moved down the Indus River towards the sea.

Continuing heavy rain hampered the rescue efforts. After visiting Pakistan, the UN Secretary-General, Ban Ki-moon, said that this disaster was worse than anything he'd ever seen. He described the floods as a slow-moving tsunami.

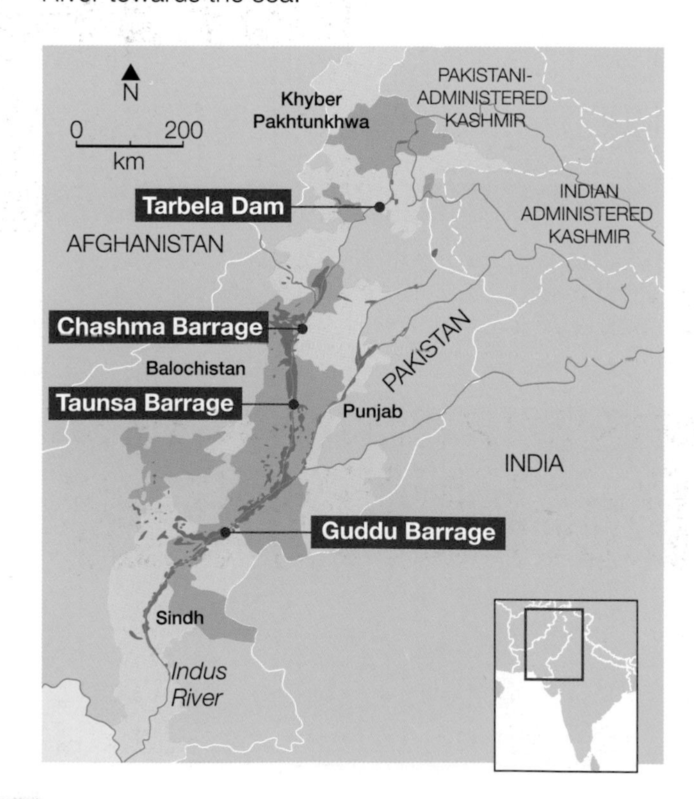

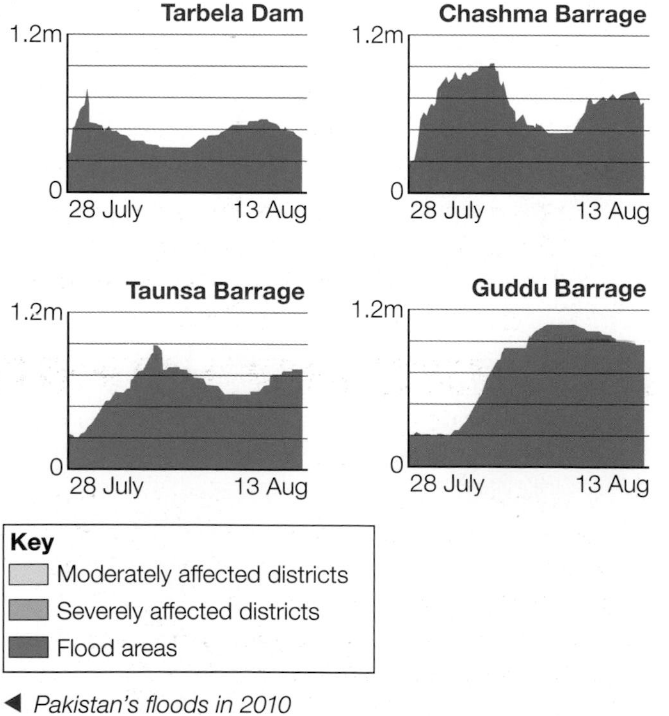

◀ *Pakistan's floods in 2010*

The effects of the floods

- At least 1600 people died.
- 20 million Pakistanis were affected (over 10% of the population). 6 million needed food aid.
- Whole villages were swept away, and over 700 000 homes were damaged or destroyed.
- Hundreds of thousands of Pakistanis were displaced, and many suffered from malnutrition and a lack of clean water.
- 5000 miles of roads and railways were washed away, along with 1000 bridges.
- 160 000 km^2 of land were affected. That's at least 20% of the country.
- About 6.5 million acres of crops were washed away in Punjab and Sindh provinces.

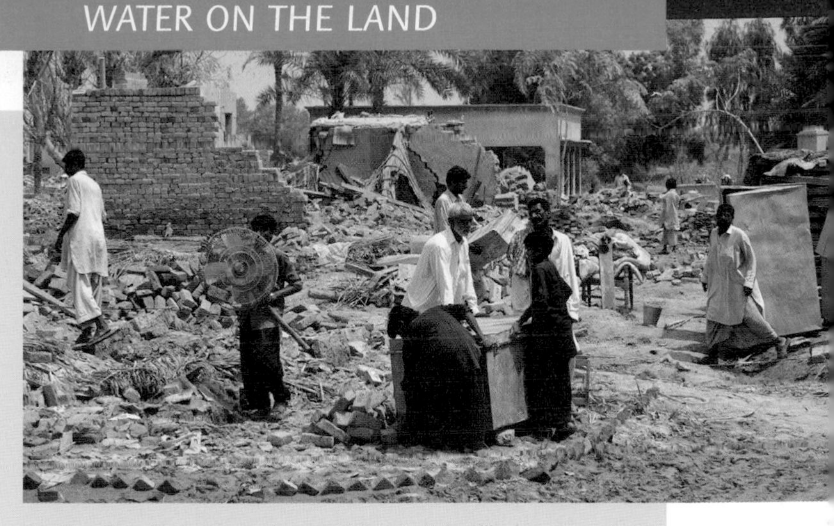

▲ *The floods destroyed everything in this village*

- Punjab, alone, needed US$1.3 billion for immediate relief and the short-term repair of roads, dykes, the electricity network and the irrigation system.
- But, three months after the flood, 7 million Pakistanis still had no proper shelter.

Responses to the floods

- Appeals were immediately launched by international organisations, like the UK's Disasters Emergency Committee – and the UN – to help Pakistanis hit by the floods.
- Many charities and aid agencies provided help, including the Red Crescent and Médecins Sans Frontières.
- Pakistan's government also tried to raise money to help the huge number of people affected.
- But there were complaints that the Pakistan government was slow to respond to the crisis, and that it struggled to cope.
- Foreign governments donated millions of dollars, and Saudi Arabia and the USA promised $600 million in flood aid. But many people felt that the richer foreign governments didn't do enough to help.
- The UN's World Food Programme provided crucial food aid. But, by November 2010, they were warning that they might have to cut the amount of food handed out, because of a lack of donations from richer countries.

▲ *Pakistani soldiers loading up relief supplies*

YOUR QUESTIONS

1 Look at the top photo on this page. List as many words as you can to describe the situation the people are in.

2 Draw a spider diagram on a sheet of A4 paper about the Pakistan floods. Cover what happened, where it happened, why it happened, when it happened, who was affected, and who responded. Use the spider diagram on page 107 as a model.

3 Complete a table like the one on the right to compare the Cumbrian floods (on pages 104-106) with the Pakistan floods.

	Cockermouth flood (richer part of the world)	Pakistan flood (poorer part of the world)
Effects		
Responses		

5.8 » Managing river flooding

On this spread you'll find out about the options we have for managing floods.

What's more important?

In the aftermath of the 2009 Cumbrian floods (see pages 104-106), local people were angry that more hadn't been done to prevent them. They accused the authorities of 'putting salmon before people', after their earlier request to lower the river bed by 3 metres in Cockermouth had been turned down, because it might harm fish stocks.

The cost of protection

Professor Samuels advises the government on managing rivers. He said 'It is technically possible to defend places like Cockermouth against extreme events, but only by building huge walls and embankments along the river, which would cost billions and alter the character of the town. For most people, that would be as unacceptable as the floods.'

No-one wants to go through what the residents of Cockermouth and Workington experienced. But as the likelihood of flooding increases, what options do we have to try to prevent flooding – or at least minimise its effects? And what are the costs?

Hard engineering

Hard engineering involves building structures to defend places from floodwater.

▶ *New flood walls built next to the River Rother, at Rotherham, to protect homes and businesses there*

Method	How it works	Comment
Build flood banks	Raise the banks of a river, so it can hold more water.	Relatively cheap, one-off costs. Looks unnatural. The water moves quickly and increases the flood risk downstream.
Straightening and deepening the river	Straighten the river channel to speed up the flow, or line it with concrete. Dredge the river to make it deeper and able to carry more water.	Dredging needs to be done every year. Lining with concrete is expensive. Speeding up the flow increases the flood risk downstream.
Dams and reservoirs	Trap and store water, and release it in a controlled way. Can be multi-purpose and generate electricity.	Very expensive. The creation of a reservoir means huge changes to the ecosystem. Dams trap sediment.
Flood walls	Built around settlements, industry or roads.	Expensive and looks unnatural. Effective if the flood isn't too extreme.
Storage areas	Water is pumped out of the river and stored in temporary lakes. It's then pumped back when the water level in the river has dropped.	Effective, but it needs large areas of spare land.
Barriers, e.g. Thames Barrier	The barrier is raised when a high tide or flood is forecast.	Very expensive and, in the case of the Thames Barrier, a new one may be needed in future to protect London from higher flood surges.

Soft engineering

Soft engineering involves adapting to flood risks, and allowing natural processes to deal with rainwater.

Method	How it works	Comment
Flood abatement	Change the land use upstream, e.g. by planting trees.	This slows down the flow of rainwater into rivers, e.g. by increasing interception.
Flood proofing	Design new buildings, or alter existing ones, to reduce the flood risk.	It's expensive to alter existing buildings.
Flood plain zoning	Different uses are allowed, depending on the distance from the river (see the diagram below).	Land close to the river might only be used for grazing. The land furthest from the river is used for hospitals, old people's homes, etc.
Flood prediction and warning	The Environment Agency monitors river levels and rainfall. They use this information, plus weather forecasts, to predict flooding.	They issue warnings (see page 107) and produce Flood Maps, which show areas at risk from flooding.
Washlands	These are parts of the flood plain that are allowed to flood.	Washlands can't be built on. They're usually used for sports pitches or nature reserves.

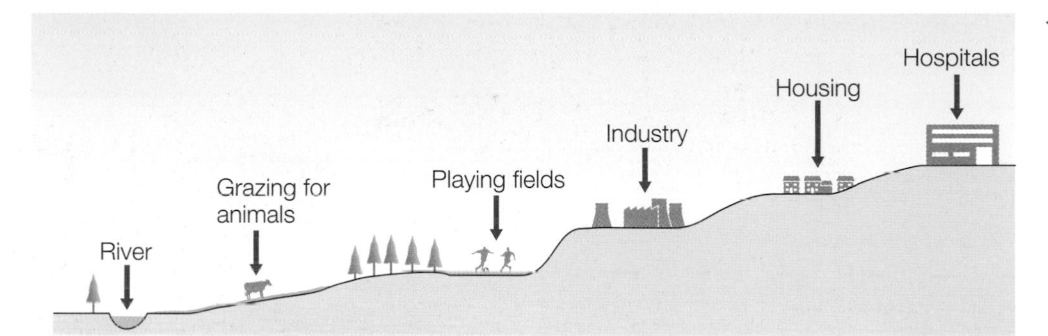

◀ *Flood plain zoning*

Decisions ...

Flooding will keep on happening. In 2007, not only Gloucestershire flooded (see pages 58-59), but Sheffield did as well. Since then, the Sheffield authorities have debated what they can do to avoid serious flooding again. They have a choice of hard or soft engineering solutions. However, any decision has to take into account the costs involved, and the impact on people and the environment.

The Environment Agency believes that hard engineering is not the answer, and that soft engineering will make managing floods more sustainable. And that applies to other places too, not just Sheffield.

Phil Rothwell is the Head of Flood Strategy at the Environment Agency. After the Cumbrian floods, he said: 'The future isn't about hard engineering. We can't stop floods, and, in any case, people don't want us to turn our rivers into canals hidden behind huge embankments.'

Of course, there is one more option – and that is to do nothing.

YOUR QUESTIONS

1. Explain the difference between hard engineering and soft engineering options for managing floods.
2. Work in small groups.
   - **a** Go to the Environment Agency website www.environment-agency.gov.uk and find the Flood Map for Cockermouth. (Follow the link 'Prepare for flooding'.)
   - **b** Go to www.geograph.org.uk and look for photos of the river (either the Cocker or the Derwent) flowing through Cockermouth.
   - **c** Use the Flood Map, the photos, and all the information on this spread to make a decision about the best way to protect Cockermouth from flooding in the future. You need to think about the costs involved, and the impact on people and the environment.

Hint: There is no one 'right' answer. But you must be able to justify the decision you make.

On this spread you'll learn about water use, water stress, and the increasing demand for water in the UK.

Enough water?

The next time you head for the power shower, think again. While sometimes it seems like it never stops raining and we shouldn't be short of water in the UK, consider this. In England and Wales there is only 1334 cubic metres of water available per person per year. That's about half the amount that hotter countries like Italy and Spain have. And in the Thames Valley, in southern England, there are only 266 cubic metres per person – that's just 20% of the average for England and Wales.

Water facts

In England and Wales:

- the amount of water used per household has risen by 70% over the last 30 years – mainly due to the introduction of appliances that use a lot of water, like washing machines and dishwashers.
- only 66% of households owned a washing machine in 1972 but, by 2010, that figure was 94%.
- water is mostly used for washing and for flushing the toilet, but also for drinking, cooking, washing the car and watering the garden.
- the average person currently uses 150 litres of water every day (but this ranges from 107-176 litres, with the highest use in south-east England) – by comparison, someone in Africa uses on average 47 litres a day, while someone in the USA uses on average 578 litres.

Water stress in the UK

The UK is a crowded island, and we're not evenly spread out. The **population density** (the number of people per km squared) is highest in the south-east and in large cities like Birmingham and Manchester (see the map on the right). Rainfall (which is where our freshwater comes from) also doesn't fall evenly across the UK – the west is wetter, and the south and east drier (see opposite).

These facts help to explain why the Environment Agency has described most of south-east England as suffering from 'serious water stress' (see opposite). **Water stress** happens when the amount of water available isn't enough to meet the demand, or is of poor quality. And differences in the distribution of rainfall and population can lead to areas of **deficit** and **surplus**.

- Areas of deficit do not have enough rainfall, or water.
- Areas of surplus have more water than they need.

West Wales, for example, has a high rainfall but a low population density, and therefore it has a water surplus.

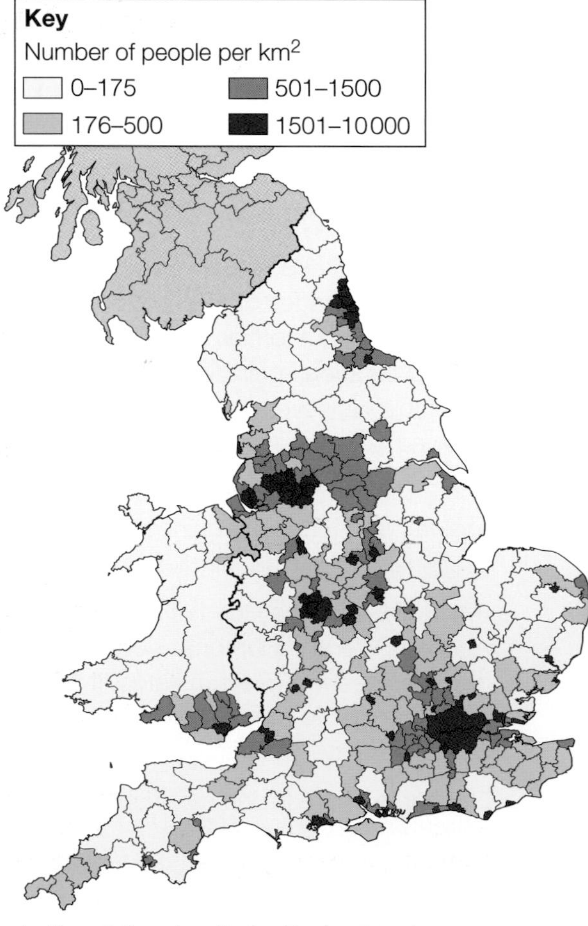

▲ *Population density for England and Wales in 2006*

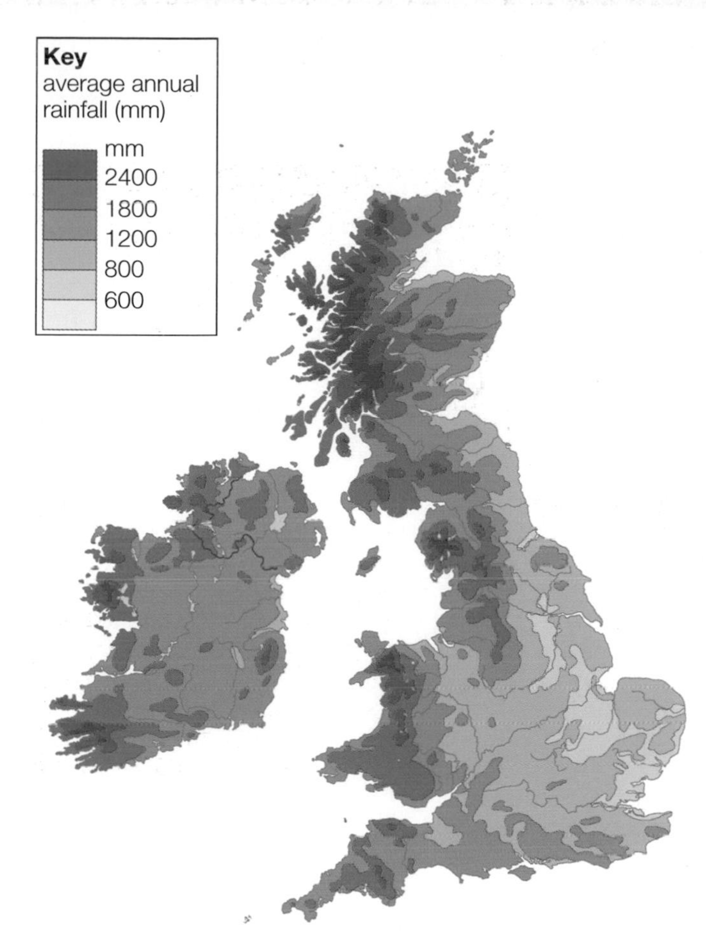

▲ *Average annual rainfall in the UK*

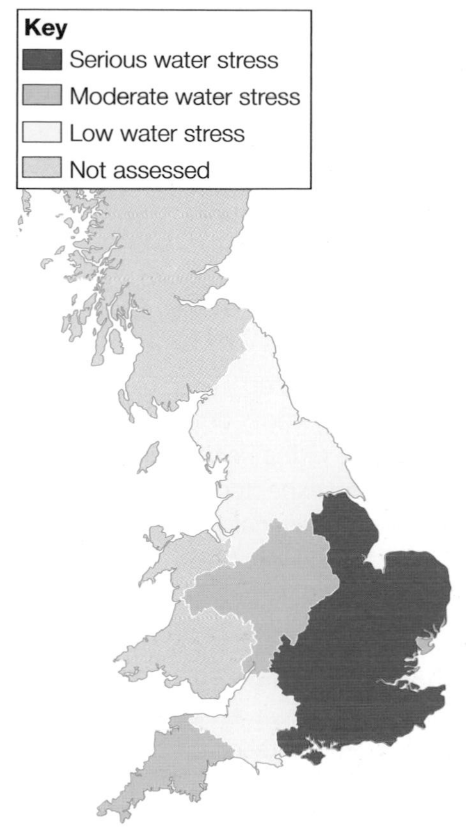

▲ *Water stressed areas in England and Wales in 2007*

Increasing demands

The UK's population is growing – in 2008, it was estimated to be 61.4 million, and was expected to increase by 10 million by 2033 and by a further 10 million – to over 80 million – by 2050.

More people means more housing – and increasing demands for water. The largest increase in demand is expected to be in regions of the UK where water use is already high (like the south-east).

By 2020, the demand for water could be about 5% higher than it is today. That's a staggering 800 million litres of water a day. Not only will individual and household water use increase, but so will industrial use (e.g. in the building industry), and agricultural use (e.g. for irrigating crops).

And, as temperatures rise – because of global warming – drought and water shortages could become more common, especially in the south of the UK.

YOUR QUESTIONS

1 Explain what these terms mean: population density, water stress, water deficit, water surplus.

2 Look at the maps showing population density, rainfall in the UK, and the areas that are currently water-stressed. Annotate a blank map of England and Wales to indicate which areas you think are likely to have a water deficit in the future, and those that are likely to have a water surplus.

3 Write a one-minute radio broadcast to describe the increasing demands for water in the UK.

4 Work in a small group. Think of as many solutions as possible to the problem of water stress. Record your ideas as a spider diagram. (You won't find the answers on this spread, but it will start you thinking about some of the ideas you'll meet on the next spread.)

5.10 » Managing our water supplies – 2

On this spread you'll find out about Kielder Water, and investigate the need for sustainable water supplies.

Kielder Water fact file

- Kielder Water in Northumberland is the biggest man-made reservoir in northern Europe.
- It is 12 km long and up to 52 metres deep.
- It cost £167 million to build, and was completed in 1982.
- It was built to meet an expected increase in water demand from north-east England – because of the rising population there and the expected growth of the steel and chemical industries. However, these industries haven't grown as expected; in fact they've declined.
- It is a **water-transfer scheme (**water is transferred from one area to another). The water from the reservoir is released into the Rivers Tyne, Derwent, Wear and Tees. This helps to maintain river flows when levels are low. Extra water can also be released for household and industrial use.
- Kielder Water can provide up to 909 million litres of water a day (almost as much as all the other sources in the region added together).

▼ *Kielder Water*

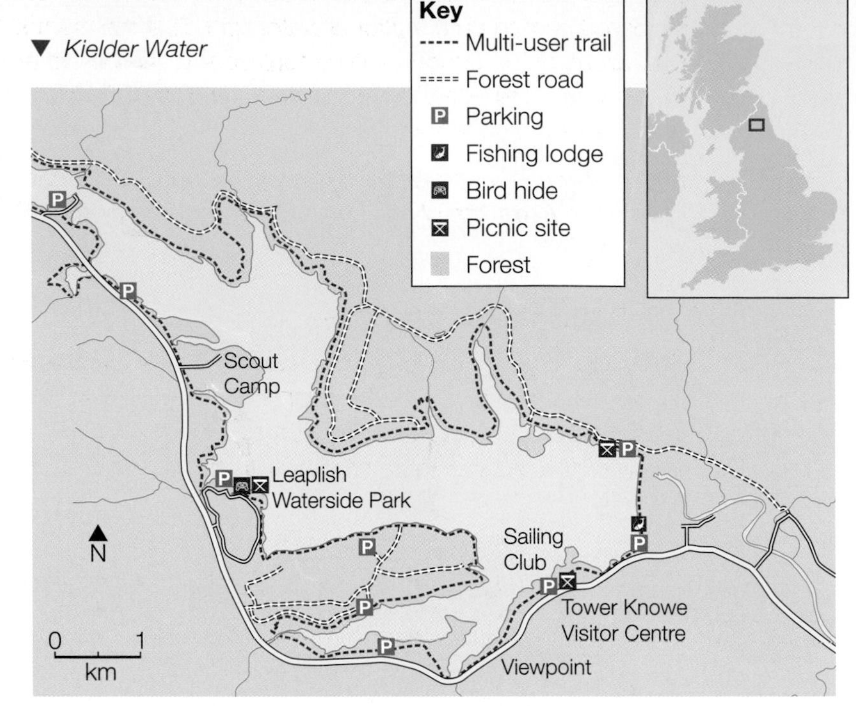

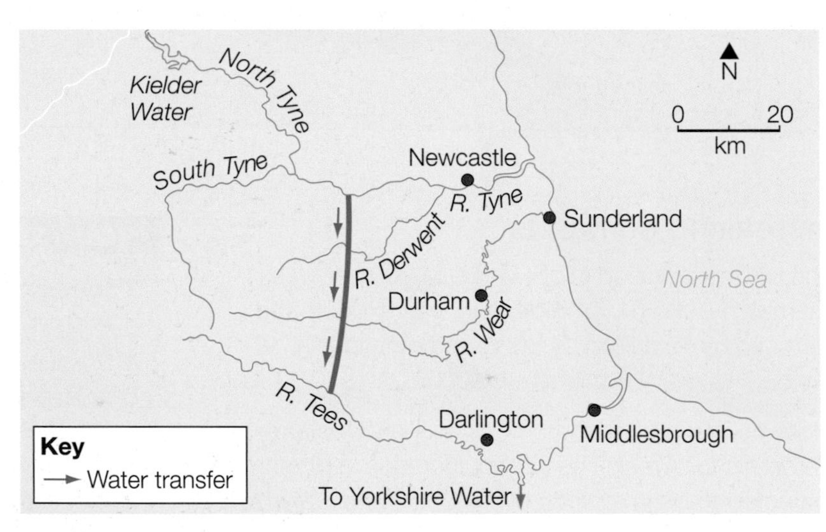

▲ *Kielder Water transfers*

Kielder Water – benefits and disadvantages

Kielder Water has brought both benefits and disadvantages to the area.

- Kielder Water has become a major tourist attraction. This has created jobs and benefited the local economy.
- One and a half million trees were cut down to build the reservoir.
- The north-east now has the most reliable water supply in England.
- Only a few families had to be moved and re-housed when the reservoir was built.
- The release of clean water into the River Tyne has encouraged salmon and sea trout to migrate upriver to breed.
- Forest Park, surrounding Kielder Water, is harvested for timber and employs about 200 people.
- If pollution occurs downstream, clean water can be released to dilute it and flush it out to sea.
- The water is used to generate hydroelectric power at Kielder Dam.

The need for water transfer

Kielder Water is a major water-transfer scheme. Before it was completed, the British government considered setting up a 'national water grid' (like the electricity grid) to transfer water from areas of the country with a surplus (e.g. Wales), to areas with a deficit (e.g. the south-east). This idea became popular in the mid 1970s, because of drought, but it didn't happen in the end.

By 2006, a national water grid was being talked about again – after some lengthy dry periods, and when the impacts of climate change became clearer. So far, it still hasn't happened.

The need for sustainable supplies

The Environment Agency's view about water is that: 'We need to plan so that there are sustainable, reliable water supplies for people and businesses'. It is also concerned about protecting the environment. Some of the ways in which the Environment Agency thinks our water supplies can be made more sustainable are shown below.

YOUR QUESTIONS

1 a Explain what the term 'water-transfer scheme' means.

b Use the map and text opposite to explain how Kielder Water acts as a water-transfer scheme.

2 The Environment Agency wants to reduce water use from 150 to 130 litres per person per day by 2030. Which of the suggestions given above do you think are most likely to make this happen? Explain your answer.

3 a Use the information on 'Kielder Water – benefits and disadvantages' to draw three spider diagrams. One should show the economic issues created by Kielder Water, another the social issues, and a third the environmental issues. Highlight the benefits in one colour and the disadvantages in another colour on each diagram.

Hint: Some issues might appear on more than one diagram.

b Use your spider diagrams to decide whether the benefits outweigh the disadvantages. Explain your answer.

6 » Ice on the land

What do you have to know?

This chapter is from **Unit 1 Physical Geography Section B** of the AQA A GCSE specification. It is about ice – changes in the amount of ice, and the landforms and landscapes that ice has created. The table shows how the pages in this chapter match the content in the specification.

Specification content	Pages in this chapter
The amount of ice on the land has changed over time. Evidence for changes in ice cover.	p118-119
The glacial budget, and why glaciers advance and retreat.	p120-121 Case study of the Mer de Glace p120-121
Ice shapes the land as a result of weathering, erosion, transportation and deposition.	p122, 126-127
Different landforms are created as a result of erosion, transport and deposition.	p123-129
Landscapes affected by snow and ice attract tourists, and this has a range of impacts. Tourism needs to be managed.	Case study of the Alps p130-133
Avalanche hazards.	p134-135
Retreating glaciers can have economic, social and environmental impacts.	p136-137

Your key words

Ice sheet, glacier

Pleistocene Era, Ice Age, glacial periods, interglacial periods

Firn

Glacial budget, accumulation, ablation

Erosion (plucking, abrasion)

Weathering (freeze-thaw)

Corries (cirques, cwms), tarn (corrie lake), pyramidal peak, arêtes, hanging valleys, glacial troughs, truncated spurs, ribbon lakes

Rotational slip, bulldozing

Boulder clay (till)

Moraine (lateral, medial, terminal, ground)

Drumlins

Avalanche

Exam help ...

Advice See pages 297-299 for information on how to be successful in your exams.

Practice See page 305 for exam questions on this chapter.

What if ...

- we were plunged into an ice age?
- all the glaciers melted?
- you got hit by an avalanche?
- you could go inside a glacier?

On this spread you'll learn about glaciers and ice sheets, and how the amount of ice has changed over time.

What's so special about ice?

People will travel thousands of miles to see the kind of scenery shown in the photo on the right.

- Places like Greenland and Antarctica are cold and desolate, but they have an outstanding beauty that attracts tourists – as well as wildlife and adventure sports enthusiasts.
- They are also important study areas for scientific researchers, who use them to gather evidence of changes in global climate over thousands of years.
- These ice-bound regions have their own special ecosystems and wildlife.

▲ *The Greenland ice sheet*

your planet
The ice in Antarctica is over 5 kilometres thick in some places!

What's the difference between a glacier and an ice sheet?

An **ice sheet** is a huge mass of ice that covers a vast area of land (see the photo above). The two largest ice sheets today are in Greenland and Antarctica. Between them, they account for 96% of the ice-covered land on Earth.

A **glacier** is like a river – only it's made of ice and flows much more slowly. Glaciers begin high up in the mountains, where snow collects in hollows on the mountainside and compacts to form ice. As the glacier begins to slide downhill – due to the weight of the compacted ice and gravity – it follows the easiest route and flows down river valleys. There are many glaciers in the Alps, like the Gornergrat Glacier in Switzerland (on the right), but glaciers are found all over the world at high altitude – even on Mt Kilimanjaro in Tanzania, which is close to the Equator!

▶ *The Gornergrat Glacier at Zermatt in Switzerland*

Ice in the past ...

For long periods in the distant past most of Britain was covered by thick ice sheets – along with about 30% of the planet! The last Ice Age began around 2 million years ago, during the **Pleistocene Era**. During this Ice Age there were cooler times (**glacial periods**), when the ice advanced, and warmer times (**interglacial periods**), when the ice retreated. Much of Britain at that time would have looked like Antarctica or Greenland does today. The maps below show how far the ice sheets extended during the last Ice Age.

An **Ice Age** is a period of time when ice sheets are found on the continents. So some scientists believe that we are still in an Ice Age – but in an interglacial period.

The **Pleistocene Era** is a period of geological time which began about 2 million years ago and ended about 10 000 years ago.

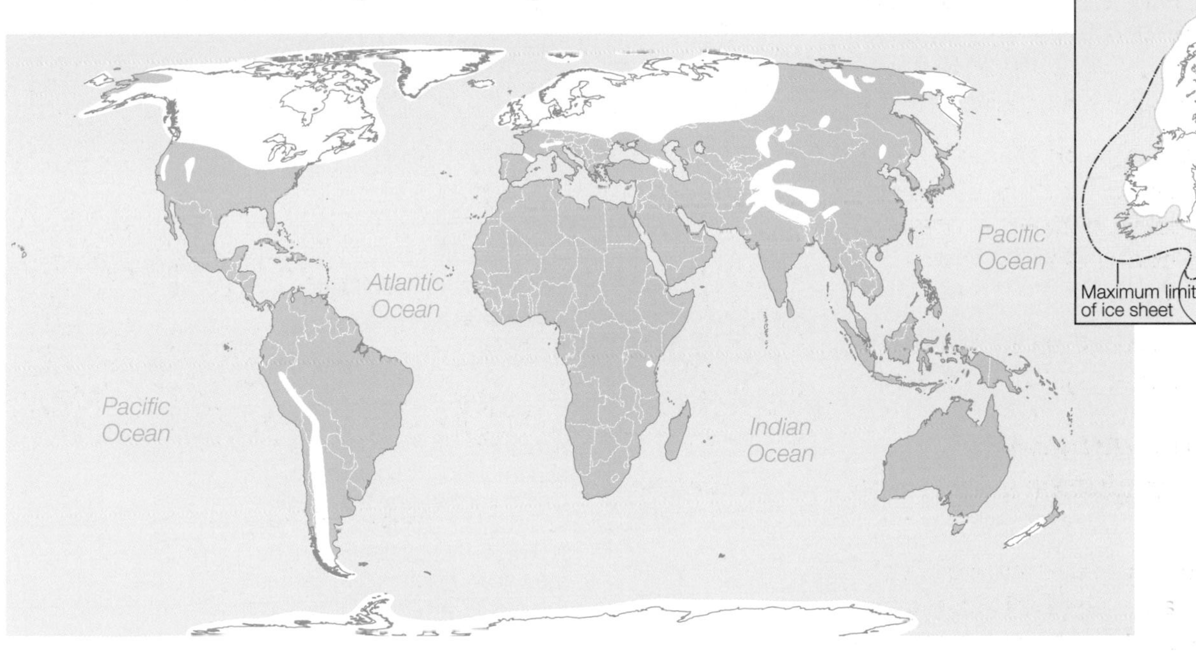

▲ *The maximum extent of the ice sheets during the last Ice Age*

How do we know what happened in the past?

- By taking core samples from existing ice sheets and looking at the chemical composition of the ice (and the marine organisms trapped in it at different depths), scientists can work out changes in temperature over the last few thousand years.
- Fossils are also good indicators of climate. Some fossilised plants and animals found in the UK could only have survived in a warmer climate. Others must have lived in a colder climate – glacial and interglacial periods.
- Some of the landforms found in Britain today, e.g. valleys in Scotland and the Lake District, could only have been formed by ice. So these can help to pinpoint where the ice sheets reached in the past.

... And ice now?

Ice now covers about 10% of the world's surface – much of it in Antarctica.

- For the last 150 years, ice sheets and glaciers all over the world have been melting and becoming smaller.
- But – in the last 20 years – the rate of melting has speeded up noticeably.
- Scientists believe that this increased melt rate is due to rising global temperatures.

YOUR QUESTIONS

1 Start a dictionary of key terms for this chapter. Begin with: ice sheet, glacier, Ice Age, Pleistocene Era, glacial period, interglacial period. Add an explanation for each term.

2 Find out the locations of Fox Glacier, the Mer de Glace, and the Athabasca Glacier. Draw up a table that shows the continent, the country and the mountain range for each one.

3 What evidence do scientists use to show how climate has changed?

6.2 » Glaciers – advance and retreat

On this spread you'll find out how, and why, glaciers advance and retreat.

From snowball to glacier

We've all made a snowball by squeezing the snow in our hands until it becomes hard and solid. Ice forms at the bottom of a glacier in exactly the same way – newly fallen snow each year compresses the previous year's snow. The individual snowflakes are compacted into smaller particles of ice, which stick together to form larger granules – called **firn**. With more snow falling each year, eventually the firn is compressed to form solid glacial ice.

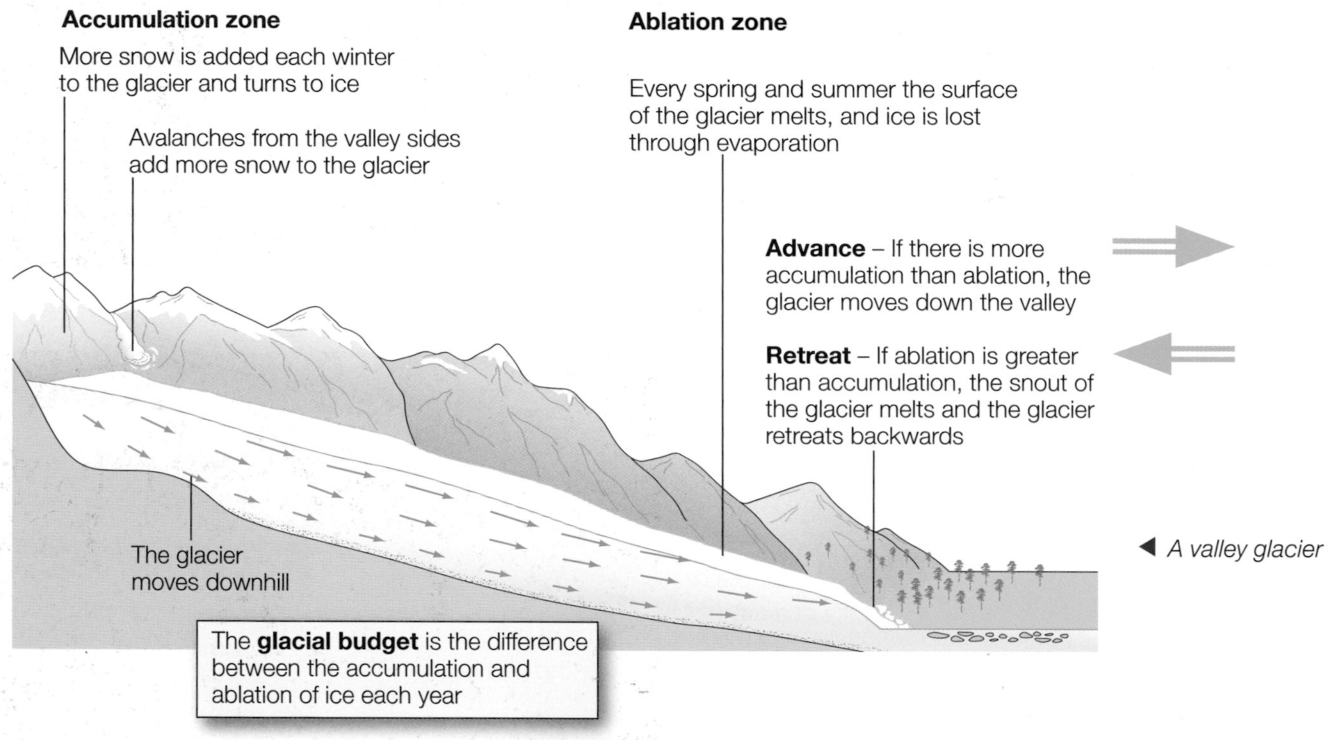

◀ *A valley glacier*

The shrinking Mer de Glace

The Mer de Glace (Sea of Ice) glacier in the French Alps is retreating. Since 1850, it has become much shorter and thinner. Although in the 1970s and 1980s it actually advanced by 150 metres, it's now retreating again by about 30 metres a year (it's already 500 metres shorter than it was in 1994).

▶ *The Mer de Glace in the French Alps*

The Mer de Glace:

- is the largest glacier in France
- is 7 kilometres long and 200 metres deep
- flows NNW from Mt Blanc towards Switzerland
- is close to Chamonix.

▲ *A painting of the Mer de Glace in 1893*

How do we know that the Mer de Glace is retreating?

There is all sorts of evidence:

- Nineteenth-century paintings, like the one on the right, show how much deeper and wider the glacier used to be. Compare that with the photo from 2008.
- Aerial photos and satellite images show recent changes in the position of the glacier.
- Old maps from a century ago show the extent of the ice at that time.
- Debris left by the glacier on the valley's floor and sides as it retreats gives an idea of how much longer and wider it used to be.

Why is the Mer de Glace retreating?

Records show that average temperatures in France have risen by 1°C in the last 100 years. But above 1800 metres (i.e. in the Alps), temperatures have risen by 3°C in the last 40 years. Not only that, but winters have become drier – especially the winters of 1989, 1993 and 2002. And in the mountains a drier winter means less snowfall. So, while a glacier would normally advance in the winter and retreat in the summer, the combination of warmer winters and normal summer melting means that glaciers like the Mer de Glace are retreating all year round. As a result, scientists believe that there is a direct link between retreating glaciers and climate change.

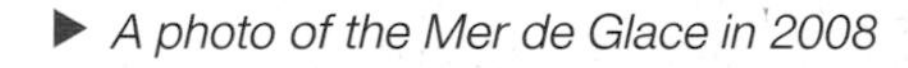

► *A photo of the Mer de Glace in 2008*

YOUR QUESTIONS

1 Write a definition for these terms: firn, accumulation, ablation, glacial budget. Add them to your dictionary of key terms.

2 Copy these phrases and choose the correct words from the right-hand side.

accumulation > ablation = glaciers advance/retreat in winter/summer

ablation > accumulation = glaciers advance/retreat in winter/summer

3 Draw a flow diagram to show how snow turns into glacial ice.

4 Write an e-mail to a friend describing how and why the Mer de Glace has changed from the time the picture above was painted to the present day.

Hint: You need to include information about how the glacier has retreated, and the evidence we can use to tell us why it is retreating.

6.3 » Glaciers at work – 1

On this spread you'll learn how glacial erosion leads to the formation of distinctive landforms.

Making the impossible happen

In 2008, Dean Potter (a 38-year-old American) scaled the 4300 metre-high Eiger mountain in Switzerland – without using any safety equipment – and then jumped off the top! Dean is famous for his extreme adventures, and he climbed the Eiger using his free-base climbing method. 'Free-base is a combination of free climbing and base jumping' said Dean. 'Free climbing is where you climb a mountain without any safety equipment, and base jumping involves diving off high mountains – flying through the air and breaking your fall with a parachute.'

(Adapted from an article on the Mail Online website)

▲ Dean Potter free climbing the Eiger in 2008

How do glaciers shape the land?

Ice is a powerful force. It shapes the land using the processes of erosion, weathering, transportation, and deposition – carving out landscapes and creating distinctive landforms. It helps to create spectacular scenery that people want to see and climb (or jump off in Dean Potter's case).

your planet

At the beginning of 2010, temperatures stayed around freezing for days on end in some parts of the UK, so lots of freeze-thaw weathering would have been going on!

Erosion

- **Plucking** – Have you ever picked up an ice cube with a slightly damp hand and found that it sticks to your finger? The same thing happens when glaciers move over bedrock – loose rocks stick (or freeze) to the bottom of the glacier. As the ice continues to move forward, it pulls them away from the bedrock ('plucks' them), and later deposits them further down the valley.
- **Abrasion** – Rocks that are stuck in the bottom of the ice act like sandpaper – grinding over the bedrock. This can leave scratches called **striations**.

Weathering

Freeze-thaw weathering happens when the temperature is above freezing during the day and below freezing at night.

Water gets into cracks in the rock and freezes when the temperature falls below 0°C. When water turns to ice, it expands by about 9% – this forces the cracks apart. The ice may melt during the day and more water may be added before it refreezes at night. If this happens often enough, the cracks will widen and large pieces of rock will eventually fall off.

Cracks and joints in rock fill with water

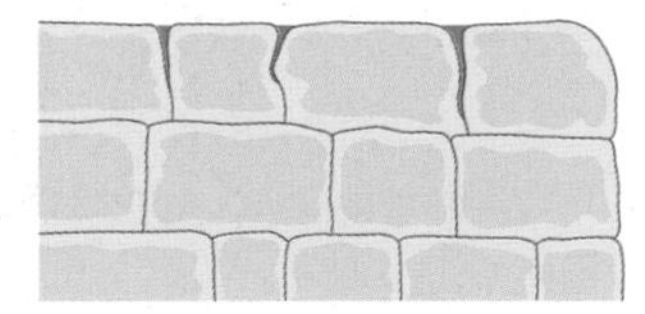

The water freezes and expands – pushing the rocks apart

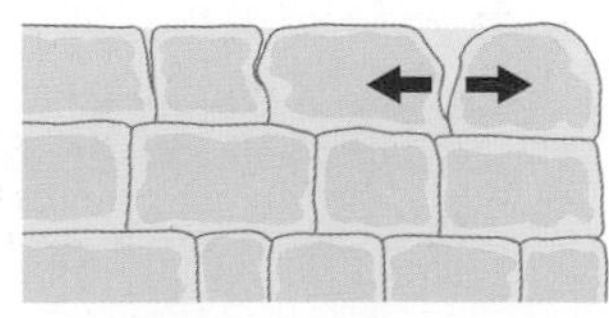

Repeated freezing and thawing widens the crack, and eventually bits of rock fall off

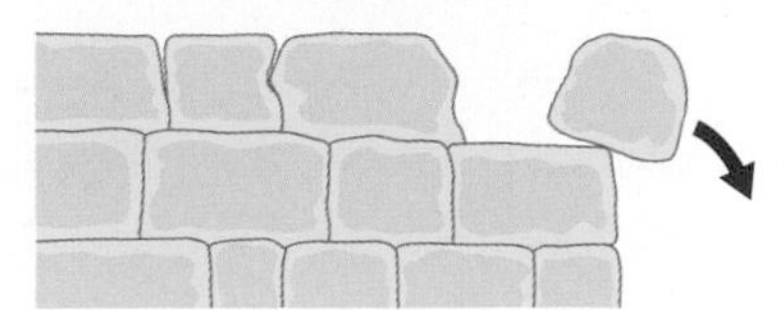

▶ *Freeze-thaw weathering*

Landforms created by erosion

Glaciers and ice create a lot of different features. Some are described here, and others on Spreads 6.4 and 6.5.

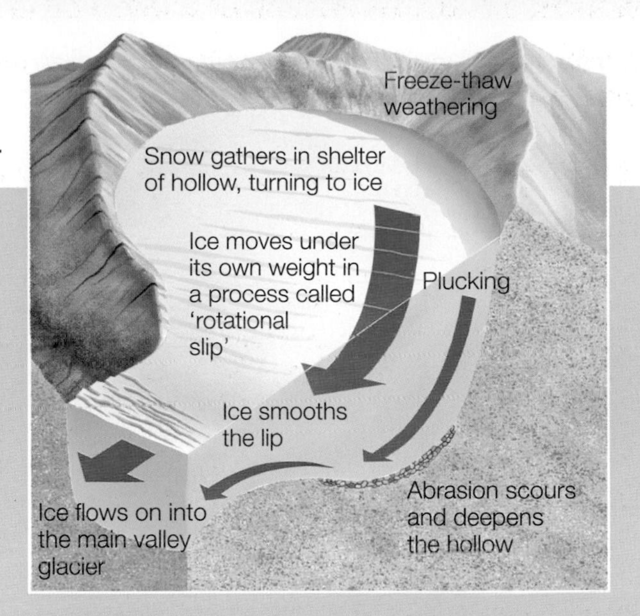

Corries, cirques or cwms

A **corrie** is a steep-sided hollow, formed by a small glacier. They're called **cirques** in France and **cwms** in Wales.

- It starts as a sheltered hollow high up on a mountainside that fills up with snow.
- The sheltered high-level snow doesn't melt in the summer. So, as more continues to collect in the hollow each winter – and becomes tightly compacted – a small glacier eventually forms.
- Plucking and freeze-thaw weathering make the corrie's **back wall** very steep.
- Rocks fall on to the ice and help to scrape away the base of the corrie.
- As the glacier moves, it deepens the hollow by plucking and abrasion.
- The glacier moves down and out of the corrie in a circular way (called **rotational slip**).
- There is less erosion at the front of the corrie, so it forms a lip or rock bar.
- When the ice melts, a small lake often forms in the hollow – called a **tarn** or **corrie lake**.

Arêtes

An **arête** is a narrow ridge that's formed when two corries cut back towards each other. The back wall of each corrie is worn away by freeze-thaw and plucking – leaving a sharp ridge between them. The photo below shows a corrie and an arête above the Gornergrat Glacier in Switzerland.

Pyramidal peaks

A **pyramidal peak** is formed when three or more corries cut back towards each other at the top of a mountain – forming a pyramid shape. The Matterhorn (below) is one of the most famous pyramidal peaks in the world.

YOUR QUESTIONS

1. Explain the terms: abrasion, plucking, freeze-thaw weathering, rotational slip. Now add them to your dictionary of key terms for this chapter.
2. Use the information above to draw a flow diagram showing how a corrie is formed.
3. Draw an annotated sketch to show how a pyramidal peak is formed. Don't forget to include the words: plucking, abrasion, freeze-thaw weathering, three corries, cutting back

   Hint: 'Annotate' means adding explaining labels.

On this spread you'll find out about other landforms caused by glacial erosion

More erosional features

Glacial troughs, truncated spurs, and hanging valleys.

When a glacier moves downhill, it usually takes the easiest route. That often means down an old river valley. Note the interlocking spurs of land and the river tributaries feeding into the main river valley in the diagram above.

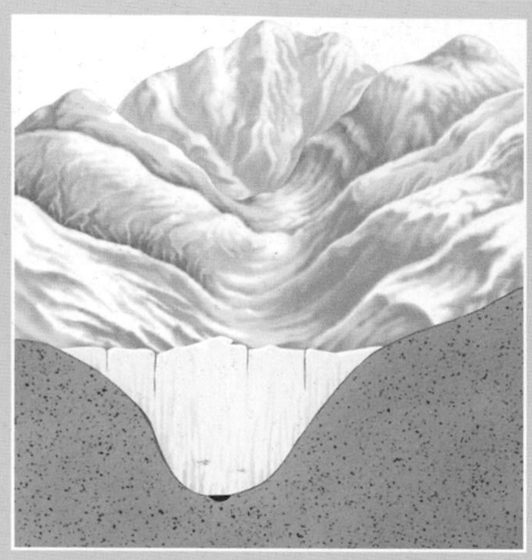

Over time, the glacier carves its way through the rock. It erodes a deeper and wider valley by **abrasion** and **plucking**.

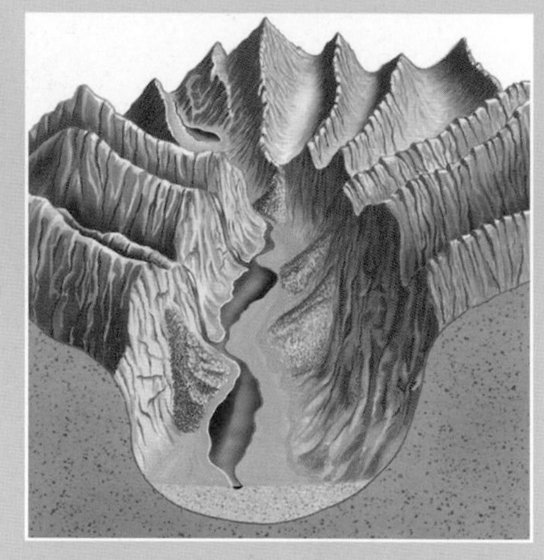

When the ice eventually melts, and the glacier retreats, a **glacial trough** is left behind. Note that the old interlocking spurs have been eroded away to leave **truncated spurs**. The river tributaries have also been left hanging high up on the valley sides – so they're called **hanging valleys**.

Hanging valleys and waterfalls

The photo on the right shows a hanging valley in New Zealand. To spot one of these, you need to look out for a tributary valley high up on the side of the main valley – with a steep vertical drop to the valley below (as in the photo). The tributary valley once contained a smaller glacier that fed into the main glacier below it. When all the ice melted, the tributary valley was left high above the heavily eroded main valley floor – just hanging.

Rivers or streams often flow through these hanging valleys. When they reach the end of the hanging valley, they tumble down the main valley side as a **waterfall**, like Stirling Falls on the right.

▶ *Stirling Falls in Milford Sound, New Zealand. This waterfall is 155 metres high.*

Glacial troughs, truncated spurs, and ribbon lakes

The photo on the right, of Wast Water in the Lake District, shows the glacial trough left behind when the glacier that formed it retreated. Glacial troughs are also called U-shaped valleys – because they're shaped like a U! They have steep sides and a flat valley floor.

When the glacier eroded the old river valley, it wore away the original interlocking spurs – leaving **truncated spurs** behind (truncated means cut short, or broken off).

In some glacial troughs, like this one, long narrow lakes fill the valley floor. These are called **ribbon lakes**. Wast Water, Windermere and Coniston are some famous ribbon lakes in the Lake District.

Why are valleys carved by glaciers called **glacial troughs**? Here's a clue!

your planet

Wast Water is 76 metres deep – the deepest lake in England. But Loch Morar in Scotland is 310 metres deep!

YOUR QUESTIONS

1 Name four features formed by glacial erosion. Add them to your dictionary of key terms for this chapter.

2 Describe the shape of a glaciated valley.

3 a Explain why hanging valleys are left above the level of the main valley.

b What happens if the hanging valley has a river or stream flowing through it?

4 a Draw a sketch of the Wast Water photo above.

b Annotate it to explain how truncated spurs, ribbon lakes and glacial troughs are formed.

Hint: To draw a sketch from a photo, first of all draw a frame the same shape. Draw in the main features. Don't spend time drawing details that you don't need (like the people in this photo)! Give your sketch a title.

6.5 » Glaciers at work – 3

On this spread you'll discover how glaciers transport material and then deposit it to form distinctive landforms.

Glaciers on the move

Most glaciers move very slowly – grinding their way down a mountain. They usually move less than 30 cm a day. As they move, glaciers don't just erode the landscape – they also transport material from one place to another:

- Rock fragments fall onto the surface of the glacier as the result of freeze-thaw weathering on the valley sides above it.
- Rocks are also plucked from the valley floor as the glacier moves over it.
- As well as transporting material on and under it, the glacier also pushes rocks and rubble in front of it as it moves down the valley – called **bulldozing**.
- All material carried or moved by the glacier is called **moraine**, but there are different sorts of moraine, as you will see below.

Glacial deposition

When a glacier stops moving, or melts, it drops the material that it's been transporting. This is called **deposition**. The deposited material can range in size from boulders the size of a bus, to very fine dust called **rock flour**. The deposited material is called **boulder clay** or **till**.

The landforms created when a glacier deposits material include: lateral, medial, terminal, and ground moraine – and drumlins. You can see these landforms in areas that have glaciers today, but you can also find them in Britain – so they're evidence that glaciers and ice sheets once covered much of this country.

Depositional landforms

Lateral moraine

This is found along both edges of a glacier, and is made up of weathered rock fragments from the valley sides above it. The photo on the right shows tributary glaciers flowing down to join a main valley glacier. Look carefully and you can see lines of rubble along the edges of the glaciers. These are lateral moraines. When the ice eventually melts, the lateral moraines will form ridges along the sides of the valleys.

Medial moraine

This is found in the middle of a glacier, and is caused by two lateral moraines joining together when two glaciers meet (as in the photo). When the ice melts, the medial moraine will form a ridge down the middle of the valley.

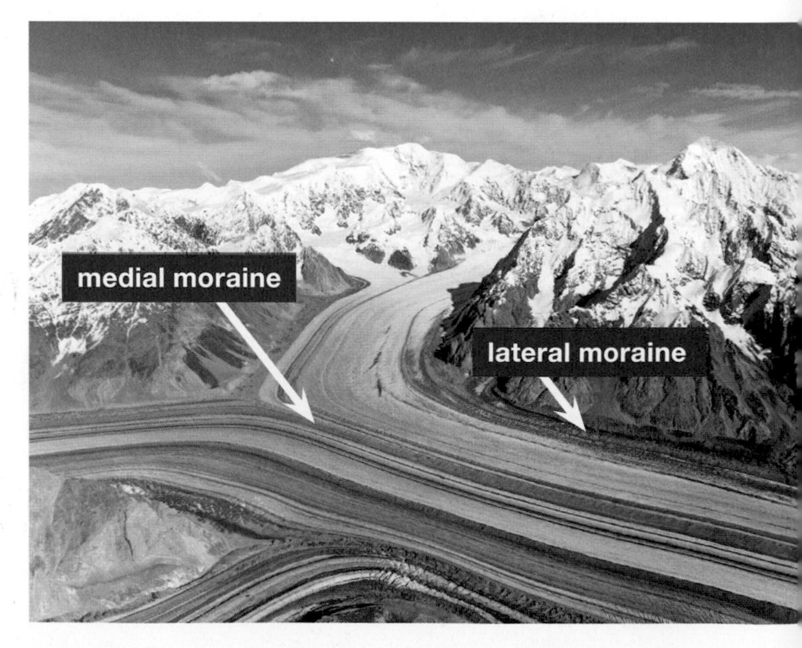

▲ *Lateral and medial moraines on the Barnard Glacier in Alaska*

Terminal moraine

This forms when material is deposited at the very end of a glacier (that's why it's called 'terminal'). It's a ridge of deposited material that runs across the glacier's snout. Terminal moraine shows exactly how far the glacier managed to move down the valley at its greatest extent. It can be tens of metres high and is often a semi-circular shape.

Ground moraine

This is a thin layer of material left under the glacier on the valley floor as the ice melts.

Drumlins

Glaciers can shape the moraine on the valley floor into small hills – called **drumlins**. These can be up to 50 metres high and 1000 metres long. Drumlins have one steep, or blunt, side and one gently sloping side. They are often found in groups, or 'swarms', on the valley floor. This is called a 'basket of eggs' topography or landscape. A very good example of a swarm of drumlins can be seen in the Yorkshire Dales, near Hellifield.

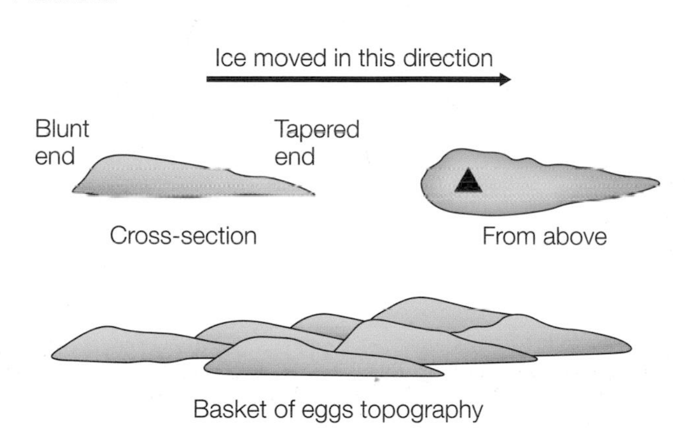

▲ *Drumlins*

▲ *Terminal moraine left behind after the retreat of the Miage Glacier in the Italian Alps*

▲ *A drumlin field at Ribblehead in the Yorkshire Dales*

YOUR QUESTIONS

1 Define the terms: moraine, boulder clay, and drumlin. Add them to your dictionary of key terms for this chapter.

2 Copy and complete the table to explain where the different types of moraine are found.

	Lateral moraine	Medial moraine	Terminal moraine	Ground moraine
Location				

3 Find and download a photo of a drumlin from the Internet. Annotate your photo to explain how drumlins are formed.

6.6 » Glacial landforms on maps

On this spread you'll learn what glacial landforms look like on an OS map.

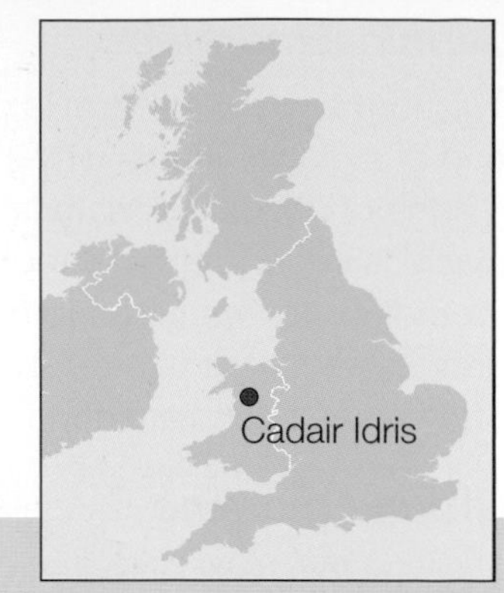

Cadair Idris

Cadair Idris was formed as glaciers ground their way through Europe in the last Ice Age. It's the second most popular mountain in Wales, after Snowdon. This is because it has many outstanding examples of glacial landforms located very close together. And, because it's in Wales, some of these features have Welsh names. So, for example, a corrie is called a cwm on the OS map.

Recognising glacial features on an OS map

Corries

Find these grid squares on the OS map extract opposite: 7011 and 7111. You can see that the contours (the brown lines) lie in a semi-circular pattern. They're close together, indicating the steep back wall of the corrie. The stream Cwm Amarch flows south out of the corrie. The black dots show the loose rock or scree that has broken off the back wall due to freeze-thaw weathering.

▲ *Craig Cwm Amarch*

Arêtes

Find this grid reference on the map – 710120. You can see three narrow ridges, or arêtes, radiating out (Craig Cau, Craig Cwm Amarch and Craig Lwyd). These arêtes have steeply sloping sides, which are shown by the black cliff symbol. The black dashed line of the Minffordd Path shows the line of one of the arêtes.

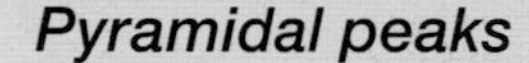

Pyramidal peaks

You find these where three or more corries cut back and meet. At grid reference 710120 three corries meet at the spot height 791. This is as close to a pyramidal peak as you can see in Britain.

Tarns

Llyn Cau on the right is a tarn inside a deep corrie. You can find it in grid square 7112. The black symbols on the map show the steep rocky cliffs around the tarn.

▲ *Llyn Cau*

Truncated spurs

Look at grid reference 733106. This shows a truncated spur. Look at the shape of the contours as they change from semi-circular ones at the top of the slope, to straighter ones at the bottom (near the road). This shows that the old interlocking spur has been truncated.

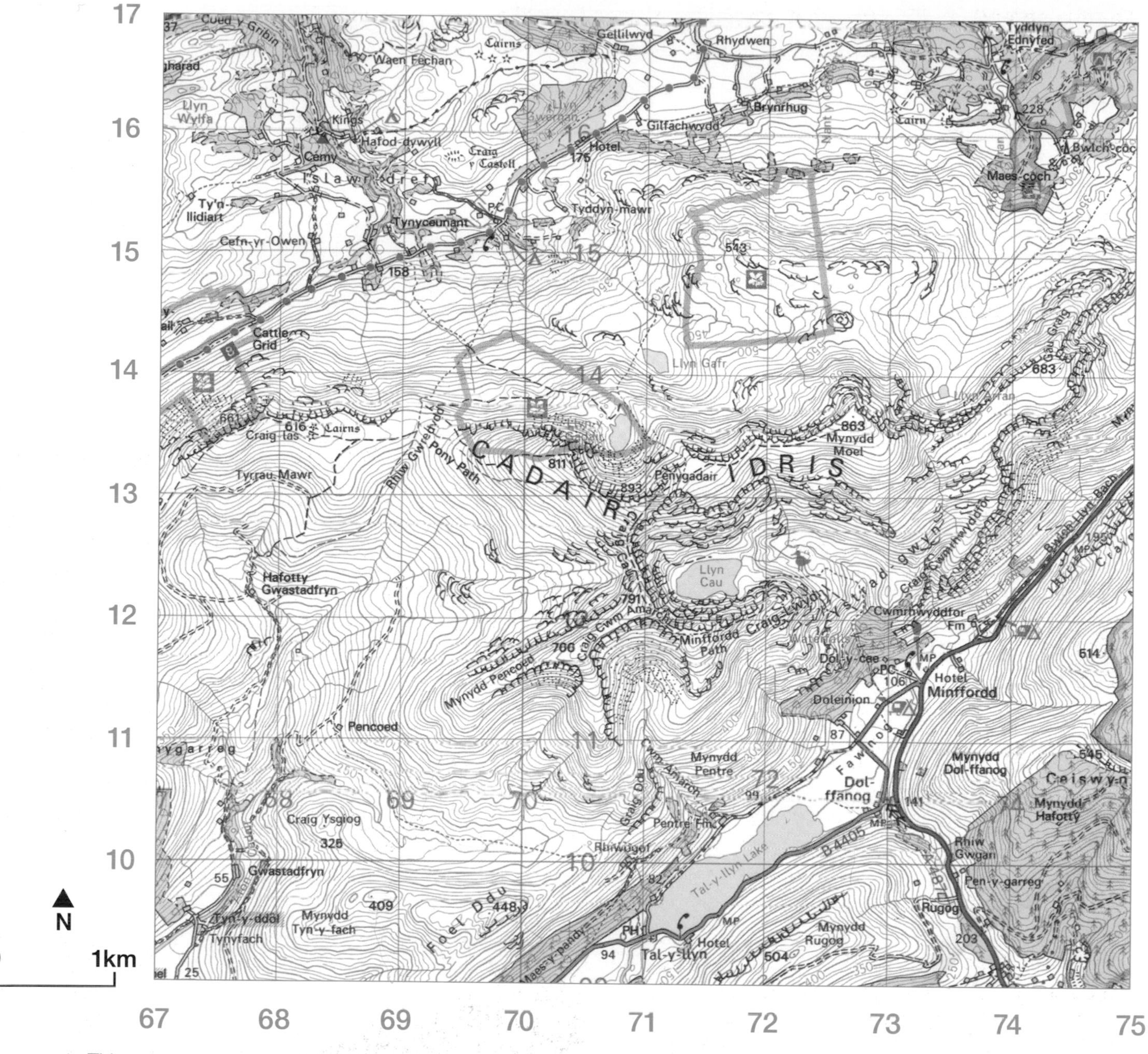

▲ *This map extract is from the Ordnance Survey Landranger series, map 124, 1:50 000 scale*

YOUR QUESTIONS

1 Find the summit of Cadair Idris in grid square 7113. (It's marked by a triangulation point.) What's the height in metres?

2 What's the height of the road that runs alongside Tal-y-llyn Lake at grid reference 719106?

Hint: There are contour lines every 10 metres, and every fifth one is a darker brown.

3 Match the following glaciated landforms with the grid references on the right:

corrie/cwm, ribbon lake, corrie lake/tarn, truncated spur, pyramidal peak, glacial trough, arête.

a 720100 **b** 727108 **c** 710118

d 715124 **e** 733110 **f** 709121

g 710125

4 a Look at the photo of Llyn Cau opposite and find it on the map.

b Name two glacial landforms in the photo.

c In which direction do you think the camera was pointing when the photo was taken?

d At about what height was the photo taken?

6.7 Tourism in the Alps

On this spread you'll learn how landscapes affected by snow and ice attract tourists.

Winter tourism

Fancy going snowboarding or skiing this winter? Millions of people do. The Alps are Europe's main area for winter tourism – attracting 120 million visitors a year. The tourists come for winter sports like skiing and snowboarding, as well as ice climbing, cross-country skiing and bobsleighing. But people don't just go to the Alps for fun in the snow. More-extreme places, like Antarctica, are now attracting tourists (see pages 284-287). Even the world's highest mountain, Mt Everest, now attracts its share of 'normal' tourists, who go trekking as far as Everest Base Camp (see page 14).

There are a lot of lively resorts in the Alps, with many of them developing and expanding in the last 50 years. Although many tourists also visit the Alps in the summer, for walking holidays, it's the winter snow that attracts most people and has led to the widespread resort development.

La Plagne

La Plagne is a typical example of a modern Alpine ski resort in the French Alps. It was built in the 1960s at a height of 2000 metres. That's a high enough altitude to guarantee snow from December to April. It also has a glacier (Glacier de Bellecote), which can be used for skiing in the summer.

▼ *La Plagne*

The resort has a large selection of bars, restaurants and nightclubs.

There are a variety of slopes for all levels of skiing – from beginners to advanced.

You can ski from your doorstep.

For non-skiers, there's paragliding, husky sledding, snowshoeing and sleigh-riding – as well as the Olympic bobsleigh run for the more adventurous!

- Four airports serve the French Alps, so the resorts there are easy to reach from other European countries. Plus there are links to Europe's motorway network, and the TGV (the fast train from Paris).
- La Plagne consists of six high-altitude villages and four lower-village resorts – all linked up by a system of lifts. Visitors to La Plagne can also ski in Les Arcs – a large resort connected to La Plagne by cable car. This link has created one of the largest ski areas in the Alps – Paradiski.
- Paradiski has 425 km of pistes (ski runs), and endless off-piste opportunities. There are also snow parks and plenty of opportunities for snowboarders.
- La Plagne expanded in the 1970s with high-rise apartment blocks, but some of the villages (such as Belle Plagne and Plagne Soleil) are built in a traditional style and are traffic-free, with underground parking.

▲ *The station for the Montenvers train is perched on a rocky ledge. It has a restaurant and viewing platform with breathtaking views of the mountains and the glacier.*

Chamonix and the Mer de Glace

Chamonix is a town in the Alps on the border between France and Italy. Mt Blanc (Europe's highest mountain) towers above the town. Chamonix has a permanent population of 10 000, but this can increase to 100 000 on a winter's day and 130 000 on a summer's day.

The town heaves with skiers in the winter, and with mountaineers and tourists in the summer. The tourists come for the scenery, walking and mountain biking – as well as the shopping and restaurants. People have been coming to Chamonix to see the nearby Mer de Glace glacier since the Montenvers Railway was first opened in 1908. This mountain railway takes tourists to the edge of the glacier. They can then take a cable car down to a cave carved right into the heart of the glacier. This cave contains a collection of ice sculptures. However, because the glacier is constantly moving, the cave has to be recut every year!

▲ *Inside the ice cave in the Mer de Glace*

YOUR QUESTIONS

1. Name four sporting activities associated with snow.
2. Prepare a poster for La Plagne. Include the reasons why people go there, what activities they can do, how they get there, and some information about the resort itself.
3. Write a postcard to a friend about a visit to the Mer de Glace ice tunnel.
4. Work with a partner to create a spider diagram showing the impacts that tourism might have on a place like La Plagne.

   Hint: You won't find the answers to this question on this spread. This question is meant to make you think, and prepare for the next spread.

6.8 » Is tourism good for the Alps?

On this spread you'll find out about the impacts of tourism in the Alps, and some of the management strategies used to deal with them.

The impacts of tourism

Winter sports have changed many Alpine villages over the last 50 years, but are all the changes good?

Economic impacts

- Alpine tourists put money into the local economy by spending in resort shops, ski schools, hotels and restaurants. Before tourism arrived, the lower slopes of the Alps were a farming area that provided few jobs. Now a thriving economy exists.
- Tourism also provides jobs and a demand for services that doesn't stop when the snow melts. In the summer, outdoor enthusiasts like walkers, sightseers and nature lovers ensure that the demand for services continues.

▲ *Tourism provides a range of employment in the Alps*

Social impacts

- Many local young people now stay in their villages when they leave education, rather than migrating to urban centres. This is because tourism provides them with employment – but not just in shops and hotels. For example, skilled people like engineers are needed to keep equipment like ski lifts working.
- However, there have been negative social impacts as well. The traditional way of life in the mountains has changed, and there has been a decline in local crafts and skills as a result of tourism.

Environmental impacts

- Large amounts of water and energy are needed to keep the Alpine resorts running, and for making artificial snow. The Olympic Bobsleigh run in La Plagne uses millions of gallons of water to maintain its icy surface.
- Noise from machinery like ski lifts scares wild animals and disturbs the peace.
- Forests are cleared to make room for more tourist accommodation or ski slopes, and this can lead to soil erosion and changes in the water cycle.

▲ *This is what skiing can do to the environment*

- Alpine vegetation is destroyed by the building of access roads and other infrastructure.
- Skiing over thin snow can damage the fragile vegetation underneath and leave scars of bare earth in the summer.
- An increase in vehicles, such as transfer coaches from airports, leads to more air pollution. This pollution often becomes trapped in the Alpine valleys, which affects the villages – and is thought to be the main cause of damage to trees.

Conflict

Many people welcome the benefits that tourism brings, but not everyone is happy with it. Conflicts can arise between different groups of people. For example, some people come to the Alps for the peace and quiet of the mountains, while others come for active sports like skiing in winter and mountain biking in summer – and partying all night! Conservationists can see the damage that tourism is doing, and they're unhappy about further Alpine development. But many local people feel that tourism is good for the local economy and they want much more development. With all these different views, Alpine tourism has to be managed carefully.

Managing tourism

Belle Plagne

Belle Plagne is the highest of the villages in the La Plagne resort. It's careful development has minimised its environmental impacts.

◀ *Belle Plagne*

New buildings are made of natural materials, like wood, and built in traditional styles.

It has underground parking, so it's traffic-free all year. This means that you can ski all the way to your apartment or hotel – and the village has a clean, pollution-free atmosphere.

Avalanche fences have been built on the hillside above the village – to stop any avalanche caused naturally, or by the activities of tourists, from threatening lives and buildings.

Belle Plagne is above the tree line (the height at which most trees grow), so no trees have to be cut down to meet the demands of tourism, e.g. for ski runs.

La Plagne is part of the Vanoise National Park, which is promoting conservation.

Chamonix – responsible tourism

As well as being a major ski resort in winter, Chamonix attracts thousands of visitors in the summer. Its two great attractions are the Mer de Glace (page 131) and Mont Blanc. The local council has developed a policy of 'integrated tourism', which educates and involves visitors and local people in the protection of the Alpine environment. It:

- provides free public transport, to minimise the use of cars by visitors
- encourages local farming to maintain a traditional way of life alongside tourism
- provides education about the glacier and the environment
- has links with nearby Italy and Switzerland to manage the international flow of traffic through the valley.

▲ *Chamonix wants to protect the mountains that people visit*

YOUR QUESTIONS

1 Give two advantages and two disadvantages of the impacts of tourism in the Alps.

2 Describe the ways in which Belle Plagne is managing tourism. In your opinion, how successful have they been?

3 Use the website www.chamonixgoesgreen.org to find out the different ways in which ecotourism is being promoted in Chamonix.

4 a Think about different groups of people who might have an interest in an area like La Plagne.

b Which groups are likely to agree with each other, or come into conflict about the development and use of the area for tourism?

c Suggest why they are likely to agree or disagree.

6.9 » Avalanche!

On this spread you'll learn what causes avalanches, and why they're such a hazard.

Alpine resorts fear killer avalanche season

Thousands of winter sports enthusiasts are packing their bags and heading for the mountains – from the wilds of Scotland to the resorts of the snow-capped Alps. But amid this flurry of activity, mountain-rescue authorities are urging tourists to be careful. As a reminder of the lethal power of snow, just last week dozens of people were caught in the paths of Alpine avalanches. While some of them escaped, others – including five Britons – weren't so lucky.

Daniel Goetz, an avalanche forecaster for Météo France, warned that skiing off-piste – the choice for many adrenaline junkies – should be avoided at all costs.

(Adapted from an article in *The Guardian* on 2 January 2010)

After heavy overnight snow in the Alps, skiers can be woken up by the sound of blasting, as ski patrols deliberately set off controlled avalanches to make the area safer. It's a warning to skiers to be careful. But, even so, around 20 people die each year as a result of avalanches in the Swiss Alps alone.

What is an avalanche?

An avalanche is a rapid flow of snow down a slope. It can be set off by a natural event or by human activities.

Natural causes

When a lot of fresh snow falls on top of frozen snow, and the top layers are softer than the lower layers, they begin to slip. A block of snow will break off and start flowing downhill – pushing more and more snow in front of it.

Human causes

If people are skiing off-piste, or walking or climbing across a snow-covered slope, they can cause the top layer of snow to start moving – and so set off an avalanche.

Snowfall Avalanches usually happen within 24 hours of heavy snow. The risk is greater with over 30 cm of snowfall.

Temperature A large rise in temperature can weaken the upper layers of snow and cause an avalanche.

Wind direction Wind blows snow up the windward side of a mountain and drops it on the leeward side. This can cause an uneven build-up of snow on the leeward side, which is more likely to avalanche.

Steepness Most avalanches occur on slopes with an angle between 30° and 50°.

Type of slope Avalanches are more likely to occur on convex rather than concave slopes.

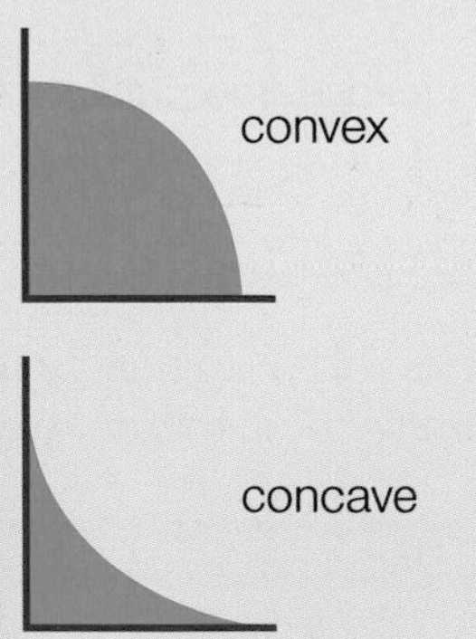

your planet

Avalanches can reach a speed of 80 mph in just 5 seconds!

The effects of avalanches

Immediate effects

There's very little warning when an avalanche starts to flow.

- If people are caught in one, they're likely to be killed straight away – or buried by snow and suffocated.
- If they survive, they'll almost certainly be seriously injured and have broken bones.
- Buildings in the path of the avalanche will be flattened or buried.
- Roads are often blocked, which slows down rescue efforts.

your planet

265 people were killed by Alpine avalanches in 1950-51. It was called 'the winter of terror'.

Longer-term effects

Avalanches kill people every year. But, once the immediate effects are dealt with, things generally carry on as normal. However, a major avalanche – like the one pictured below – can have a big impact on the tourist industry and the local economy.

- The number of tourists can fall, due to fear and the loss of facilities.
- Restaurants and hotels might be destroyed or have to close.
- Local businesses, such as shops, ski hire and tour guides, will lose money.
- If businesses are forced to close, this can lead to long-term unemployment.

YOUR QUESTIONS

1 What is an avalanche? Add this term to your dictionary of key terms for this chapter.

2 Draw a spider diagram to show the reasons why avalanches happen.

3 a Put these phrases in the correct order and draw a flow chart of the effects of an avalanche on an Alpine ski resort.
loss of life, damage to economy, loss of jobs, fewer tourists, damage to village

b Use your flow chart to explain how avalanches can affect the local economy.

On 9 February 1999, after several days of non-stop heavy snow, avalanches had already cut off the road and come close to villages near Chamonix. Then, at 2pm, a huge avalanche swept into the village of Montroc – killing 12 people in their chalets and destroying 20 houses.

6.10 » Climate change and the Alps

On this spread you'll find out how climate change might affect the Alps.

your planet

In the nineteenth century, local people made a living from the Lower Grindelwald Glacier. They sent cartloads of ice to Paris, where cafes used it in cocktails!

Europe's glaciers in retreat

In 2006, the BBC reported that shrinking glaciers were causing tonnes of rock to break away from one of Switzerland's most famous mountains, the Eiger, and crash into the valley below. The rock had been held in place by the ice of the Lower Grindelwald Glacier. However, this supporting ice had now melted – leaving a mass of unstable limestone exposed. And scientists say that climate change is to blame for the ice melting.

Martin Grosjean, a specialist in glaciers and climate change, says that what's happening to glaciers across the world is an extreme reaction to extreme climate change – a response to global warming caused by increasing greenhouse gas emissions.

Climate change and the Alps

What's happening?

- Average temperatures in the Alps are rising. Above 1800 metres, the average temperature has risen by 3°C in the last 40 years.
- Alpine glaciers are retreating. For example, the Mer de Glace is 500 metres shorter than it was in 1994.
- Winter snowfall is becoming more unreliable, and the lower Alps are receiving less snow overall.
- Alpine summers are also getting drier. 2003 was the driest summer for 500 years.

Climate change could bring about a range of economic, social and environmental impacts in the Alps.

Economic impacts

- A lack of snow in villages at lower altitudes (and in the southern Alps, where warming is greater) might mean that some resorts can no longer rely on winter tourism.
- If the Alpine glaciers continue to shrink, there will be fewer sightseers and ice climbers.
- Fewer visitors will mean a loss of income for businesses that rely on tourism.
- The higher Alpine resorts, which have more-reliable snowfall, might survive because they tend to be part of larger ski areas, such as Paradiski. There are also more amenities for tourists and a greater diversification of businesses related to tourism. And they have more money to invest in artificial snow machines.
- Alpine agriculture might benefit from a higher snowline and warmer drier summers, because it means a longer growing season.

▲ *The Rhone Glacier in the Swiss Alps in 1920 and 2005. You can see how far the glacier has retreated.*

Social impacts

What will happen to the Alpine villages?

- If tourist businesses close, unemployment will rise.
- Without tourism, young people will leave the Alpine communities to find work elsewhere.
- Older people will be left in the villages and local services will decline.

Environmental impacts

Even though it looks like nothing much can live there, the snow-covered Alps have their own ecosystem of plants and animals. But as the glaciers retreat and the **snow line** (the height at which snow stays on the ground for most of the winter) gets higher, the area available for these animals and plants to exist in is shrinking.

The Alpine ecosystem is an example of a fragile environment, which means that it's easily damaged. Animals and plants survive in a delicate balance of climate and soil conditions. If that balance changes, the ecosystem will suffer.

There will be other environmental impacts too:

- Some Alpine species are already threatened with extinction because of rising temperatures.
- Forests at lower levels are also under threat.
- As the Alpine glaciers continue to retreat, there will be less meltwater in lakes and their fish might die out. Also, the summer levels of water in Europe's rivers will drop.
- The increasing use of artificial snow-making machines in Alpine resorts takes up huge amounts of water.
- As temperatures rise there will be more avalanches.
- Melting glaciers will lead to increased flooding in lower valleys.
- There will also be more rock falls as the ground thaws.

your planet

Some people predict that half of the 600 ski resorts in the Alps will close by 2050.

▲ *Under the snow there's a whole living world*

▲ *Animals such as the mountain hare will have a smaller area to live in as climate changes*

YOUR QUESTIONS

1. Why is the Alpine ecosystem an example of a fragile environment?
2. What evidence is there that climate is changing in the Alps?
3. Write a newspaper article about the environmental impacts of climate change in the Alps. Write no more than 200 words and add some photos to your article.
4. Draw a mind map to show the consequences of climate change in the Alps.

   Hint: Think about how when one thing happens it affects others, e.g. less snow in the valleys means …

7 » The coastal zone

What do you have to know?

This chapter is from **Unit 1 Physical Geography Section B** of the AQA A GCSE specification. It is about the processes and landforms found at the coast, rising sea levels and coastal erosion, coastal habitats , and how the coast should be managed. The table shows how the pages in this chapter match the content in the specification.

Specification content	Pages in this chapter
Different processes shape the coast – weathering, mass movement, erosion, transportation and deposition.	p140-149
Erosion and deposition create distinctive landforms.	p142-143, 146-147
Rising sea levels will affect people living in the coastal zone.	p152-153 Case study of the Maldives
Coastal erosion can lead to cliff collapse creating problems for people and the environment.	p148-149 p150-151 Case study of Christchurch Bay
Coastal management – hard and soft engineering.	p154-156 p157 Case study of Alkborough Flats Tidal Defence Scheme
Coastal habitats need to be conserved, and this can lead to conflict.	p158-159 Case study of Studland Bay Nature Reserve

Your key words

Swash, backwash

Fetch

Destructive waves, constructive waves

Erosion (hydraulic action, abrasion, attrition, solution)

Wave-cut notch, wave-cut platform

Headlands, bays

Caves, arches, stacks

Transport (suspension, solution, traction, saltation)

Longshore drift

Beaches, sand dunes, spits, bars

Marine processes

Sub-aerial processes

Weathering (mechanical – freeze-thaw; chemical – solution)

Mass movement (sliding, slumping)

Environmental refugees

Hard engineering (sea walls, groynes, rock armour, gabions)

Soft engineering (beach nourishment, sand dune regeneration, salt marsh creation, managed retreat)

Shoreline Management plans

More exam help ...

Advice See pages 297-299 for information on how to be successful in your exams.

Practice See page 306 for exam questions on this chapter.

What if ...

- all our cliffs collapsed?
- the Maldives disappeared under the sea?
- you landed on the beach opposite?

On this spread you'll learn what causes waves, and about the different types of waves.

Surf's up!

The Cribbar is the biggest wave in Cornwall – surfers travel hundreds of miles to ride it. But what are waves? And why do some places have bigger waves than others?

▲ *Surfing the Cribbar, off Fistral Beach, Newquay*

What causes waves?

Waves are formed by the wind as it blows over the sea, as the diagram shows.

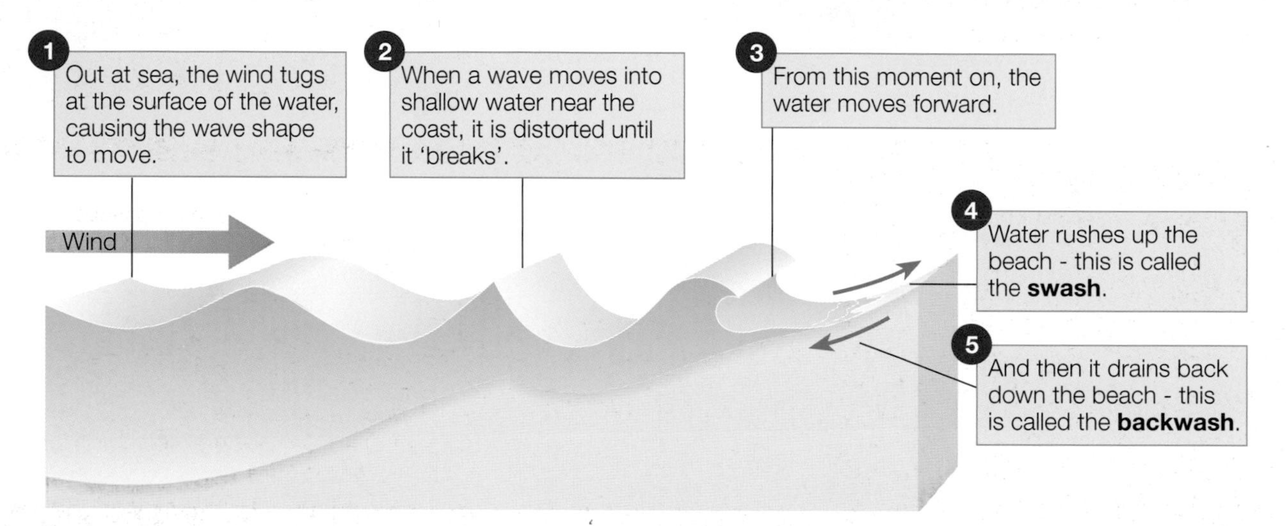

Why are some waves bigger than others?

Some days are better than others for surfing. On windy days, strong, steep waves are common, but on calm days – with little wind – the waves are gentle and quiet. You can guess which days surfers prefer!

The size and strength of the waves depends on:

- how strong the wind is
- how long it blows for
- how far it travels.

The longer the wind blows for, and the stronger it is, the higher and more frequent the waves are. The distance the wind blows over is called the **fetch**.

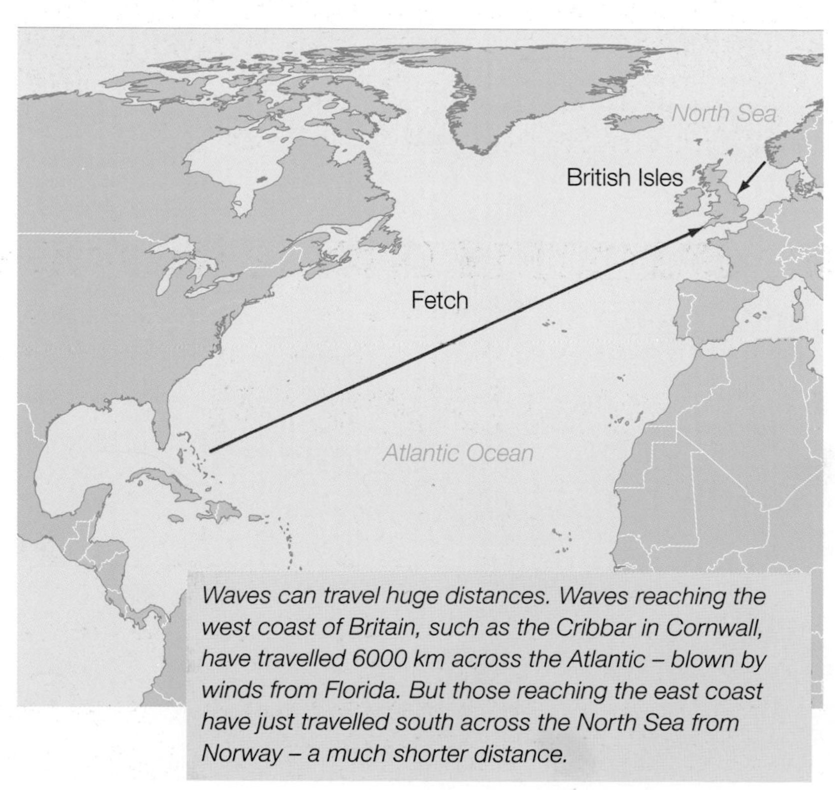

Waves can travel huge distances. Waves reaching the west coast of Britain, such as the Cribbar in Cornwall, have travelled 6000 km across the Atlantic – blown by winds from Florida. But those reaching the east coast have just travelled south across the North Sea from Norway – a much shorter distance.

Which waves are best for surfing?

Destructive waves

Waves that have a weak swash and a strong backwash pull sand and pebbles back down the beach when the water retreats. They are called **destructive waves**, because they remove material from the beach. They are often steep, high waves that are close together and crash down on to the beach. If you were counting them, they would be coming in very quickly – up to 15 every minute. Another name for them is 'plunging waves', and they are ideal for surfing!

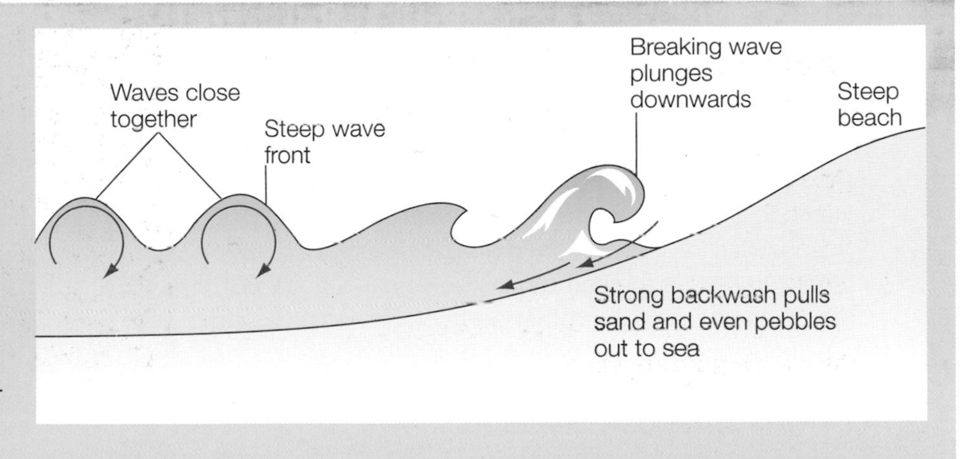

Constructive waves

Waves that have a very strong swash and a weak backwash are known as **constructive waves**, because they build up the beach. They push sand and pebbles up the beach and leave them behind when the water retreats, because the backwash is not strong enough to remove them. They are often low waves with longer gaps between them. As they break, they spill up the beach, so they're also known as 'spilling waves'. They come in at a rate of 6-8 every minute.

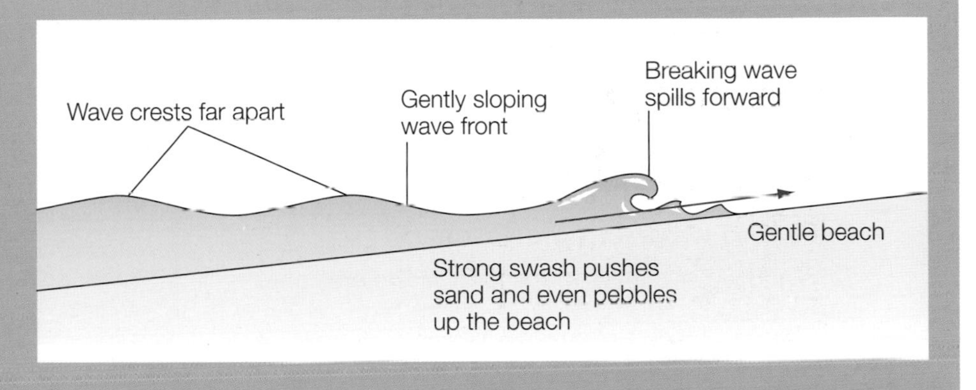

your planet

The best surfing in the world is found in Hawaii, California (USA), Durban (South Africa), Western Australia and Cornwall.

YOUR QUESTIONS

1 Explain what the following terms mean: swash, backwash, fetch.

2 Copy and complete the table below to show the differences between destructive and constructive waves.

Destructive waves	Constructive waves

3 Why are some waves stronger and bigger than others?

4 ***Either*** locate the places mentioned in the 'your planet' on a world map, and then explain why they're the best surfing locations in the world.

Or look at a surfing website for the UK, such as magicseaweed.com. Decide where in the UK would be the best place for surfing today. Explain your answer.

Hint: Look at the map for 'Latest Wind Strength'. It has a scale called the Beaufort Force, which measures wind strength. The higher the number, the stronger the wind.

On this spread you'll learn how waves erode the coast, and about some of the landforms that are created as a result.

The Jurassic Coast

Dorset, in southern England, has some of the UK's most stunning coastal scenery. In 2001, part of the Dorset and east Devon coast was made a World Heritage Site. It's known as the Jurassic Coast, because it's important from a geological point of view (and 'Jurassic' is the name of a geological period). But the Jurassic Coast, like other parts of the coast around the UK, is under threat from erosion.

▼ *Durdle Door is a famous coastal arch on the Jurassic Coast, which has been eroded by the sea*

How is the coast eroded?

Erosion is the process of wearing away and breaking down rocks. The most important feature of a coast is the type of rock in the area. Coasts made of hard rock are worn away slowly over hundreds of years, with most of the erosion occurring during storms – when waves are very powerful. Three types of coastal erosion are explained in the diagram.

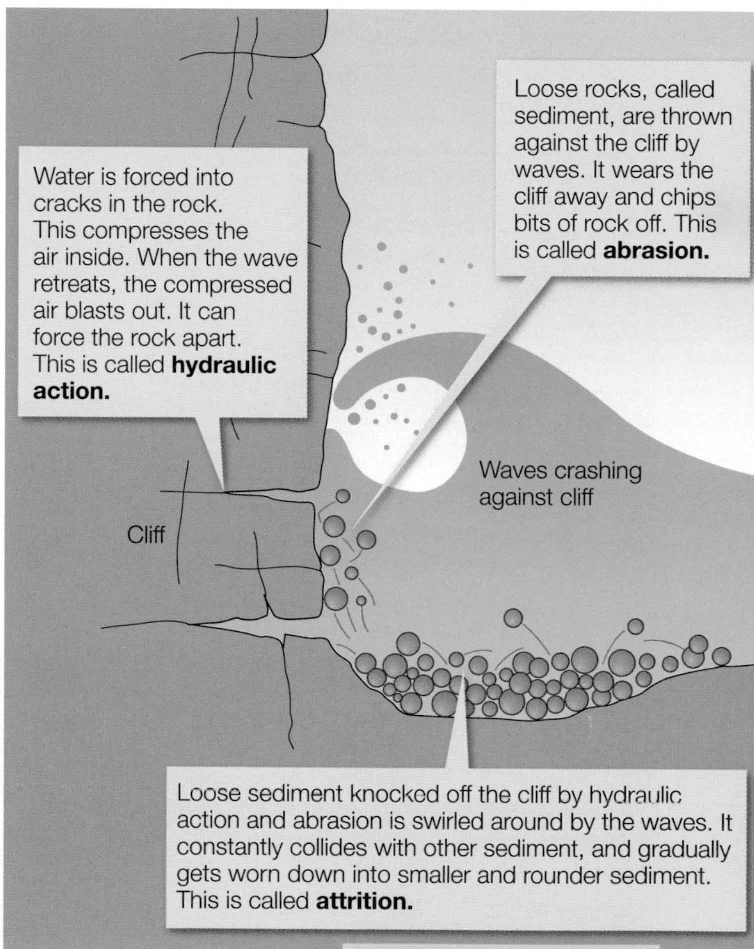

▲ *The main types of coastal erosion*

*There is another process of erosion – called **solution**. This happens when seawater dissolves material from the rock. It happens along limestone and chalk coasts, when calcium carbonate is dissolved.*

What landforms are caused by erosion?

Cliffs and wave-cut platforms

Cliffs and wave-cut platforms form where erosion is taking place.

Headlands and bays

Where coasts are formed from alternating bands of hard and soft rocks, destructive waves will erode the softer rock to form bays and coves. The more-resistant harder rock then sticks out into the sea as headlands. You can see these on the OS map extract. Look for Peveril Point at 041786 and The Foreland at 055824.

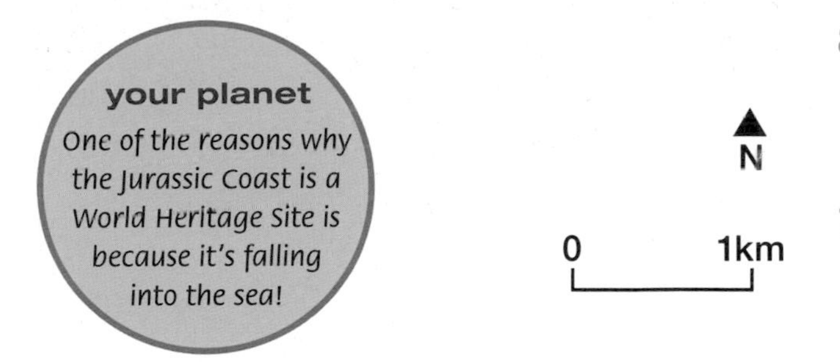

Caves, arches and stacks

Many rocks contain joints and faults. These are areas of weakness, that are eroded to form caves, arches and – in the end – stacks, as the diagram shows.

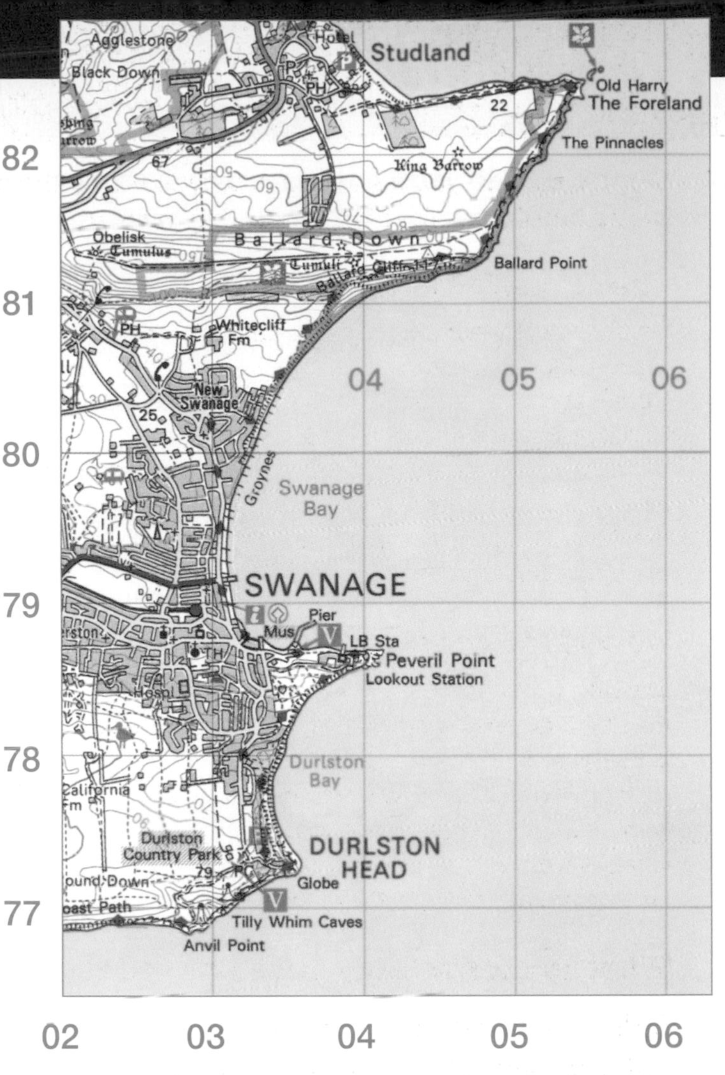

▶ *A 1:50 000 OS map extract of Swanage Bay, Dorset*

▼ *How caves, arches and stacks form*

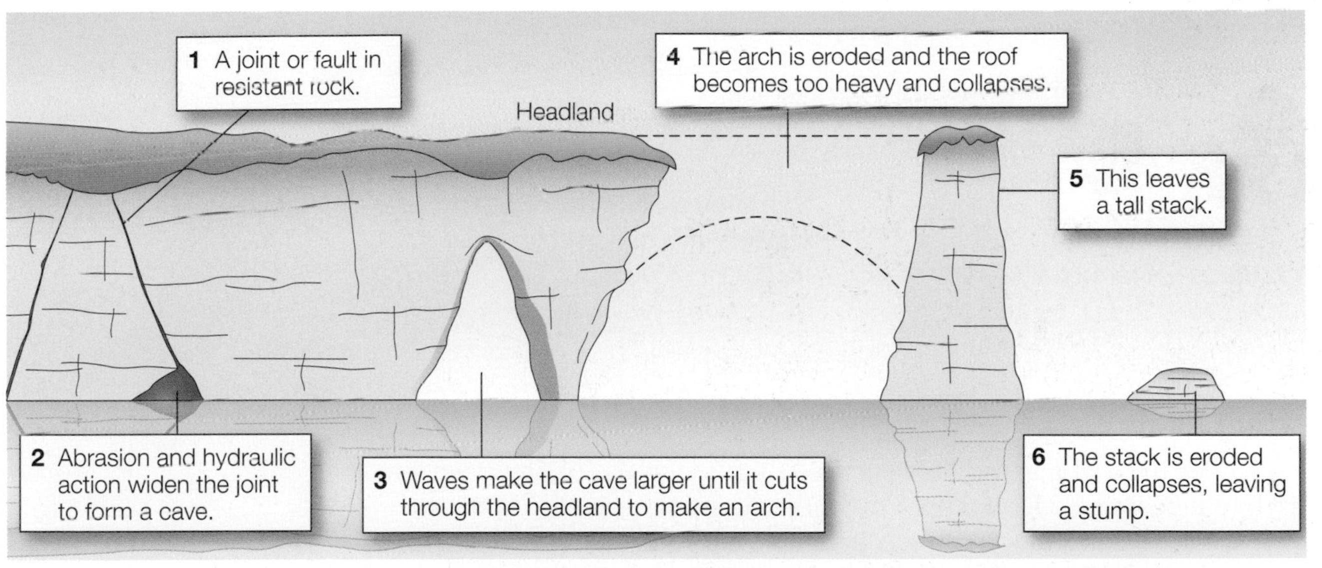

YOUR QUESTIONS

1. Define these terms from the text: erosion, wave-cut notch, wave-cut platform.
2. Describe two ways in which cliffs are eroded, using a diagram to help.
3. Find Old Harry and Old Harry's Wife on the OS map extract at 056825. Describe and explain how they became stacks.
4. Find the following on the OS map, and give six-figure grid references for them:

   Tilly Whim Caves, Anvil Point, Ballard Cliff, The Pinnacles.
5. Look carefully at the OS map. What evidence can you see that there are bands of harder and softer rocks on this bit of coastline?

7.3 » Coastal transport

On this spread you'll find out how waves move material along the coast.

On the beach ...

Have you ever run backwards as the sea raced towards you? Or felt the sand being dragged back into the sea between your toes? If so, then you already know something about how waves move up and down the beach.

How does the sea transport material?

On pages 142-143, you learned how waves erode the coast. Once material is eroded, the sea **transports** it – or carries it away. As waves move the material – or sediment – around, **attrition** makes it smaller and more rounded.

There are four ways in which sediment is transported by the sea, as the diagram shows.

▲ *It's all to do with swash and backwash*

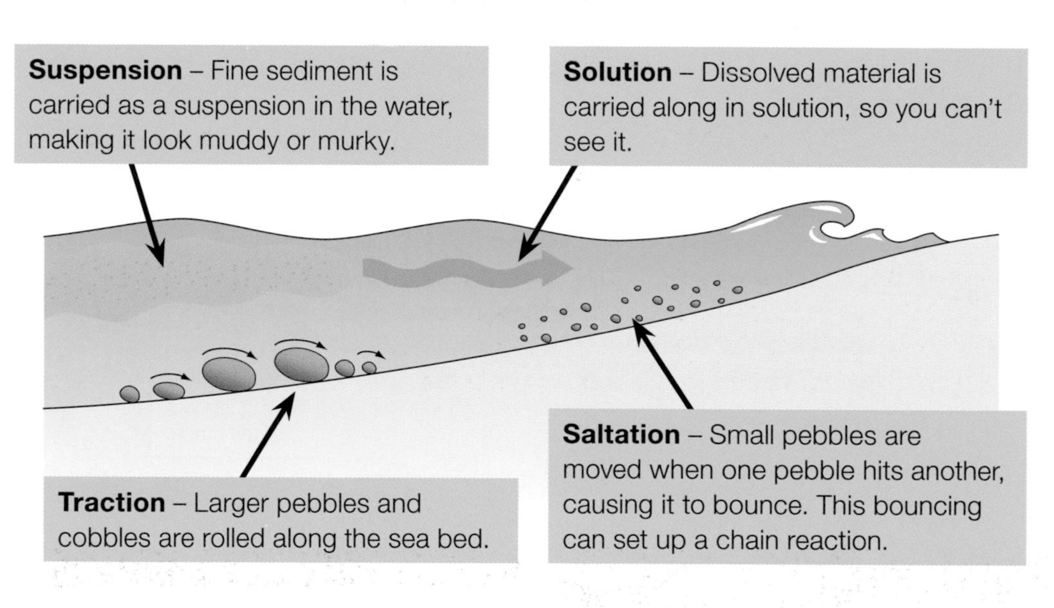

Drifting along?

Some eroded material is transported out to sea. But a lot of it is carried along the coast by a process called **longshore drift**. The diagram on the page opposite shows how this works. As you can see, the waves are approaching the beach at an angle – from the direction of the prevailing wind.

The map on the right shows that, along the south coast of England, the prevailing wind direction is from the south-west. This means that sediment is transported from west to east. However, on the east coast, the prevailing wind direction is from the north or north-east, so sediment there is generally transported from north to south.

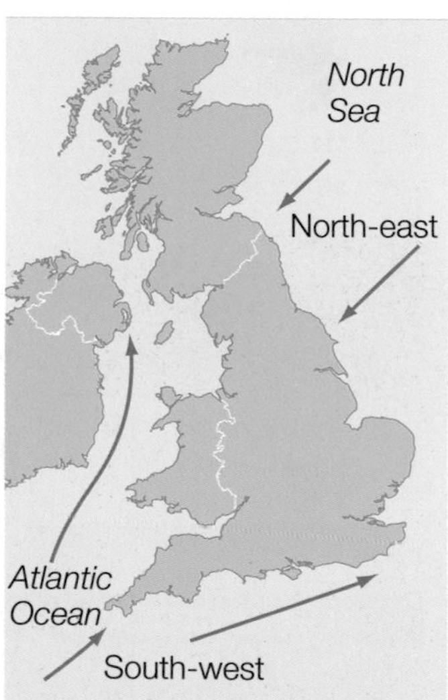

▲ *Prevailing wind directions around the coast of Britain*

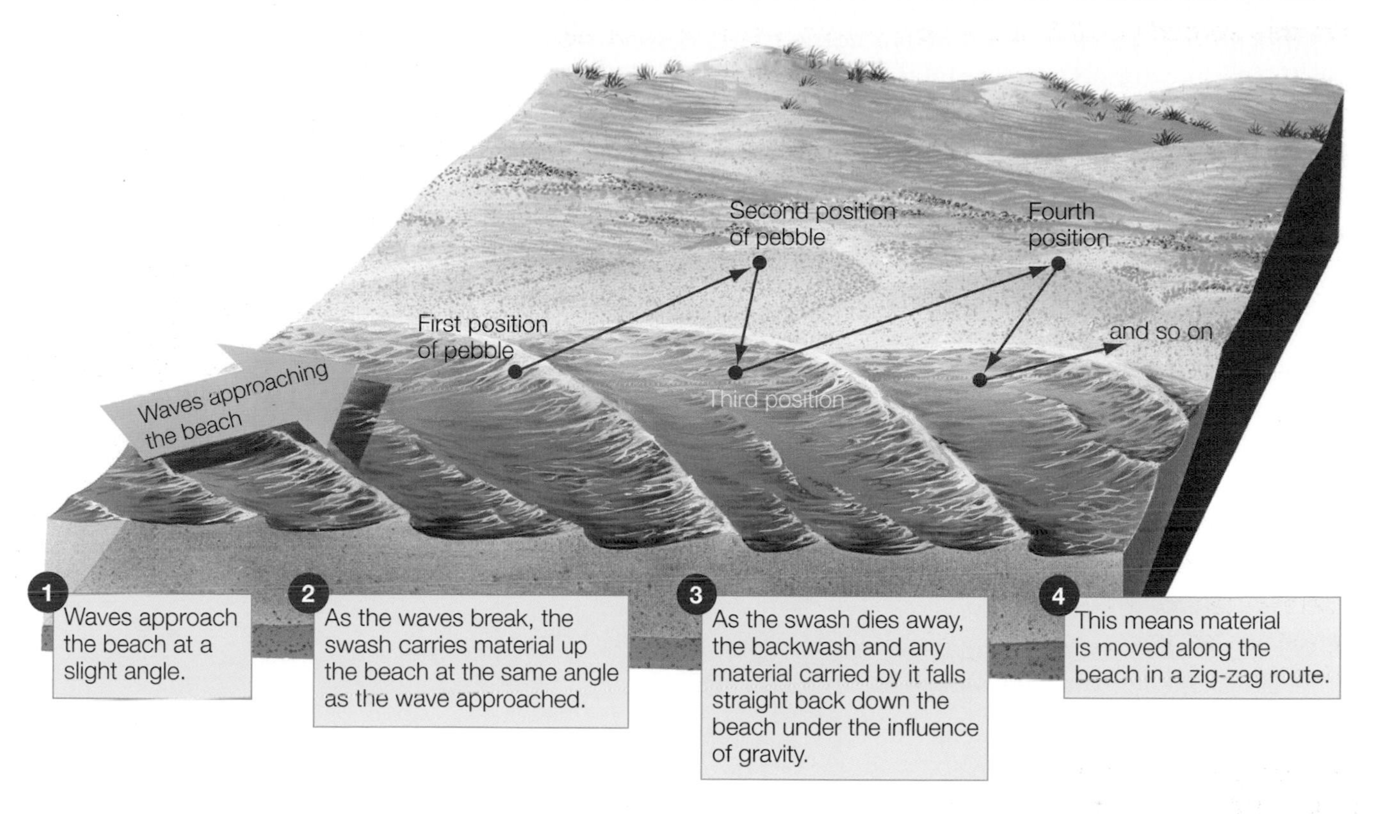

▲ *How longshore drift works*

When longshore drift transports sediment away from one part of the coast, it reduces the size of the beach and leaves the cliffs and shoreline exposed to further erosion by storm-force waves. In an effort to preserve the size of their beaches, and protect their shorelines from erosion, many local councils build wooden fences – called **groynes** – on the beach, to trap the sediment and stop it being transported along the coast. However, by solving the problem in one place, groynes just move it further along the coast instead. Because less material is transported along the coast to build up the beaches there, locations further down the coast suffer more erosion as a result.

▼ *Groynes on the beach at Swanage Bay (in grid square 0379 on the OS map on page 143)*

YOUR QUESTIONS

1 Describe the four ways in which the sea transports sediment.

2 Draw your own simple labelled diagram to show how longshore drift moves sediment along the coast.

3 Work in small groups. What problems do you think longshore drift causes for **a** the cliffs at the back of the beach **b** tourists **c** local councils?

Hint: You won't find the whole answer on this spread. But bounce some suggestions around in your group and you should come up with some ideas.

7.4 » Coastal deposition

On this spread you'll find out about some of the distinctive landforms produced by coastal deposition.

Beaches

What's your ideal beach – a gently sloping white sandy one, so you can walk into the sea without stubbing your toe, or one with boulders, rocks and rock pools? Whatever your ideal, beaches are created as a result of the processes of erosion, transportation, and deposition:

- The material forming a beach may have been *eroded* from a cliff, or removed from a beach somewhere else.
- Waves *transport* the eroded material by longshore drift and then *deposit* it.

Constructive waves deposit material when they break on the shore in a sheltered bay, because they run out of energy. The backwash is weak and sand and pebbles are left behind to form a beach.

Storm waves throw pebbles and sand as far as possible up the beach. This may form a ridge above the high tide mark. Although we think of beaches as permanent features, in fact they are temporary and can change every day.

▲ *Beaches can be mainly made up of sand, like this one at Studland Bay. Sandy beaches tend to be flat. Strong on-shore winds can also blow sand inland to form* ***sand dunes*** *at the back of the beach. They're held in place by* ***marram grass****, which is a coarse grass that helps to stabilise them and stop them from moving.*

▲ *Beaches can also be made up of shingle or pebbles, like this one at Chesil Beach. Pebble beaches tend to be steep.*

your planet

At 11 km long, Bournemouth beach is the longest unbroken stretch of sandy beach in England. But the longest beach in the world is Praia do Cassino Beach in Brazil – over 254 km long and great for surfing!

Spits

Spits are long narrow ridges of sand and shingle stretching out from the coast. They form where longshore drift moves sediment along the coast in the same direction as the prevailing wind. When the coastline changes direction, such as at the mouth of a river, the sediment is then deposited as a long ridge, which stretches away from the coast to form a spit.

Many spits develop a hooked – or recurved – end. This is caused by the wind and waves changing direction. Sand dunes are usually found on this hook. Behind the spit, a sheltered area of saltwater marshes and mudflats forms, which is covered by the sea at high tide.

Bars

Bars are narrow ridges of sand and shingle that grow across a bay as a result of longshore drift. They can trap shallow lakes – called lagoons – behind them. Lagoons don't last forever and may eventually fill up with sediment. Storm waves sometimes crash over the top or break through a bar. That's happened before at Slapton Ley, the bar in the photo.

▲ *Slapton Ley, Devon*

▲ *Hurst Castle Spit, Hampshire*

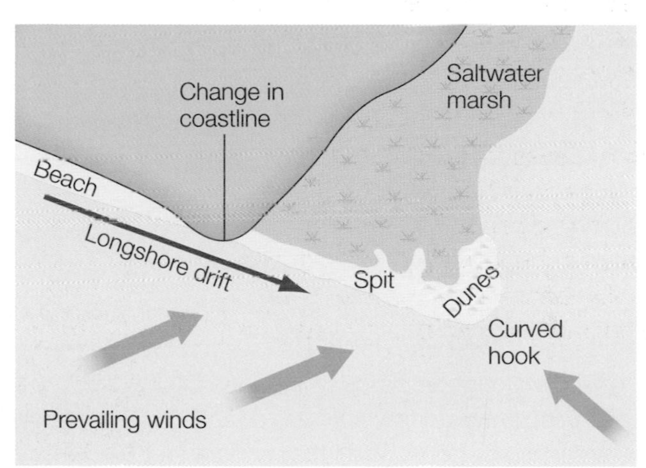

▲ *How a spit forms*

YOUR QUESTIONS

1. Why are beaches formed by constructive (and not destructive) waves?
2. Use this website www.geograph.org to find a photo of your favourite beach in the UK. Annotate the photo to explain how the beach was formed.
3. Explain how longshore drift can lead to the formation of a spit.
4. Outline the main differences between a bar and a spit.

   Hint: Make sure you understand what the command words in questions mean. 'Outline' means you need to describe and explain, but you need more description than explanation.

7.5 » Collapsing cliffs

On this spread you'll discover why some cliffs around Britain are collapsing.

The biggest jaws ever found!

The fossilised skull of a pliosaur – the largest marine reptile that ever lived – has been discovered on the Jurassic Coast. Experts believe that the head (measuring more than 2 metres long) had the biggest jaws ever seen in Britain, and possibly the world. The monster find is believed to be about 155 million years old.

The pliosaur was at least 12 metres long, and its jaws would have been powerful enough to bite a small car in half!

▲ *Stay away from the cliffs – the safest place to hunt for fossils is on the beach!*

The pliosaur mentioned in the article was found by a local fossil hunter. Hunting for fossils is hugely popular along the Jurassic Coast. The reason why they can be found there quite easily, is because the cliffs often collapse and reveal them.

Cliffs are dangerous places!

Watch out the next time you're walking along a cliff. Cliffs are slowly changing all the time – due to weathering and erosion – but in some places they can suddenly collapse without warning! In January 2010, an angler died after a cliff collapsed under him in Northumberland. Extreme snowfalls, ice and high winds are thought to have made the cliff top unstable.

The cliffs are retreating rapidly at Holderness in East Yorkshire, Barton-on-Sea in Hampshire, and Happisburgh in Norfolk (where scientists estimate that they are retreating by 8-10 metres a year).

Why do some cliffs collapse?

There are a number of reasons for this. Some are known as **marine processes** – the base of the cliff is eroded by hydraulic action and abrasion, making the cliff face steeper. Others are known as **sub-aerial processes** – the cliffs are attacked by **weathering**, such as freeze-thaw (see right). The loosened rocks then fall or slide because of gravity – called **mass movement**.

Weathering ...

... is when rocks are broken down. The rocks are weakened by mechanical and chemical processes.

- **Mechanical weathering** includes **freeze-thaw**. This happens when temperatures drop below freezing at night and then rise during the day. Any water held in cracks in the rock freezes, expands and then thaws again. This happens over and over again until the rock is weakened and fragments break away.
- **Chemical weathering** includes **solution**. This occurs when water reacts with the calcium carbonate in rocks like limestone and chalk. The calcium carbonate dissolves and is washed away in solution, weakening the rock.

Cracks and joints in rock fill with water

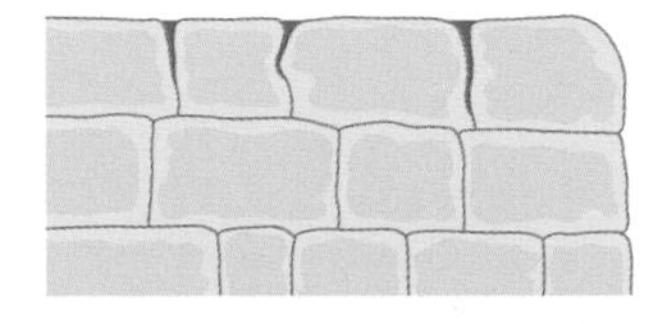

The water freezes and expands – pushing the rocks apart

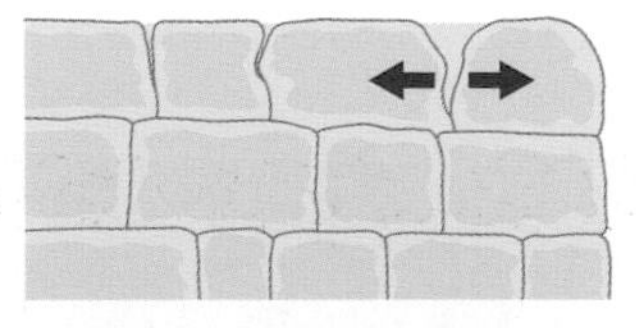

Repeated freezing and thawing widens the crack, and eventually bits of rock fall off

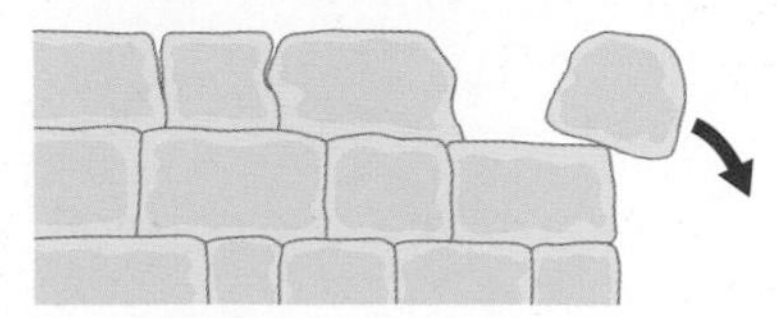

▲ *Freeze-thaw weathering*

Mass movement ...

... is when rocks loosened by weathering move down slope under the influence of gravity. The rocks can slide or slump.

- **Sliding** is when large chunks of rock slide down slope quickly without any warning. This can make it very dangerous to walk along a beach under the cliffs (see the photo on the left).
- **Slumping** is common where the cliffs are made of clay. The clay becomes saturated during heavy rainfall and oozes down towards the sea as part of a mud or debris flow (see the photo on the right).

Human actions make it worse

At Barton-on-Sea, in Hampshire, permeable sand on top of clay allows water to seep down until it saturates the clay underneath. This then slumps and is eroded by the waves.

Building on top of unstable cliffs, like those at Barton-on-Sea, can put too much pressure and weight on them adding to the chances of a cliff collapse. In the past, holiday homes were built close to the cliff edge to take advantage of the sea views. Many homes have crumbled or been demolished as the cliffs below them have collapsed. You will read more about Barton-on-Sea on pages 150-151.

your planet

The White Cliffs of Dover are 100 metres tall. That's nothing compared to St John's Head in the Orkney Islands. That's 355 metres tall – and the most vertical sea cliff in the UK!

YOUR QUESTIONS

1. Define the term weathering.
2. Explain the difference between sliding and slumping.
3. Draw a simple sketch of the photo of the slumped cliffs at Holderness. Annotate your sketch to show why the cliffs have collapsed.
4. Work in pairs to produce a PowerPoint presentation about the reasons why cliffs collapse.

▲ *Slumping of the boulder clay cliffs at Holderness, East Yorkshire*

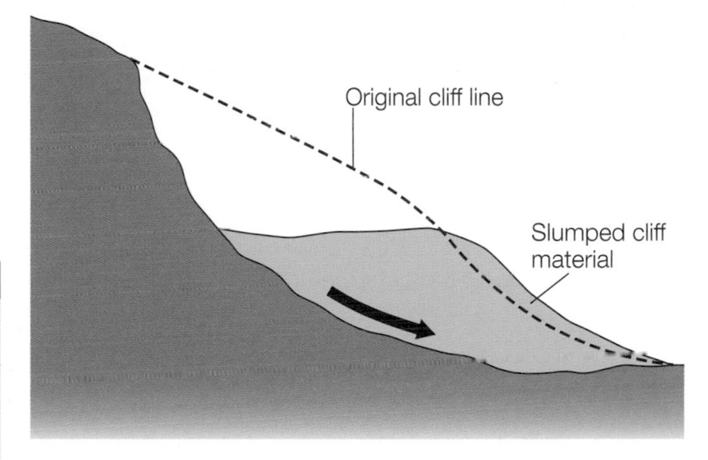

▲ *Slumping in action*

On this spread you'll investigate how cliff collapse is causing problems in Christchurch Bay.

Christchurch Bay

Christchurch Bay, near Bournemouth on the south coast of England has a big problem. It is constantly under attack from the sea.

So the cliffs there are eroding rapidly at the rate of 1-2 metres a year. Christchurch Bay is a 16 km stretch of open coastline which is exposed to waves that have a fetch of 3000 miles across the Atlantic Ocean. The area is densely populated with coastal towns and resort areas, such as Highcliffe, Barton-on-Sea and Milford-on-Sea.

Tourism is big business here, and is very important for the local economy. So the collapsing cliffs are a major issue. In 2007, the holiday beach huts had to be evacuated because the cliffs were in danger of collapsing.

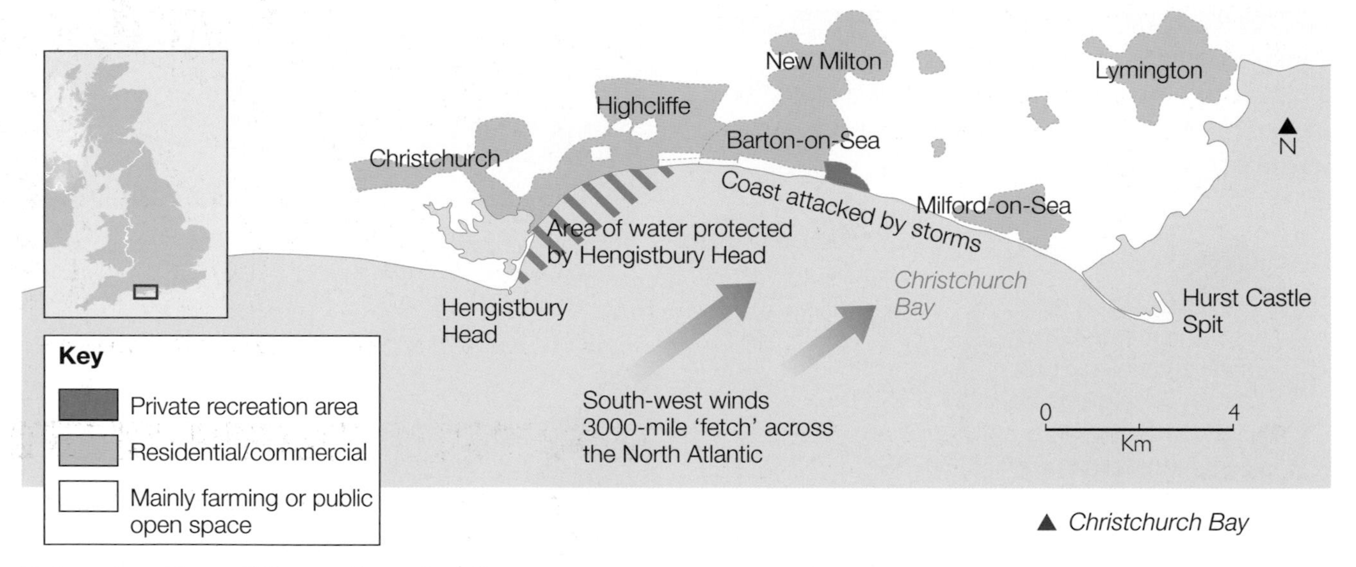

▲ *Christchurch Bay*

Why are the cliffs collapsing?

Coasts all around the UK are suffering from erosion, but weathering and mass movement are making the problem worse.

There are a number of reasons why the cliffs of Christchurch Bay are collapsing:

- **Marine processes** – The base of the cliffs is being eroded by hydraulic action and abrasion.
- **Sub-aerial processes** – Weathering is weakening the rock and then mass movement – slumping and rock fall – is leading to cliff collapse and erosion.
- **Geology** – Permeable sand lies on top of impermeable clay. During storms, heavy rain saturates the permeable sand, making it much heavier and making the top of the cliff unstable.

- **Fetch** – South-west winds, which have blown all the way across the Atlantic, create strong destructive waves.
- **Human activity** – This area is well known for its sunny climate. It's a tourist honeypot (see page 282) and there has been extensive building along the cliff top as a result. The extra weight on top of the cliff weakens it and makes it more unstable.

At Barton-on-Sea, in Christchurch Bay, mass movement is the main cause of cliff collapse (see right), but the other factors all play a part.

What impacts does cliff collapse have?

Cliff collapse affects people (social impacts), as well as the economy and the environment.

Social impacts

- People lose their homes if they fall into the sea.
- Homes close to the cliffs go down in value.
- It is difficult and expensive to insure houses close to the cliffs.
- It is dangerous for people to walk on the cliff tops and on the beach if the cliffs are likely to collapse.

Economic impacts

- Roads and railways near the coast are under threat.
- Tourists may not visit because of the danger. This affects local businesses, such as cafes, hotels, shops, taxis.
- Barton Golf Course has had to expand inland as some of its land has been lost to coastal erosion.

Environmental impacts

- Cliff collapse makes the area look unattractive.
- The cliffs near Naish Holiday Village at New Milton (on the right) are being eroded. But they have not been protected, because they're classed as a Site of Special Scientific Interest.
- Cliff collapse exposes different rock types and fossils.
- Bird-nesting sites and green land at the top of the cliffs are being lost.

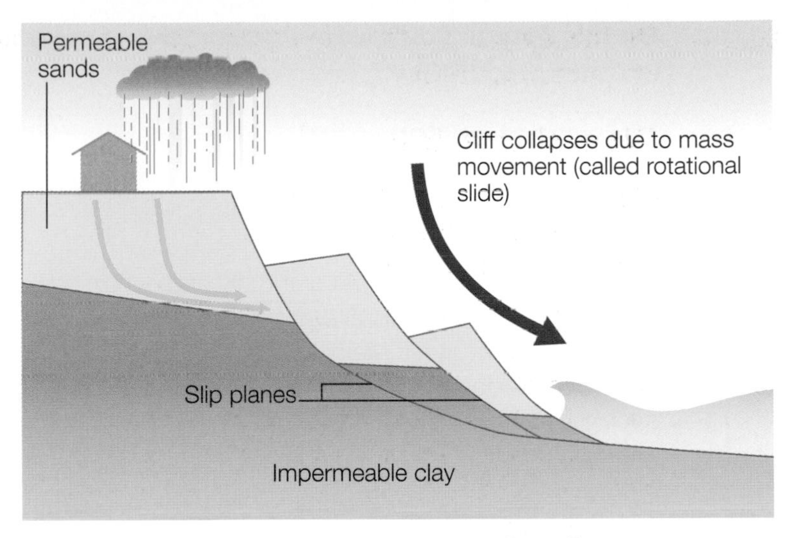

▲ *How the geology of the cliffs at Barton-on-Sea affects erosion*

▲ *Naish Holiday Village. How close to the cliff do you want to be?*

YOUR QUESTIONS

1. Draw a spider diagram to show the causes of cliff collapse at Christchurch Bay.
2. Describe, with the help of a diagram, how the geology of the cliffs at Barton-on-Sea contributes to rapid cliff erosion.
3. Use the information on this page to explain how cliff erosion affects the economy of Barton-on-Sea.
4. ***Either*** research examples on the Internet of collapsing cliffs affecting people's lives, and then write a newspaper article (of no more than 200 words) about one example.

   Or imagine that you live in Barton-on-Sea. Write a letter to the local council giving your concerns about cliff erosion, and the reasons why the coast should be protected.

   Hint: Read the question! This is an either/or. You don't have to do both parts!

7.7 » Rising sea levels in the Maldives

On this spread you'll learn about rising sea levels, and how this is going to affect people living in coastal zones.

your planet
One in ten people worldwide live less than 10 metres above sea level and near the coast.

MPs meet underwater!

The government of the Maldives held an underwater cabinet meeting in October 2009. Why? To show the world the threat they faced from global warming. President Mohamed Nasheed and his cabinet signed a document calling for global cuts in carbon emissions. The government of the Maldives says that the country will almost certainly be completely underwater if sea levels rise as predicted.

President Nasheed's plan is to buy a new homeland for his people – perhaps in India or Sri Lanka – using money from tourism, in case the entire population is forced to become **environmental refugees**.

▲ *The Maldives government makes a point*

The Maldives

The Maldives is a small country, made up of over 1000 islands in the Indian Ocean. It's the lowest country in the world – with its highest point only 2.4 metres above sea level.

The islands' stunning scenery, clear blue tropical seas and white beaches make them an ideal holiday location. Tourism is very important to the Maldives' economy – providing almost 30% of its GDP.

Environmental refugees are people forced to move away from their home area because of changes in environmental or climatic conditions, such as drought, flooding, deforestation.

▼ *Tourist paradise – but for how long? One of the many islands in the Maldives.*

What's happening?

Sea levels are rising and islands in the Pacific Ocean are already beginning to disappear. Scientists have suggested that sea levels are rising as a result of global warming, and will continue to do so for the rest of this century. But they don't know by how much – estimates range from 30 cm to 1.4 m.

As average global temperatures continue to rise:

- the polar ice sheets and mountain glaciers around the world are melting, leading to more water in the sea
- the water in the sea gets warmer and, as it does, it expands

... and that's what's causing rising sea levels.

your planet

A one-metre rise in sea level worldwide would displace about 100 million people in Asia; 14 million in Europe; and eight million each in Africa and South America.

What will the impacts be on the Maldives?

As sea levels rise, both people and the environment will be affected.

- Coral reefs will die as they're bleached and the water gets deeper.
- The ecosystem associated with the reefs will be lost.
- People will be forced to leave their homes and possibly become environmental refugees. The total population is about 400 000.
- The traditional way of life will be lost.
- Rising sea levels could put an end to the tourist industry.
- ... and, of course, the whole country might actually disappear underwater!

The Maldives' contribution to global warming is small, yet it's suffering the worst consequences. It will need international help to cope with the impacts.

And elsewhere ...?

For people living near the coast – and in low-lying areas – rising sea levels could be catastrophic if land is permanently flooded on a massive scale. Places at risk include:

- many small islands in the Indian and Pacific Oceans
- The Netherlands, where 27% of the land is below sea level
- Bangladesh, a very low-lying country with a population of 156 million
- cities like London, New York and Shanghai, which would have to spend billions on flood defences
- The Fens and Essex in eastern England.

▼ *Dhaka, the capital of Bangladesh, is used to flooding during the monsoon season. But the city, like most of the country, is very low-lying – and could be at even greater risk from rising sea levels.*

YOUR QUESTIONS

1. Why might the people of the Maldives become environmental refugees?
2. Describe and explain how rising sea levels are related to global warming.
3. Draw up a table with four columns headed Economic, Social, Political, Environmental. Classify the impacts of rising sea levels on the Maldives under each heading.

   Hint: Classify means you have to sort them into the right category – economic, social and so on.

On this spread you'll learn what coastal management is, and look at different methods of hard engineering.

Protecting the coast – hard engineering

The coast is used in many different ways – from simply enjoying the view, to industrial uses like power stations and ports. And people have been finding different ways to protect it for centuries.

For a long time, the emphasis with sea defences has been to build physical structures to stop the waves in their tracks. This approach is known as **hard engineering**. Those solid constructions that you see when you go to the seaside – the concrete **sea walls**, the **groynes** which shelter you from the wind, the **rock boulders** at the foot of the cliffs – are all examples of hard engineering. Hard engineering works, as the table below shows, but it costs a fortune.

your planet
Nearly a third of Britain's coastline is falling into the sea.

Hard-engineering method	How it works	Comment
Sea wall	Concrete structures which absorb the energy of the waves and provide a promenade for tourists, e.g. at Newton's Cove below.	Costs £2000 per metre. A permanent structure that may last for many years.
Groynes	Long wooden fences, or piles of large rocks, built out into the sea to stop longshore drift and help build up sand on one side, e.g. on the beach at Swanage (see right).	Costs £2000 per metre. They stop the transport of sediment and increase the risk of erosion along the coast.
Rock armour	Large boulders piled up at the foot of cliffs to absorb the energy of the waves and stop them eroding the cliff.	Costs £300 per metre. An effective and simple method that looks more natural.
Gabions	Rocks or boulders held in wire mesh cages and used to protect vulnerable areas from destructive waves (below right).	Costs £100 per metre. The cheapest option but it's not very strong.

▲ *The new sea wall at Newton's Cove in Weymouth, on the Jurassic Coast, cost £2 million to build. It has won awards for sympathetically protecting the cliffs from erosion. It also protects a Site of Special Scientific Interest and allows access to the beach.*

Gabions protecting the coast at Chesil Beach

Protecting Swanage Bay

Swanage Bay is one of the most developed areas on the Jurassic Coast and, consequently, has had the most money spent on protecting it from erosion. Coastal defence works have been carried out in the area since the nineteenth century.

▲ *Sea defences in Swanage Bay – groynes and the sea wall*

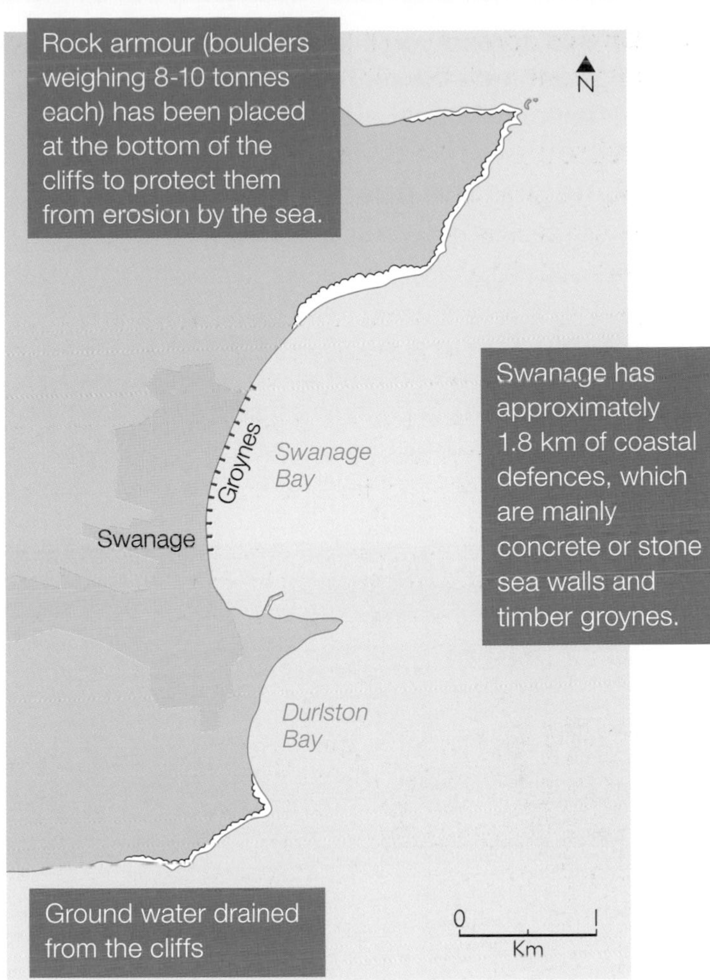

Conflict at the coast

Many people who live near coasts experiencing significant erosion are in favour of hard-engineering solutions. It looks like something 'serious' is being done to protect the coast. But it's not quite that simple, because other people have different views about protecting the coast. This can cause conflict (see the table).

In favour of hard engineering	Against hard engineering
Local people whose homes are in danger.	Local taxpayers who don't live on the coast.
Local tourist businesses, like caravan parks and hotels, which are situated right on the coast and are in danger.	Environmentalists, who fear that habitats and natural beauty will be affected.
Local politicians, who want the support of residents and businesses.	People who live down drift and might lose their beach.

However, it can be even more complicated than that, because some local businesses which rely on tourism might actually think that hard engineering is ugly and will reduce their visitor numbers, so they might decide to oppose it. It's hard to please everyone!

YOUR QUESTIONS

1 Explain what the term hard engineering means in your own words.

2 Look at the OS map extract on page 143. What will the impact of the groynes in square 0379 be **a** on the beach in Swanage Bay **b** on Ballard Cliffs in square 0481.

3 Copy and complete the table to show the costs and benefits of hard-engineering methods of coastal protection.

Method	Cost	Benefits
Sea wall	Very expensive	

7.9 » Managing the coast – 2

On this spread you'll find out about soft engineering – the more modern and sustainable way of managing the coast.

Managing the coast – soft engineering

On pages 154-155 you looked at hard-engineering methods of managing the coast. Increasingly, however, people feel that a better way of protecting the coast is to use nature itself and to build very little. This is called soft engineering and is less expensive, will last longer, and is more environmentally friendly than hard-engineering methods. The table below describes some soft-engineering methods and how they work.

Soft-engineering method	How it works	Comment
Beach nourishment	Building up beaches by adding more sand in front of cliffs.	It's natural protection, because the beach absorbs wave energy, but the sand has to come from somewhere else.
Sand dune regeneration	Allowing sand dunes to build up around wooden structures.	Sand dunes absorb wave energy and create new habitats.
Salt marsh creation	Allowing the sea to flood and spread over a large area, creating salt marshes.	New habitats are created and it reduces the risk of flooding along the coast.
Managed retreat	Abandoning the existing sea defences and building new ones further inland, creating a salt marsh which also floods in storm conditions.	Some people will lose land, houses or businesses, but new habitats are created and flooding is reduced in other areas.

▲ *Beach nourishment. 90 000 cubic metres of sand was added to the beach at Swanage Bay in 2006. The sand came from nearby Poole Harbour. At the same time, the old groynes were replaced with new tropical hardwood groynes.*

▲ *Avocets feed on tidal mudflats and benefit from managed retreat and the creation of salt marshes, such as at Alkborough Flats (opposite)*

Shoreline management plans

The modern way of managing coasts is to think about a whole stretch of coastline, like all of Christchurch Bay (see page 150) and not just one place, like Barton-on-Sea. So local councils work together to discuss the issues and protect and conserve their combined coastlines. In fact, the coastline of England and Wales has been divided up into sections, and a **shoreline management plan** has been drawn up for each one by the relevant local councils – working in partnership with the UK government. The SMPs identify the most sustainable ways of managing each coastline in the future.

There are four main options:

- Advance the line – build new, higher and better defences, and only protect valuable land.
- Hold the line – keep up and improve the existing defences.
- Do nothing – let nature take its course, so erosion takes place but new land is also built up elsewhere.
- Managed retreat – allow certain areas to flood, so that some areas are protected and some areas are not.

The Alkborough Flats Tidal Defence Scheme

Alkborough Flats lie on the south bank of the Humber Estuary, in north-east England. The tidal defence scheme completed there in 2005 is an example of managed retreat. The affected area consisted of 440 hectares of farmland, which was protected from flooding by an embankment built in the 1950s.

Who was involved?

The Environment Agency worked with local people on the project to change the flood defences. They held public meetings and sent out newsletters so that people knew what was happening. Other organisations involved included: North Lincolnshire Council, Associated British Ports, the Department for Environment, Food and Rural Affairs, and the EU.

What did they do?

They:

- broke through the existing flood embankment, creating a breach
- constructed channels to distribute the water, plus a spillway
- built a new flood bank to protect a sewage treatment plant and riding stables.

The creation of the breach and spillway means that the sea can now flood the area at high tide.

Why did they do it?

The Alkborough Flats Tidal Defence Scheme was adopted for a number of reasons:

- It provides a place to store floodwater during extreme weather and high tides. It could reduce floodwater by 1.5 metres.
- It will reduce the risk of flooding for 300 000 properties.
- It has created a new wildlife habitat.
- A new visitor centre will increase tourism.
- It will help the area to adapt to climate change and rising sea levels.

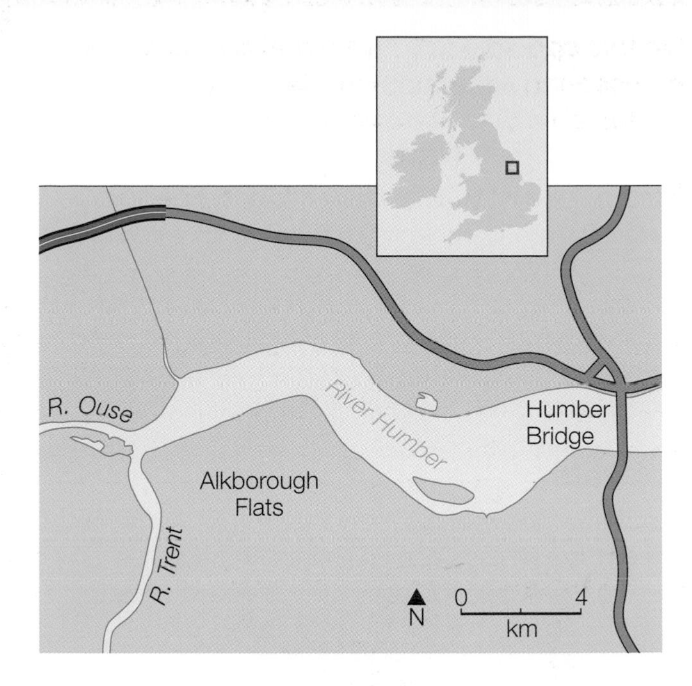

▲ *Alkborough Flats*

YOUR QUESTIONS

1 Look back at pages 154-155 and describe the differences between soft and hard engineering.

2 a Draw a spider diagram to show the options involved in shoreline management plans.

b Explain how the options involved could lead to conflict.

3 a Describe how the Alkborough Flats Tidal Defence Scheme will reduce the risk of flooding.

b Complete a copy of the table below to show the benefits and disadvantages of the scheme.

Benefits	Disadvantages

7.10 » Coastal habitats – Studland Bay Nature Reserve

On this spread you'll find out about a coastal habitat and how it's managed to allow sustainable use of the area.

Studland Bay Nature Reserve

Studland Bay in Dorset is very popular with tourists. Up to 1.5 million people visit each year, attracted by the 5 km sandy beach and shallow sea. It's only a few minutes' drive from Swanage, and most visitors arrive by car. Behind the beach is an area of grass-covered dunes, heath and scrub – an internationally important conservation area managed by the National Trust.

▲ *Sand dunes and marram grass at Studland Bay Nature Reserve*

Habitats

The nature reserve includes dunes and heath, which you can see on the OS map extract. Sand dunes develop when the wind deposits sand as low hills behind a beach on a low flat area. Dunes grow around obstacles like driftwood. In this environment there is a unique **ecosystem**. A special grass called marram grass grows on the sand and holds the dunes together. More vegetation and soil then develops, creating a natural heath.

Ecosystem – the relationship between plants, animals and the environment.

The low shrubs and small trees in the nature reserve allow butterflies, insects, small animals and sea birds to survive. It's the richest 1000 hectares for wildflowers, e.g. marsh gentian, in the country. It also supports many rare birds, such as the nightjar.

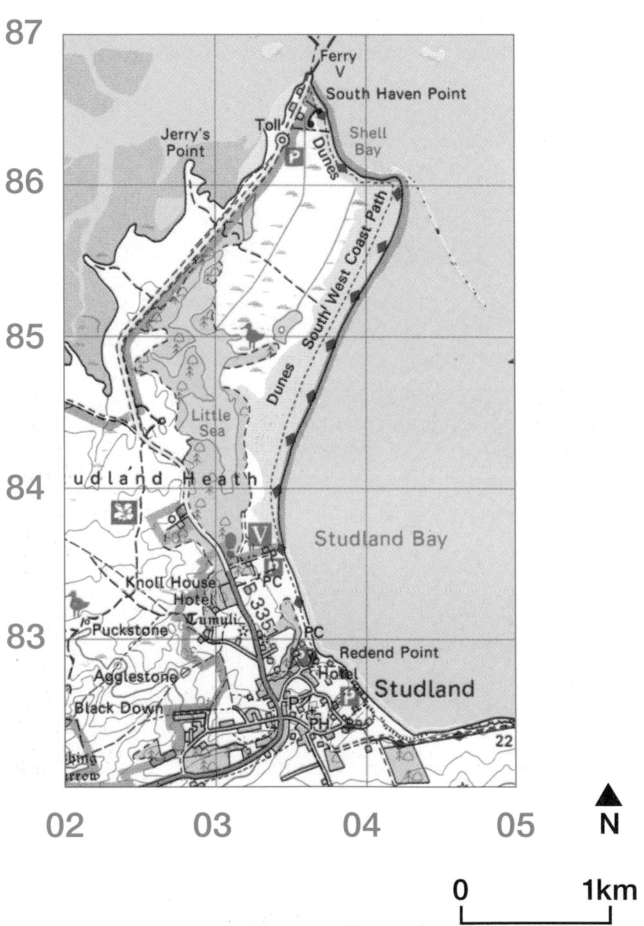

▲ *A 1:50 000 OS map extract of Studland Bay and the nature reserve*

▲ *The sand lizard is one of the native British reptiles found in the Studland Bay Nature Reserve*

Nature reserve – issues and solutions

Studland Bay attracts many different groups of people. Holidaymakers come to bathe, jet ski, sail, picnic and walk – and they want parking spaces, a café and a shop. Bird-watchers, wildflower enthusiasts and conservationists come to enjoy the quiet and natural aspects of this unique area. It can add up to a recipe for conflict.

Issues

- The nature reserve with its sand dunes and heath is a vulnerable environment. The vegetation takes many years to establish, and if the habitat is destroyed the ecosystem will break down.
- The nature reserve is home to rare species of plants and birds, and all six native British reptiles.
- The area attracts many tourists and the beach can get very crowded in summer.
- Visitors need somewhere to park, plus other facilities like paths and public toilets.
- Visitors bring problems such as litter, and create fire hazards (e.g. from barbecues and cigarette ends).

Solutions

The following solutions have been implemented so that the environment is protected to allow for sustainable use:

- Vulnerable areas and those recently planted with marram grass (to stabilise the dunes) have been fenced off to limit access and damage.
- Bird-watching hides and guided walks help visitors to enjoy the wildlife properly.
- Car-parking spaces are limited and people are not allowed to drive on the beach.
- Boardwalks have been laid through the dunes to keep tourists on specific paths.
- Fire beaters are placed in the dunes in case of fire.
- Jet skis are not allowed to be launched from the beach, and there is a 5 mph speed limit to reduce the noise.
- Dogs aren't allowed on Middle and Knoll beach from July to September.
- Facilities including a shop, café, toilets and litterbins are provided near the car parks, to focus tourists into one area.
- Information boards educate visitors about the environment and how they can help to protect it.

◀ *An aerial view of the Studland Peninsula. You can see the heath and the Little Sea.*

Families	Families			
Jetskiers		Jetskiers		
Birdwatchers			Birdwatchers	
Conservationists				Conservationists

YOUR QUESTIONS

1. Describe the ecosystem found in the Studland Bay Nature Reserve.
2. Use the OS map extract to give a six-figure grid reference for the following tourist amenities: the ferry, the Visitor Centre, a car park, and a bird reserve.
3. Explain why the car parks (labelled as a white P on a blue background on the OS map extract) are located in the north and south of the peninsula.
4. ***Either*** outline two problems caused by visitors and explain solutions to them.

   Or, describe the ways in which Studland Bay Nature Reserve is managed to ensure that it is conserved but also used sustainably.
5. Make a large copy of the conflict matrix on the left. In each box linking different users, write down whether you think the two groups will agree or disagree about how to use the Studland Bay Nature Reserve. Explain your answers.

8 » Population change

What do you have to know?

This chapter is from **Unit 2 Human Geography Section A** of the AQA A GCSE specification. It is about how population changes, policies to control population growth, ageing populations and migration. The table shows how the pages in this chapter match the content in the specification.

Specification content	Pages in this chapter
How the global population has increased.	p162-163
Population change and the demographic transition model.	p164-165
How the population structures of different countries change.	p166-167
Rapid population growth, and the need for sustainable development.	p168-169
Different strategies to control rapid population growth.	p170-171 Case study of China's population policy
	p172-173 Case study of population control in Kerala
The problems and impacts of an ageing population, and strategies to cope.	p174-177 Case Study – the UK's ageing population
Migration – push and pull factors, the impacts on source regions and receiving countries, and economic migrants within the EU.	p178-181

Your key words

- Birth rate
- Death rate
- Natural increase
- Population growth rate
- Demographic transition model
- Population pyramids
- Sustainable development
- Young dependents
- Working population (economically active)
- Elderly dependents
- Dependency ratio
- Pro-natalist strategy
- Migration
- Migrant
- Source country
- Host country
- Economic migrants
- Push and pull factors

Exam help ...

Advice See pages 297-299 for information on how to be successful in your exams.

Practice See page 307 for exam questions on this chapter.

What if ...

- the world's population fell?
- we all lived to be 100 years old?
- no-one migrated?
- we all lived to be as fit and healthy as the man opposite?

8.1 » World population growth

> **your planet**
> If the world had 100 people, 20 would be children, and 80 would be adults.

On this spread you'll find out about world population growth.

Meet Lucy

Lucy was born in Guangzhou in southern China in 2006, and spent the first two years of her life in an orphanage. Now she lives in England.

This chapter looks at population, but it's not about meaningless numbers – it's about real people – you and me … and Lucy.

In 1950, there were 562 million people living in China (that's just over half a billion). By 2009 (when Lucy was three), China's population had more than doubled to 1.3 billion. It had the largest population of any country in the world, and one in five people (20% of the world's population) lived there.

How fast is the population growing?

In 2009, the US Census Bureau estimated that the global population was 6.79 billion (that's 6 790 000 000). The table on the right illustrates how fast the world's population was growing in 2009. But has it always grown so quickly? The graph below shows how the world's population has grown over the last 1000 years.

By 1804, it had taken humans just over 300 years to double in number from half a billion to 1 billion people. Yet, by 1999, the population had reached 6 billion – and it had taken just 39 years to double from 3 billion. In other words, the bigger the population, the faster it has grown. This is called the **exponential growth**.

Additional people every …	
… year	77 760 000
… month	6 480 000
… day	216 000
… hour	9000
… minute	150

▲ *Global population growth in 2009*

▼ *Global population growth since 1000*

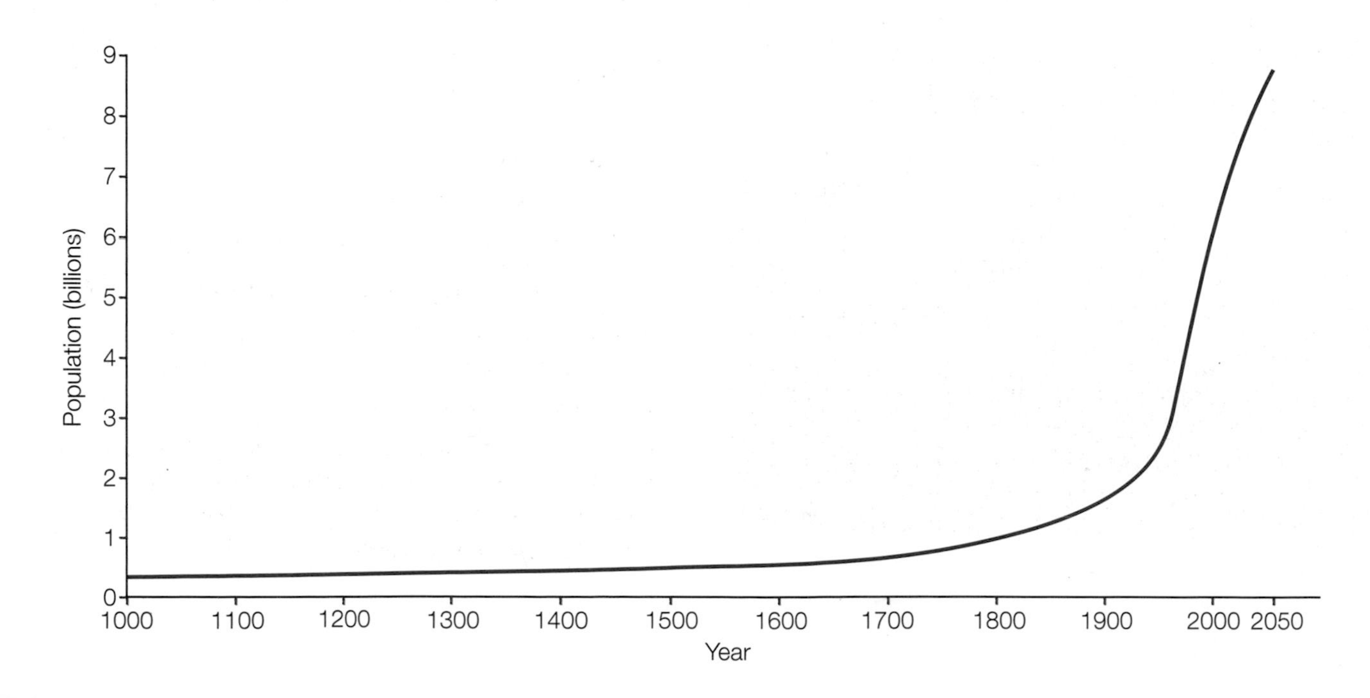

Population slowdown

In the 1960s, nearly every country on the planet had a growing population. However, from the 1970s onwards, the number of babies being born in developed countries began to drop – for reasons such as women working longer before starting a family. Some developed countries (such as Sweden and Italy) are now seeing their populations actually begin to fall, in spite of increasing life expectancy.

In addition, some developing countries (such as China and India) introduced controls to limit their populations and keep them sustainable. Now, the overall number of babies being born in developing countries is also starting to level off.

With fewer babies being born, the United Nations now expects the global population to peak at 10 billion (see the table). It may then slowly begin to drop after 2200.

Year	Estimated global population
2013	7 billion
2028	8 billion
2054	9 billion
2183	10 billion

The future

As the table shows, by 2050, it's estimated that the global population will be about 9 billion. The map on the right shows their projected distribution. It looks so strange because the countries have been drawn in proportion to their estimated populations. Look how big Asia and Africa are – it's thought that, by 2050, 62% of all people will live in Africa, and South and East Asia.

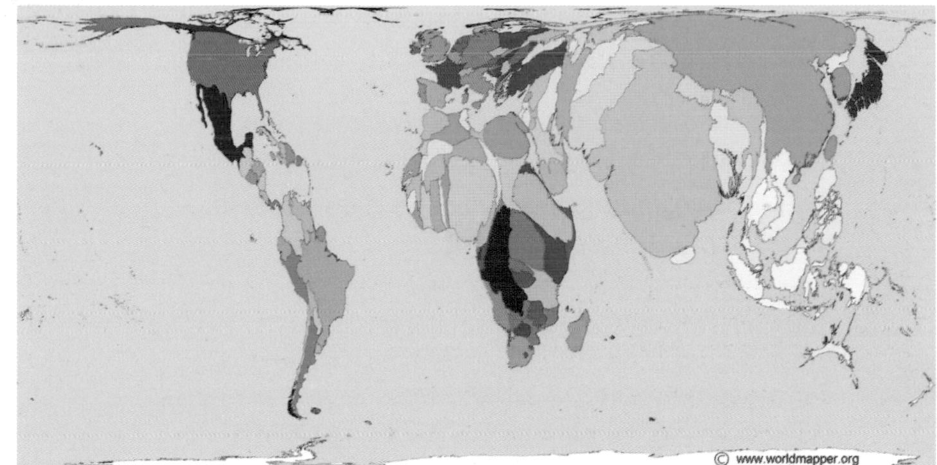

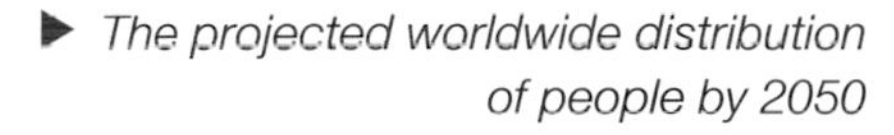
▶ *The projected worldwide distribution of people by 2050*

Measuring population

The following measures are used to work out whether a population is growing or falling.

Birth rate – the number of babies born per year, for every 1000 people.

Death rate – the number of people who die per year, for every 1000 people.

Natural increase – the number of people added to, or lost from, the population per year, for every 1000 people.

Population growth rate – the number of people added to, or lost from, a population each year, as a result of natural increase and net migration. It is given as a percentage.

YOUR QUESTIONS

1 Use the writing frame below to make some notes about world population growth.

> Birth rate is …
> Death rate is …
> Natural increase is …
> In 2009, the global population was …
> The global population is expected to peak at …

2 Look at the table showing how the population is expected to grow.

a Starting with 2013, work out how long it takes to add each extra billion people.

b What do you notice?

3 Use the following statement from the UN to hold a class discussion about future population growth: 'The choices that today's young people make about the size of their families will determine how many people there will be in 2050'.

On this spread you'll learn about some reasons for the growth in the global population.

The demographic transition model

On pages 162-163, you found out what's happening to the size of the global population. But why is it happening? Babies are born – and people die – but other things happen too.

A lot of countries have had similar patterns of population change over time. So, geographers have devised a model to illustrate and explain this. It's called the **demographic transition model** – but what does that actually mean?

- *Demographic* is to do with population, or people.
- *Transition* means change.
- A *model* is a simplified version of something that happens in real life.

So, the demographic transition model is a simplified way of looking at population change.

The model has worked quite well for countries that have gone from a rural, poorly educated society to an urban, industrial, well-educated one. So it fits what happened in the UK, the rest of Europe, and other richer countries like Japan and the USA. But poorer countries might not follow the same pattern.

▲ *Japan's population is getting older, and also starting to decrease in size, because it's reached Stage 5 of the demographic transition model. These elderly Japanese are are doing some communal exercise.*

▼ *The demographic transition model*

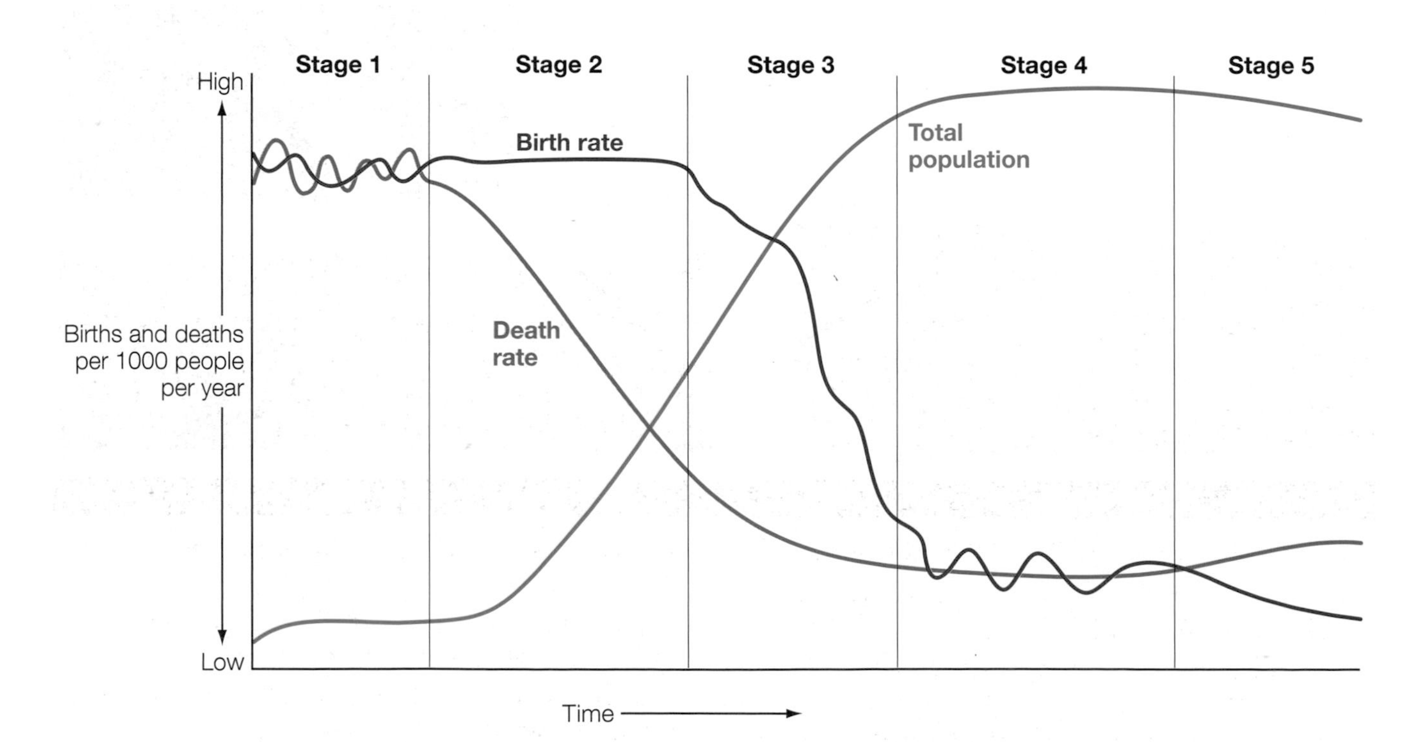

What happens at each stage?

	Stage 1	Stage 2	Stage 3	Stage 4	Stage 5
Death rate	High because of disease, famine, lack of clean water, lack of medical care.	Starting to fall because of improved medicine, cleaner water, more and better food, improved sanitation.	Still falling, for the same reasons as Stage 2.	Remains low.	Goes up slightly because more of the population is elderly.
Birth rate	High, due to a lack of birth control; women also marry very young; children are needed to work in the fields to support the family's income.	Still high, for the same reasons as Stage 1.	Starting to fall, because fewer people are farmers who need children to work; birth control is now available; numbers of infant deaths are falling; women are staying in education longer and marrying later.	Low, because of birth control – people are now having the number of children they want.	Remains low, and can fall below the death rate; changes in lifestyle mean people have fewer children later.
This means that ...	... natural increase is low; population doesn't increase much.	... natural increase is high; population increases quickly.	... there's still some natural increase, but it's lower than it was; overall population increase is slowing down.	... there is little or no natural increase, so population doesn't increase much.	... if more people die than are born, the total population will probably fall (depending on migration patterns).
Places at this stage today	Perhaps just a few remote tribes in tropical rainforests, isolated from the rest of the world.	Poor countries with low levels of economic development, such as Nigeria and Afghanistan.	Poorer countries where economic development is improving, like India and Brazil.	Richer countries which are more economically developed, such as the UK, USA and France.	A few richer countries, like Japan, Italy and Germany.

The demographic transition model shows how population patterns can change over time. The diagram provides more detail about some of the reasons, or factors, which help to explain this.

Factors affecting population change

Urbanisation

As farming methods change, and fewer people are needed to work on the land, many rural people move to urban areas to work. They need fewer children there, so they have smaller families.

Education and women

As the society and economy develop, women tend to stay in education longer. This means that they get married and start having children later, and usually have fewer children as a result.

Educated women also know more about birth control, and so can limit their families more effectively.

Changes to farming methods

If people rely on farming, and there is little technology, they often have large families to provide extra workers.

As technology increases, and countries develop, fewer people are employed in farming and the need for large families declines.

YOUR QUESTIONS

1 Explain, in your own words, the demographic transition model.

2 a Which stage of the demographic transition model are most richer countries at?

b Explain what is happening to the birth rate, death rate and natural increase at that stage.

3 a Which stage of the demographic transition model are most poorer countries at?

b Explain what is happening to the birth rate, death rate and natural increase at that stage.

4 Why do you think some poorer countries might not follow all the stages of the demographic transition model?

Hint: You won't find the answer on this page. This question is meant to make you think.

On this spread you'll look at how the population structures of two contrasting countries are changing.

Population pyramids

Geographers don't just look at total population numbers – they also look at the structure of a population. That means thinking about how many babies are being born and how many people are dying – and how the number of people in different age groups is changing. This is done using **population pyramids**.

A population pyramid is a type of graph which shows the percentage, or number, of males and females in each age group – the number aged 0-4 years, 4-9 years, and so on. The trick is to know how to 'read' the pyramid.

- Firstly, look at the overall shape. For example, if the pyramid is wide at the bottom – like the two for Mexico below – it means that there are lots of young people in the population.
- Then look for details – like bars that are longer or shorter than those above and below them. Shorter bars could indicate high death rates in those age groups – perhaps through war or famine.

Why is Mexico's population structure changing?

Mexico has a large youthful population. Under-15s currently make up 31% of the population, and just over 5% are over 65. The average age is 26.

However, Mexico's population structure is slowly starting to change, because:

- it has now managed to achieve a much lower death rate – just 4.78 deaths per 1000. Not only are more babies being added to the population, but people are living longer as well! This is due to more childhood vaccinations being introduced in an effort to reduce infant mortality, and improved healthcare generally.
- although the birth rate is starting to fall, it remains over 20 per 1000. Therefore, Mexico still has a large percentage of young people. Even if they have fewer children than their parents, the population of Mexico will continue to rise for some time to come.

It is expected to take at least 50 years before Mexico's population levels out. Today's young people will then be moving into old age.

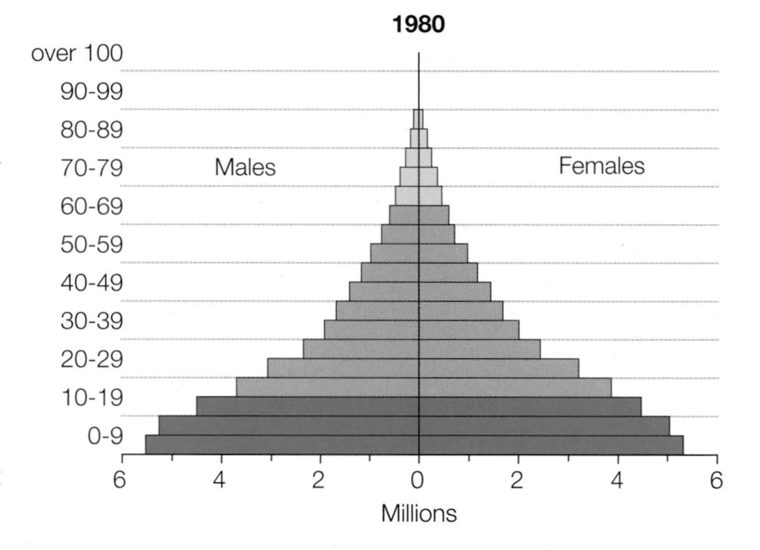

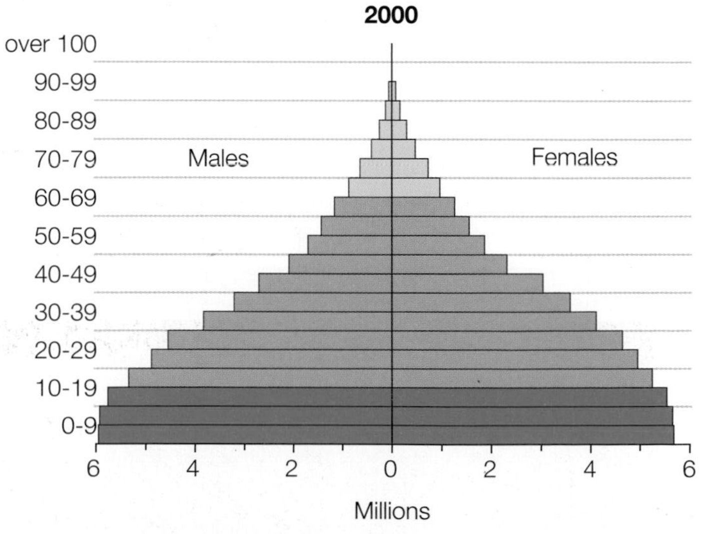

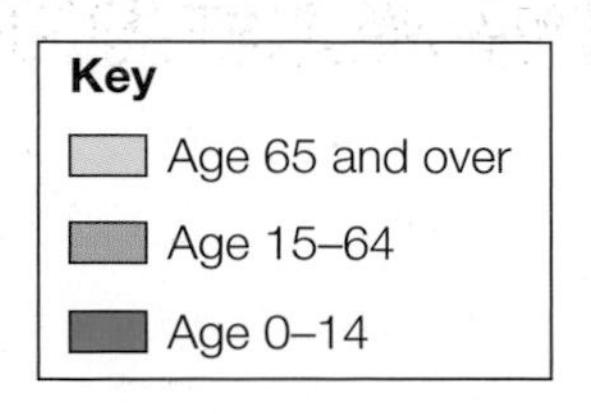

▲ *Mexico's changing population structure*

Why is Japan's population structure changing?

By contrast with Mexico, Japan has a population that is ageing – and starting to get smaller. Japan has the oldest population in the world – over 65s make up nearly 21% (with under-15s just 13.6%); the average age is 44 (the highest of any country).

Japan's population structure is changing because:

- people are living longer. The average life expectancy in Japan is 79 for men and 85 for women. This is due to a healthy diet (low in fat and salt) and a good quality of life. Japan is one of the richest countries in the world and has good healthcare and welfare systems. There are 210 doctors for every 100 000 people (compared with 190 in the UK).
- the birth rate in Japan has been declining since 1975. This is partly due to the rise in the average age at which women have their first child. This rose from 25.6 years in 1970 to 29.2 in 2006. Throughout this period, the number of couples getting married fell, and the age at which they got married rose.

Population pyramids and the demographic transition model

Countries at different stages of the demographic transition model have different-shaped population pyramids. If you can recognise the different basic shapes, and understand what they're showing, then you can tell which stage of the model a country is at (see below).

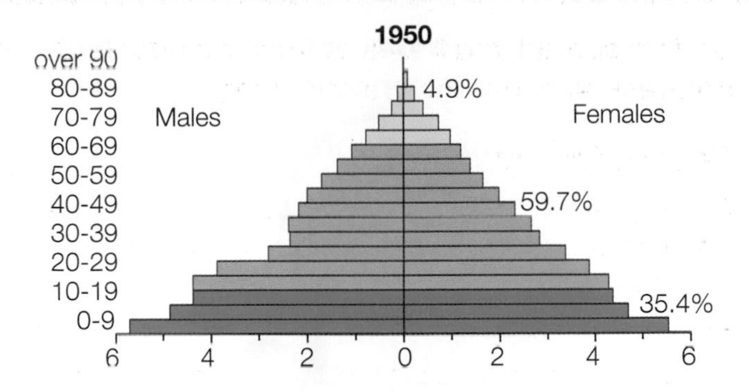

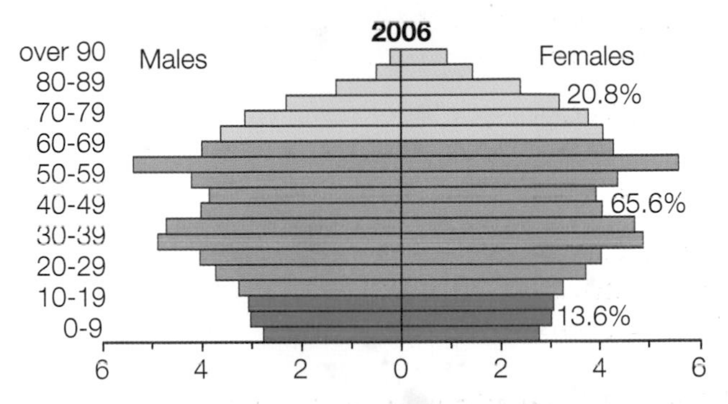

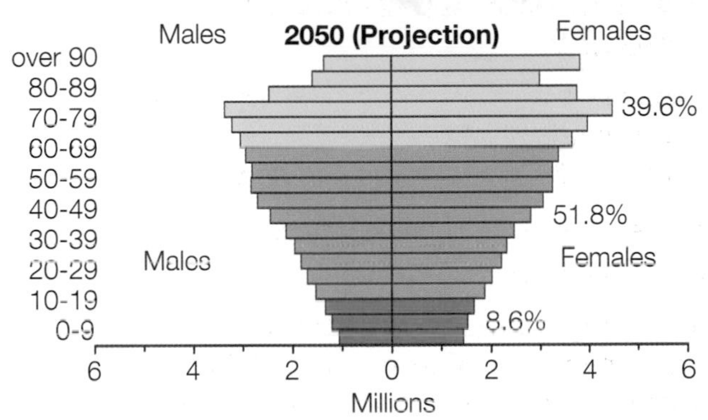

▲ *Japan's changing population structure*

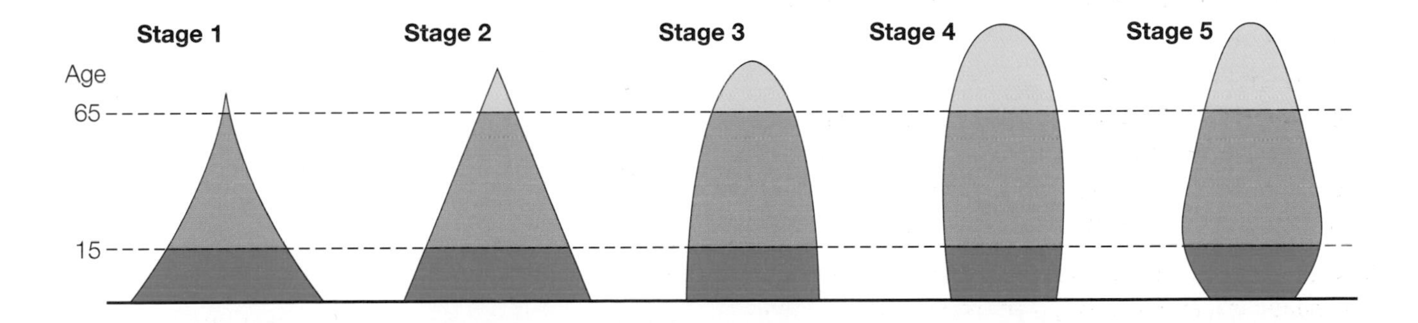

YOUR QUESTIONS

1. Explain what population pyramids show.
2. Look at the two population pyramids for Mexico.
   - **a** Describe each pyramid's shape.
   - **b** Explain the changes between 1980 and 2000.
3. Use the population pyramid and information for Japan to explain why the country's population is getting older, and is declining.

8.4 » Rapid population growth

On this spread you'll find out about some of the problems caused by rapid population growth.

What's happening where?

Worldwide, the population is still growing – but is it the same everywhere? The short answer is no, as the map below shows.

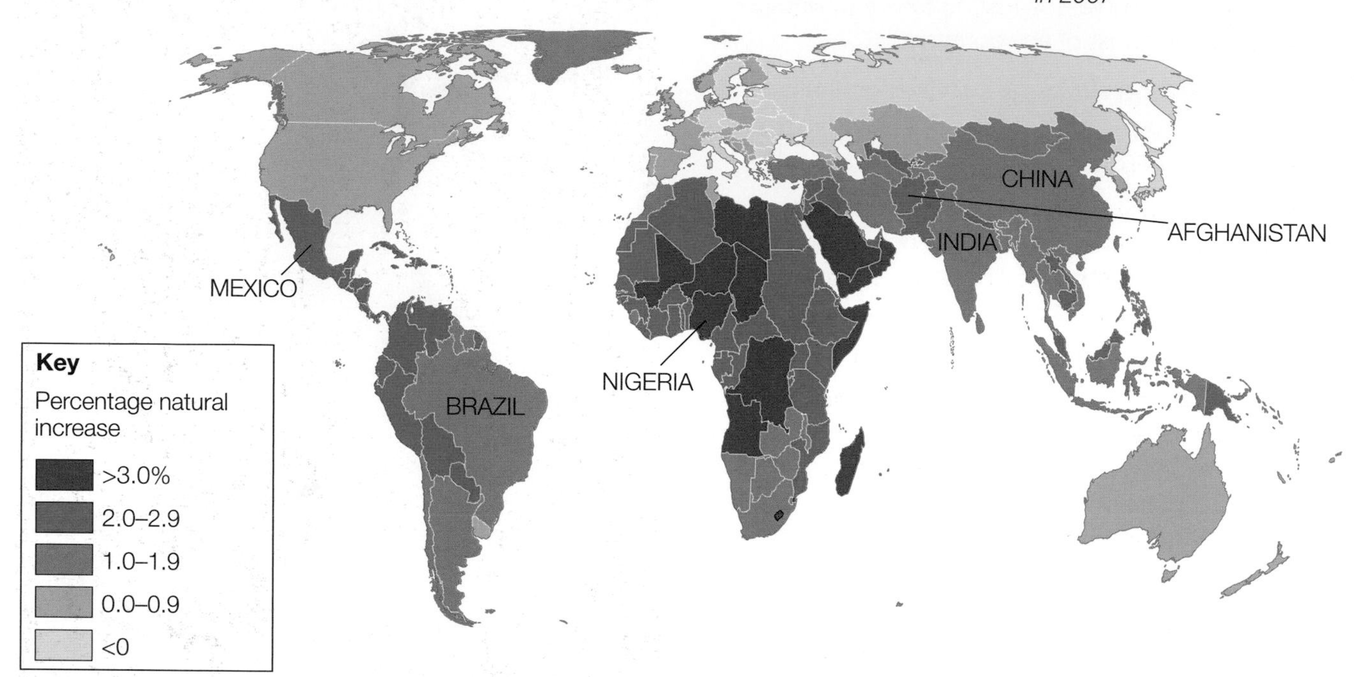

▼ *The natural increase in population around the world in 2007*

Generally, higher levels of population growth are happening in developing or poorer countries, and lower levels of growth, population balance – or even decline – are happening in developed or richer countries. Look at the table, and concentrate on the columns for population growth rate and GDP per capita (GDP shows how wealthy a country is). You should notice that there's a link between them.

Country	Infant mortality	Population growth rate	Fertility rate	Life expectancy (years)	GDP per capita (US$ ppp)
Afghanistan	151.95	2.63%	6.53	44.6	$700
Nigeria	94.35	1.99%	4.91	46.9	$2300
India	30.15	1.55%	2.72	69.9	$2900
China	20.25	0.66%	1.79	73.5	$6000
Brazil	22.58	1.20%	2.21	72.0	$10 200
Mexico	18.42	1.13%	2.34	76.1	$14 200

▲ *Indicators of population change (all figures for 2009, except GDP which is for 2008)*

Infant mortality – the number of babies dying before they reach the age of one, per 1000 births.

Fertility rate – the average number of children a woman will have in her lifetime.

Life expectancy – the average number of years someone can expect to live.

GDP (Gross Domestic Product) – the value, in dollars, of the goods and services that a country produces in a year.

- GDP is divided by the country's population to give **GDP per capita**.
- **ppp** means **purchasing power parity**. GDP is adjusted because a dollar buys more in some countries than others.

Problems of rapid population growth

For many poorer countries, rapid population growth is slowing down their development. They're struggling to earn enough money from farming and basic industry to provide for more and more people. The ever-growing population puts too much pressure on their resources. Some countries:

- find it difficult just feeding everyone – but the population keeps on growing. *The result:* millions of people go hungry.
- can't afford to provide enough schools and teachers. *The result:* millions of people don't get the education and skills that would help to raise them out of poverty – and help their countries to develop.
- can't afford to provide good basic healthcare, with enough doctors and hospitals. *The result:* millions of people suffer and die from illnesses and diseases that could have been cured or prevented.

Population growth and sustainable development

Sustainable development is defined as: 'meeting the needs of the present without compromising the ability of future generations to meet their own needs'. But what is the link between population growth and sustainable development?

For a population to be sustainable, the rate at which it grows must not threaten the survival of future generations. You can probably see that a population that is growing too rapidly, or one that is falling, won't be sustainable.

Take the example of Afghanistan. It might not be the first place you'd think of if you're talking about rapid population growth, but it's got the fourth highest birth rate in the world.

your planet

The average age in Afghanistan is 17. Only 13% of Afghan women can read and write.

Afghanistan

Only Niger, Mali and Uganda (all in Africa) had higher birth rates than Afghanistan in 2009. Not only that, but its population is growing faster – in percentage terms – than countries like China and India, as the table opposite shows.

But Afghanistan is a dangerous place in which to be born. More than 150 babies out of every 1000 born will die before they reach their first birthday. As a comparison, in the UK fewer than 5 babies out of every 1000 will die before their first birthday.

YOUR QUESTIONS

1 Explain in your own words what these terms mean: GDP per capita ppp, fertility rate, infant mortality, life expectancy, sustainable development.

2 Look at the map opposite.

a Describe where population is growing fastest (i.e. countries where the natural increase is over 2%).

b Describe where population is declining (i.e. countries where the natural increase is below 0%).

3 a Create a mind map to show how rapid population growth can affect a country's development.

b Underline the social impacts in one colour, and the economic impacts in another colour.

Hint: Social means to do with people, economic means to do with money. Some impacts might be social and economic.

8.5 » China's population policy

On this spread you'll explore how China has tried to control its population growth, and the impacts this has had.

The Beijing Ren Ai Geracomium (old people's home) is in a drab, dusty village just outside Beijing. It's an unusual place, mainly because it exists at all. Old people's homes are rare in China – most elderly people live with their families. However, in future there will be a much greater need for old people's homes in China, because its strict population control policy means that there are now too few young people being born to take care of all of their elderly relations.

Adapted from 'China's predicament', an article in *The Economist*, 27 June 2009.

▲ *One of China's problems – there are now fewer young people to support a growing elderly population*

Population policies

Many countries around the world have introduced population policies to influence population growth. These policies encourage people to have more or less children, depending on the country's circumstances.

The best-known population policy is China's one-child policy. During the 1950s and 1960s, China's population grew rapidly and was seen as unsustainable. China didn't have enough food, water and energy to provide for such a rapidly growing population. Therefore, in 1979, the Chinese government introduced rules to limit population growth – its one-child policy. Couples who only had one child received financial rewards and welfare benefits. Those who had more than one child were fined – and there were also reports of forced abortions and sterilisations.

Has China's policy led to sustainable development?

China's one-child policy has prevented around 300 million babies being born, so its population now – and going into the future – is lower than it would have been. However, by controlling one problem, has China just succeeded in creating other problems? Look at the box opposite.

▶ *China's population growth since 1950*

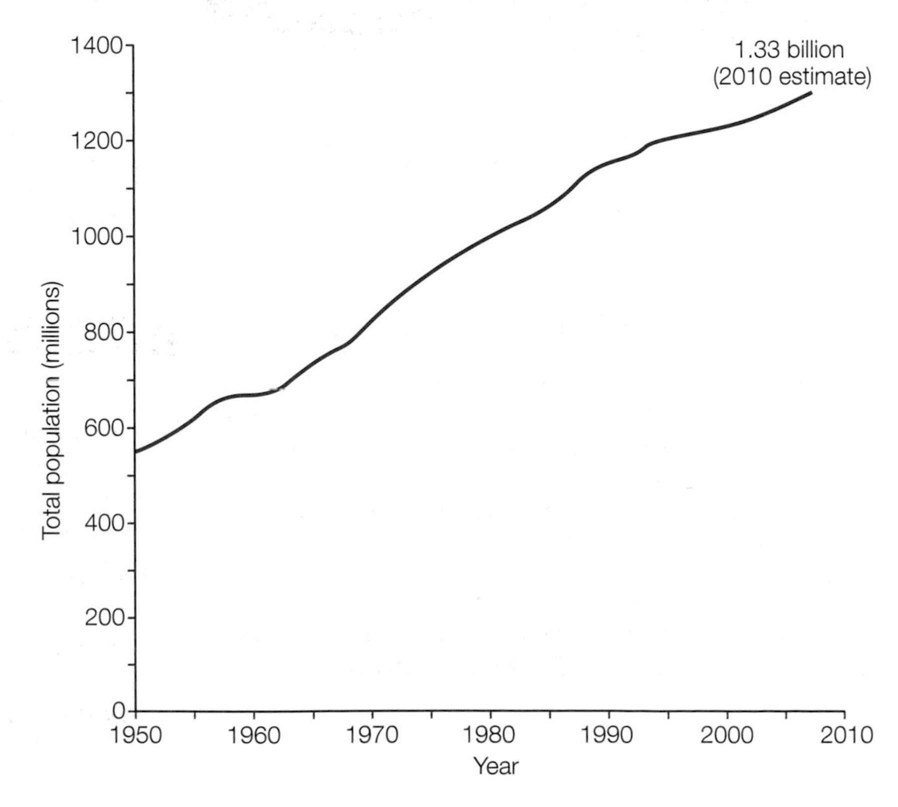

The impacts of China's one-child policy

China's chosen method of population control has had a range of social and economic impacts.

Social impacts

- A typical Chinese child today will have 2 parents and 4 grandparents to look after when they reach old age (a married couple might have up to 4 parents and 8 grandparents to look after). So more old people's homes, like the Beijing Ren Ai Geracomium, will be needed.
- Chinese society traditionally prefers boys, especially if couples are only allowed to have one child. So, baby girls have often been abandoned – with many ending up in orphanages. The lucky ones are adopted.

- By 2020, it is estimated that men in China will outnumber women by 30 million, which might lead to social tension and unrest as more and more men find themselves unable to get married.

Economic impacts

- China's population is ageing rapidly. About 22% of Shanghai's residents are over 60, and that's expected to rise to 34% by 2020. They will all need supporting financially in their old age, which includes an increasing need for expensive healthcare.
- The percentage of people aged over 65, compared to people of working age, is going to increase rapidly – from 10% in 2009, to 40% by 2050. And, from 2025, China is expected to have more elderly people than children (see below).

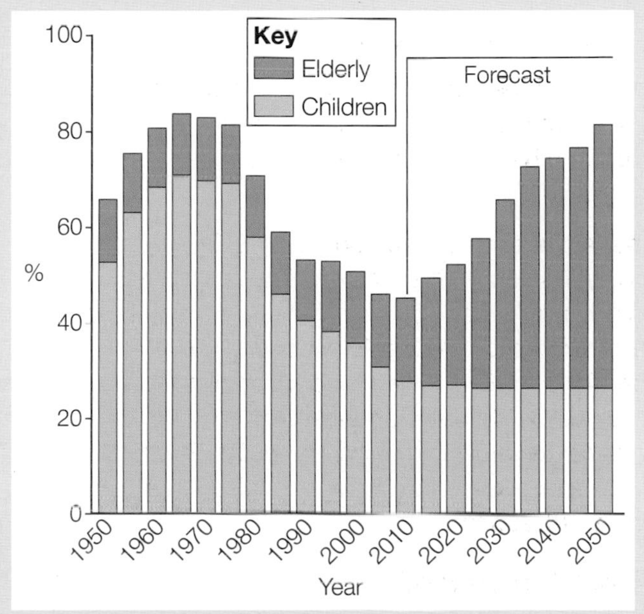

- Many experts feel that China's growing economy won't have enough workers in the future to keep it expanding, while also supporting the growing number of non-workers in the population. The number of young people starting work between the ages of 20 and 24 will drop by half from 2010 to 2020.

Will the one-child policy change?

Reports in 2009 suggested that China's one-child policy was changing. In Shanghai, couples were being encouraged to have two children (if they were single children themselves). But Xie Lingui, a Chinese family-planning official, said that this had been the case for many years, and it wasn't a sign that the policy was changing.

However, the Chinese government may have to relax the policy in future to address the problems it has created.

YOUR QUESTIONS

1. Why do countries have population policies?
2. Write a newspaper report about China's one-child policy. You should include sections about why it was implemented, what the impacts have been, and what you think might happen in the future. Write no more than 250 words.
3. Has China's one-child policy led to sustainable development? Explain your answer.
4. Hold a class debate on the statement 'China's one-child policy should be abandoned'. Some students should be in favour of abandoning the policy, some should be in favour of keeping it.

8.6 » Population control – Kerala

your planet

If it's crowded where you live, think about this. Kerala is three times more densely populated than the UK!

On this spread you'll find out how Kerala has managed to control its population.

Introducing Kerala

The south Indian state of Kerala is about twice the size of Wales, but with a population of 32 million (Wales has a population of about 3 million). Kerala is one of India's most densely populated states, but it has the country's lowest birth rate (see the table below). Its population growth rate (9.4% per decade) is less than half the Indian average (21.3%). So, what's Kerala's secret?

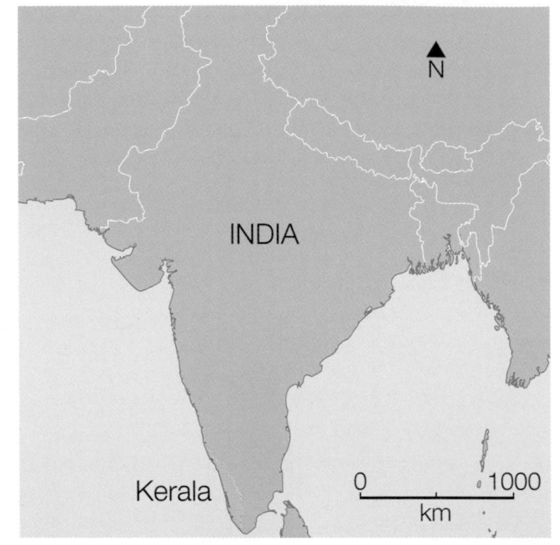

Kerala's approach

What makes Kerala different from the rest of India is its focus on healthcare and education. Kerala's levels of both are the highest in India. For instance, its literacy rate is 91%, compared with 61% in India as a whole.

Although Kerala is one of India's poorest states – its 32 million people have an average income of US$293 per year, and its GDP per capita is 90% lower than the USA's – Keralans can expect to live nearly as long as Americans.

This success story is the result of two things:

- Political decisions to invest in education and women's health. Almost all villages have access to a school and a modern health clinic within 2.5 km.
- Economics. Kerala relies less on farming and more on service industries than other Indian states, especially tourism.

▲ *Secondary school in Kerala. Most children complete 10 years at school.*

How Kerala compares with India as a whole

- From the late 1970s, Kerala has led India in public services – building roads, post offices, primary and secondary schools, medical facilities, and banks.
- Rural poverty in Kerala is the lowest in southern India.
- Women's health and education are the best in India. Food programmes focus on mothers and children, using ration cards and free school lunches.
- Attitudes toward women are positive. There are more girls than boys in higher education, and women hold some of the top jobs.
- Women in Kerala marry on average 4 years later, and have their first child 5 years later, than other Indian women. They have only 2 children on average, and experience very low infant mortality (see the table).
- Over 95% of babies are born in hospital.

Quality of life indicator	Kerala	India	Low-income countries	USA
Adult literacy rate (%)	91	61	39	96
Life expectancy in years (males)	69	67	59	74
Life expectancy in years (females)	75	72	n/a	80
Infant mortality per 1000	10	33	80	7
Birth rate per 1000	17	22	40	16

Has Kerala's approach led to sustainable development?

Kerala has managed to control its population growth by investing in healthcare and education – while still allowing people the freedom to choose their own family size. However, it looks as if Kerala's population could stop growing altogether within 30 years. The projected changes to Kerala's population structure, as illustrated in the two population pyramids, could then create new problems.

Kerala's population greying fast

A steady rise in the percentage of the population over the age of 60, combined with a low population growth rate, will have social and economic effects, says a study by Sabu Aliyar. With an increasing number of couples having only one child, or no children, Kerala's age ratio will alter dramatically. 'As Kerala is 25 years ahead of the rest of India, and in the final stages of demographic transition (low fertility and mortality), the ageing of the state is an important issue' said Sabu Aliyar.

Adapted from an article in the *Hindustan Times*, 4 August 2009.

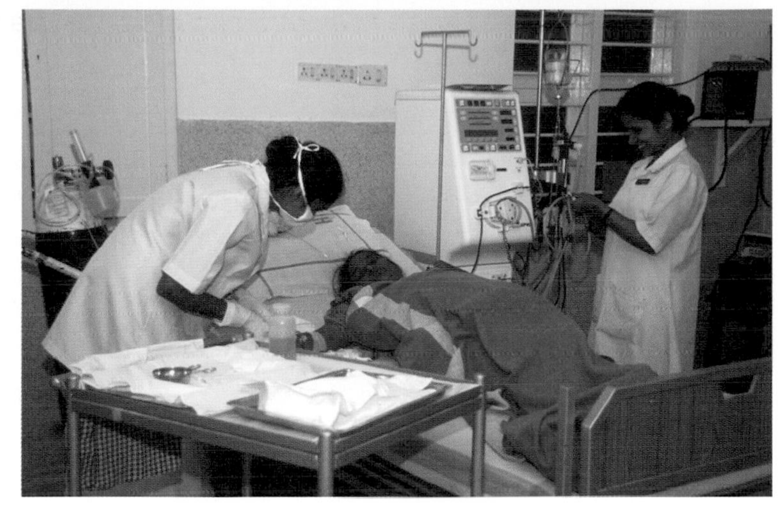

▲ *Medical care in Kerala*

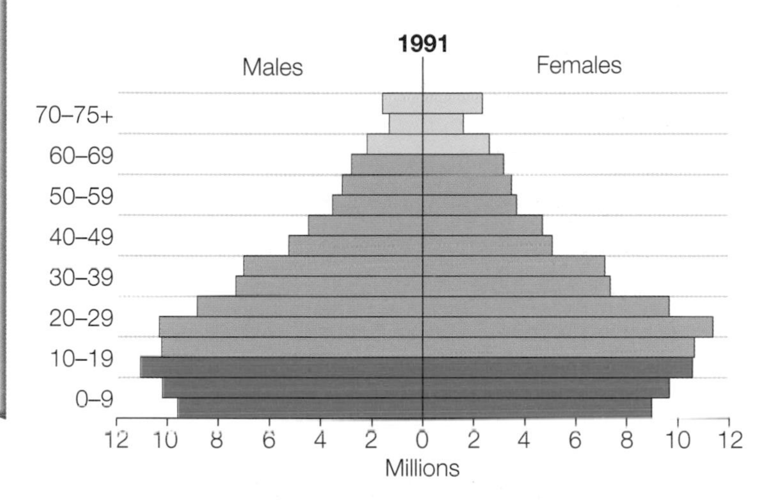

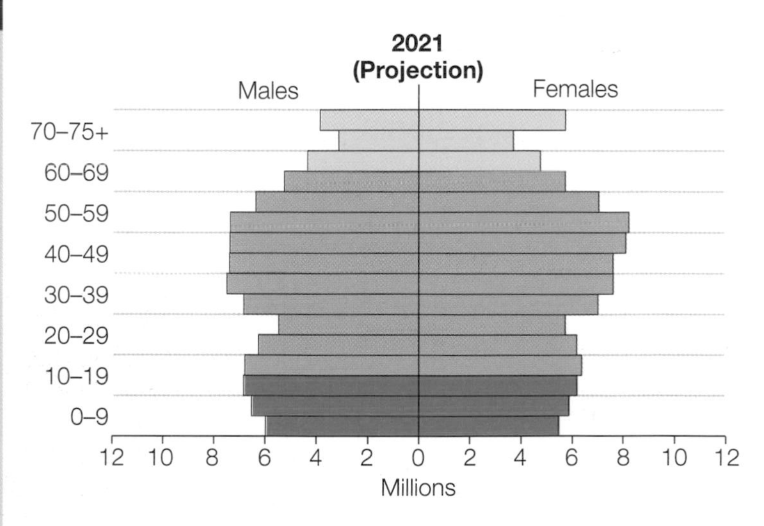

▲ *Projected changes to Kerala's population structure between 1991 and 2021*

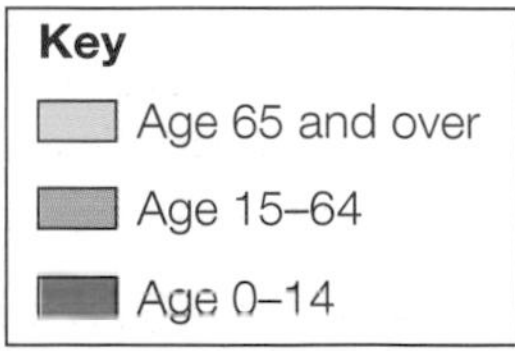

YOUR QUESTIONS

1 a Use the population pyramids to describe how Kerala's population structure is projected to change between 1991 and 2021.

b What problems might Kerala face in the future as a result of these changes?

2 Has Kerala's approach to population control led to sustainable development? Explain your answer.

3 Look back at pages 170-171 and compare the different approaches to population control adopted by China and Kerala.

Hint: In this question you are being asked to 'compare'. That means you need to identify the similarities and differences between the two approaches. You need to look at how population has been controlled, how successful the approach has been, and the impacts it has had.

8.7 » The UK's ageing population

On this spread you'll find out about the UK's ageing population.

Britain's oldest man

Henry Allingham was Britain's oldest man. Born in 1896, when Queen Victoria was on the throne, his life spanned three centuries. He fought in the First World War, was 73 when Neil Armstrong first set foot on the moon, and was 105 when 9/11 happened. He put his long life down to good times and 'a good sense of humour'!

▶ *Henry Allingham died in July 2009, aged 113 years and 42 days.*

The UK is getting older

Henry Allingham was exceptionally old – but he wasn't alone in living longer. The graph on the right, and the population pyramids opposite, demonstrate how the UK's population is ageing. And the fastest increase has been in the size of the group aged 85 and over – called the 'oldest old'. In 2008, there were 1.3 million people in the UK aged over 85 (twice the number in 1983). That number is projected to rise to 3.2 million by 2033.

▼ *The UK's changing population by age band*

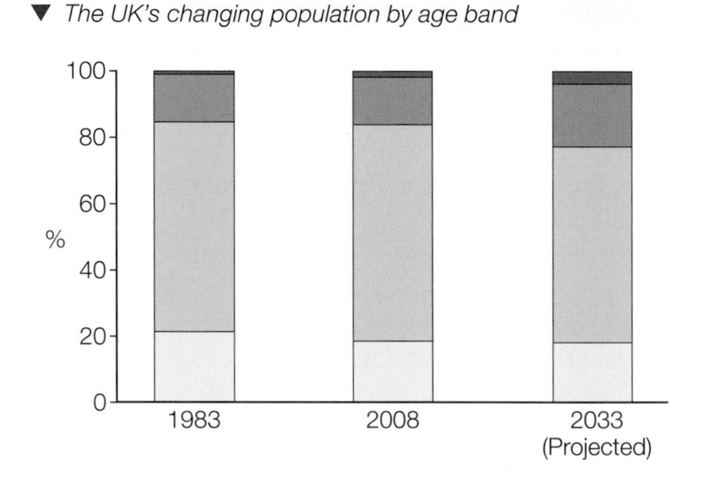

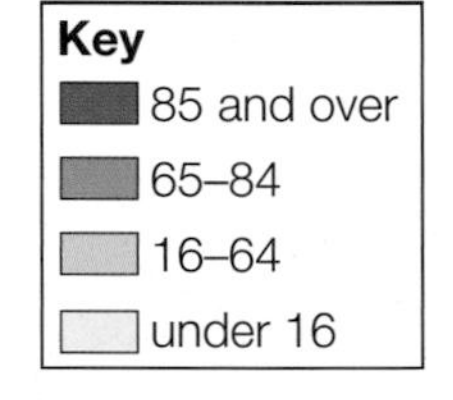

Not only that, but:

- over the last 25 years, the percentage of the UK's population aged 65 and over has increased from 15% to 16%, which might not sound much – but that's an extra 1.5 million pensioners.
- the average age of the UK's population increased from 35 in 1983, to 40 in 2009.
- by 2034, there are expected to be around 100 Britons aged over 110 – known as 'supercentenarians'!

The UK is not alone

It's not only the UK that has an ageing population. Across Europe, birth rates and population growth rates are low, and the average age and life expectancy in most European countries is increasing (see the table).

Country	Average age (years)	Population growth rate (%)	Population growth rate (rank*)	Male life expectancy (years)	Female life expectancy (years)
UK	40.2	0.28	176	77	82
Sweden	41.5	0.16	186	79	83
France	39.4	0.55	152	78	84
Germany	43.8	-0.05	210	76	82
Italy	43.3	-0.05	207	77	83
Poland	37.9	-0.05	208	72	80

**Population growth rate is ranked out of 233 countries. 1 is the highest, 233 is the lowest.*

The UK's changing population structure

The three population pyramids show the UK's actual population structure in 2000, together with the projected population structures for 2025 and 2050. Notice how the pyramids are split into three categories:

- Those aged under 15 are called **young dependents**.
- Those aged 15-64 are called the **working population** (also known as **economically active**).
- Those aged 65 and over are called **elderly dependents**.

Those in the working population are able to earn money and pay taxes, which help to support those who are too young or too old to work (the young and elderly dependents).

Look at how the shape of the pyramid changes between 2000 and 2025, and then again by 2050:

- It gets narrower at the base (fewer babies are being born).
- By 2050, there are fewer people in the working population.
- The number of over 65s has increased.

If the UK's population structure continues to change, the shape of the population pyramid will as well. In future, if the UK's population starts to decline (such as in Italy or Germany), the pyramid will become even narrower at the base.

The dependency ratio

The **dependency ratio** is a measure of the number of working people and those dependent on them. You can work it out like this:

$$\frac{\text{children (under 15)} + \text{elderly (65 and over)}}{\text{working population}} \times 100$$

In 2007, the UK's dependency ratio was 61, which meant that for every 100 people of working age, there were 61 people dependent on them.

YOUR QUESTIONS

1. Explain these terms to your neighbour: elderly dependents, working population (or economically active), young dependents, dependency ratio.
2. Look at the three population pyramids for the UK. List the projected changes to the population structure by 2050.
3. How do you think the UK's dependency ratio will change by 2050? Explain your answer.

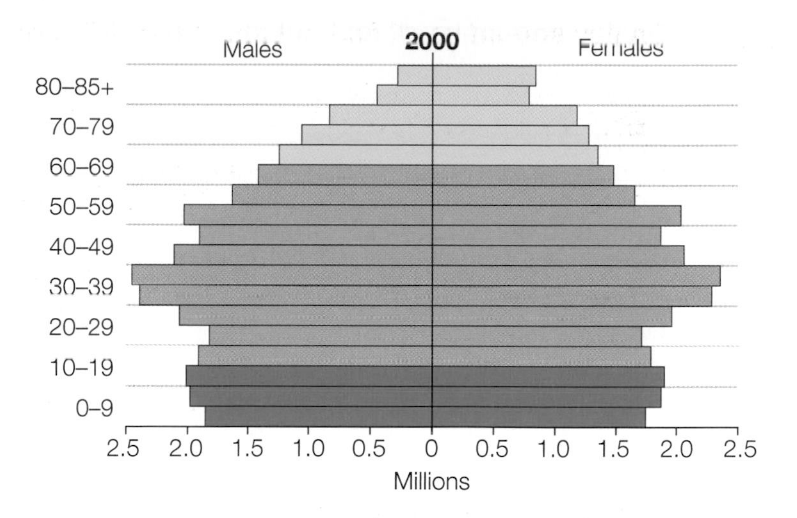

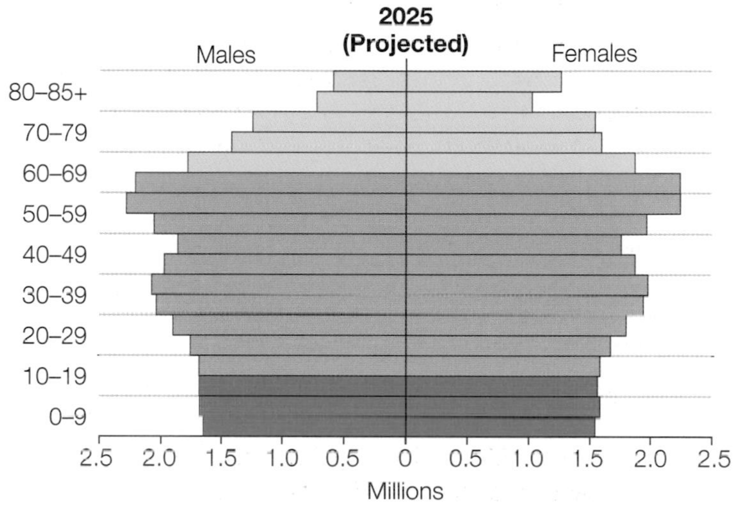

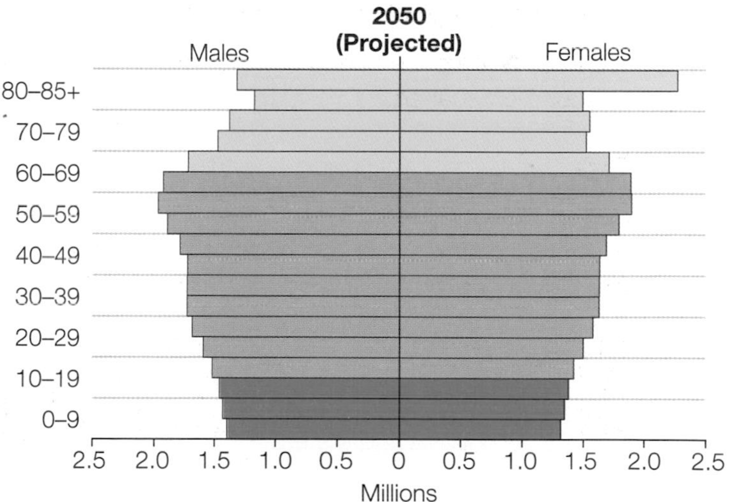

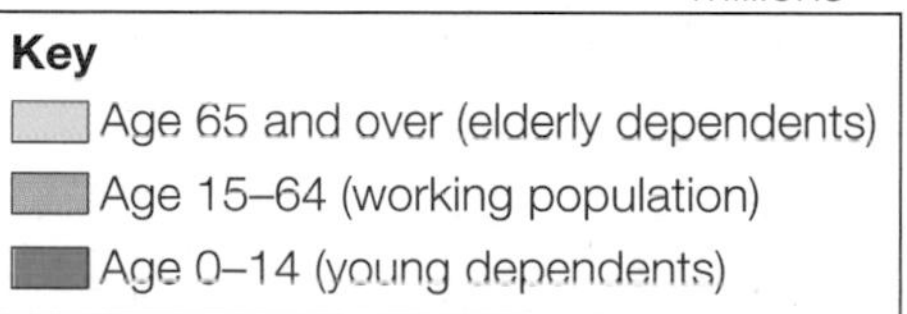

▲ *The changing population structure of the UK*

8.8 » Coping with an ageing population

This spread looks at some of the problems associated with an ageing population, and how governments respond to them.

Ageing brings problems

For some people, living longer is a good thing – but, for others, old age is no joke. Growing old can bring with it problems that affect both the individual citizen and society as a whole:

- **Health and fitness**. An ageing population inevitably leads to an increase in degenerative diseases like cancer, heart disease, diabetes, arthritis, and dementia (e.g. Alzheimer's).
- **Housing**. Many elderly people have specific housing needs, such as homes without stairs, or wider doorways and lower kitchen units for people with limited mobility (such as those in wheelchairs).
- **Increasing care needs**. Most elderly people need increasing levels of care over time. They may eventually have to move into **sheltered accommodation** (where they still have their own flat, but there are carers on site for specific needs), or **nursing homes** (where all their needs are looked after).
- **Pensions**. As more and more people live longer lives, they will claim their state pensions for longer.
- **Fewer workers**. As the UK's population continues to age, there will be a smaller and smaller working population and a larger dependent one.

▲ ***Lack of money*** *can be a problem for a pensioner on a low fixed income. Rising bills for unavoidable expenses, such as food, fuel and council tax, can hit older people hard.*

Impacts on the economy

The problems of an ageing population will increasingly affect the UK's economy and its future development. And a lot of it comes down to money:

- The higher levels of health and personal care required as the population ages will lead to increasing costs. The annual cost of caring for people with dementia is expected to double to £47 billion within 20 years.
- The cost of providing state pensions for more people for longer periods will also increase dramatically.
- A smaller working population means that less income tax and national insurance will be paid to the government. That will reduce its ability to pay for the increasing demands of healthcare, pensions, etc.

▲ *Charities, such as Age UK, run programmes to help older people stay fit and healthy for longer*

Strategies to cope with an ageing population

In 2009, the British Labour government launched a strategy called 'Building A Society For All Ages'. Its intention was to help the UK to adapt to its ageing population. The strategy covered six key areas (see the table).

Another way of coping with an ageing population is to try to persuade people to have more babies! This will eventually change the country's population structure, increase the working population, and increase the country's ability to pay for pensions, etc. Sweden has adopted this approach, which is called **pro-natalist**.

Sweden's total population is only 9 million. It is ageing and its growth rate is almost zero. Sweden's fertility rate was 2.1 in 1989, but by 1999 had fallen to 1.5. Since then the Swedish government has introduced a range of benefits to encourage couples to have more children:

- 13 months' paid paternity leave for fathers – at 80% of their salary.
- Extra money for couples if there is less than 30 months' gap between children.
- Child benefit is paid for each child.
- Sick child care – 120 paid days per child per year.
- All-day childcare and all-day schools for all.

YOUR QUESTIONS

1. Define these terms from the text: sheltered accommodation, nursing homes, pro-natalist.
2. How will an ageing population affect the UK's economy?
3. Draw two spider diagrams to show the problems or impacts that an ageing population brings. Draw one for social impacts and the other for economic impacts. (You might need to put some things on both diagrams.)
4. **a** Describe ***either*** the UK's strategy to cope with an ageing population, ***or*** Sweden's.

   b What are the benefits of your selected strategy?

Part of the strategy	Example of how it could be achieved
Having the later life you want	Offer free NHS health checks to people in England aged 40-70, to encourage them to maintain and improve their health and fitness.
Older people at the heart of families	Provide financial help for grandparents who care for grandchildren.
Engaging with work and the economy	Review the age at which people retire.
Improving financial support	Provide extra winter fuel payments for the elderly.
Better public services for later life	Introduce a health 'prevention package', to bring together things like flu vaccinations, cancer checks, etc.
Building communities for all ages	Working with developers and architects to build homes suitable for older people, e.g. with doorways wide enough for wheelchairs.

▲ *The six key areas of the Labour government's strategy: 'Building A Society For All Ages'*

▶ *Part of the Labour government's strategy was to encourage more people aged over 50 to become teachers*

8.9 » Migration – from Poland to the UK

This spread looks at why so many people migrated from Poland to the UK.

Piotr's story

Piotr Dobroniak is 29. Before leaving Poland, he was the manager of a cash-and-carry chain in Wroclaw. Now he lives in London – working as a labourer earning £50 a day (before tax). One of his first jobs was as a builder on the new Wembley Stadium. He said 'There was a lot of tension between the British and foreign workers'.

Piotr is happy in London and – unlike many other Polish migrants – may never go home. Things are also looking up for him – he's just been interviewed for a manager's job at the Lidl supermarket chain.

Piotr is not the only one

In 2004, Poland and seven other Eastern European countries joined the EU. As a result of this, a wave of migration hit the UK. No one really knows how many Eastern Europeans arrived but, by 2008, over 850 000 had registered to work in the UK – and, like Piotr, many came from Poland. They were living all over England and Wales, as the map shows.

A BBC survey of Polish migrants showed that:

- 60% came for financial reasons, and because of unemployment in Poland
- 85% were young (under 34)
- 31% came for personal and professional development
- 18% wanted to get away from the political and economic situation in Poland
- only 15% wanted to stay permanently – the remainder were seasonal workers, or had come to the UK for a few years to earn and save money
- nearly 30% had a degree or similar qualification.

Migration is the movement of people from one place to another.

The **source country** is the country that a **migrant** like Piotr comes from – in his case Poland.

The **host country** is the country that a migrant goes to, such as the UK.

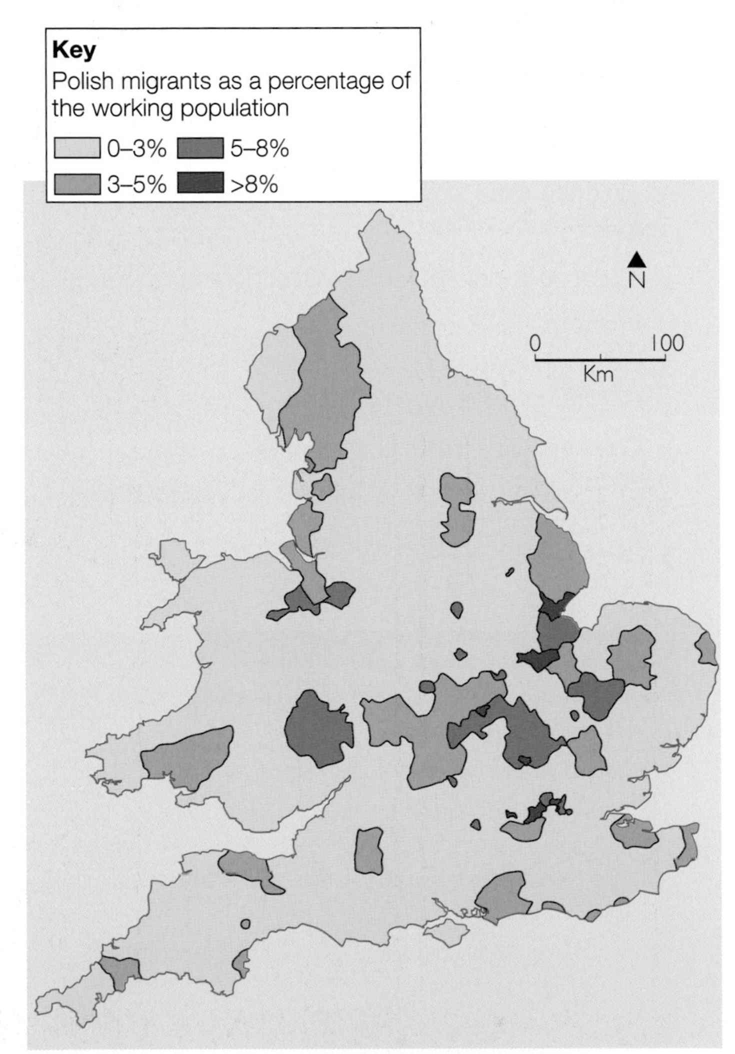

▲ *The main locations of Polish migrants in England and Wales in 2006*

your planet

In 2006, a Polish migrant could earn an average of £20 000 a year in the UK, compared with £4000 in Poland.

Most of the Polish migrants came to work. Radek, a 25-year-old electrician, said 'I am not here to claim benefits. I am here to earn money. As much as I can make, and then go home.' He, and others like him, are called **economic migrants**. Migration can be divided into different categories, as the spider diagram shows.

Migration

- **Voluntary** – people choose to move for better jobs and higher wages. They are called **economic migrants.**
- **Forced** – people have to move, or they'll face extreme hardship, persecution, and even death. They are called **refugees.** (**Environmental refugees** are those fleeing **environmental disasters.**)
- **Temporary** – e.g. **seasonal migrants**, like those who come to pick fruit and vegetables in the UK and then go home when the picking season ends.
- **Permanent** – people move and don't return home, e.g. many from the Caribbean who migrated to the UK in the 1950s and 1960s.

Push and pull

The reasons why people migrate are often described as **push** and **pull factors**.

Push factors that can force people to leave their own country include:

- not enough jobs
- low wages
- poor educational opportunities
- poor healthcare
- war with another country
- civil war and lawlessness
- drought and famine.

Conflict, or the threat of conflict (as here in Afghanistan), often forces people to flee their homes. They may never return.

Pull factors that can attract people to a new country include:

- hope of finding a job
- higher wages than at home
- better healthcare
- chance of a better education
- a better standard of living
- family and friends may have moved there already
- lower levels of crime, and safety from conflict.

Seasonal work, such as cutting celery, attracts migrants for a short period of time

YOUR QUESTIONS

1 Define these terms from the text in your own words: migrant, migration, source country, host country, economic migrant, refugee, seasonal migrant, push and pull factors.

2 Look at all of the material on this spread. Then complete a table to show some push factors (for leaving Poland) and pull factors (for moving to the UK).

3 Build your own set of case-study notes about Polish migrants.
- Start with **who** they are, and how many.
- Include **where** they went, **why** and **when**.
- You'll complete your case study on page 181.

On this spread you'll learn about the impacts of migration, and the movement of Afghan refugees to France and the UK.

What are the impacts of migration?

Piotr Dobroniak's migration from Poland to the UK (together with many other Poles) has had impacts on both countries.

Impacts on the UK

Migrants:

- provide a hard-working, motivated, workforce
- fill skills shortages
- contribute to the local and national economy, e.g. their taxes help to support the UK's ageing population (see pages 174-177)
- tend to be young, so they help to balance the UK's ageing population
- can put a strain on local services, such as schools and housing.

Impacts on Poland

- In 2005/2006, almost £4 billion was sent back to Poland from abroad (that's around a third of Poland's economic growth)
- Fewer unemployed people have been left looking for jobs in Poland.
- However, labour shortages have been caused in the service, building and science industries.
- Therefore, people from other countries, such as the Ukraine and Belarus, have been invited to work in Poland to help fill those shortages.

Population growth

Migration has had another important impact on the UK. Its population is now projected to rise from 61 to 71.6 million by 2033. Just over two-thirds of that rise is expected to be down to migration – either directly or indirectly. In other words, the migrants themselves count in the population figures and, if they then go on to have children in the UK, they will count as well.

▶ *Wroclaw, Poland – Piotr Dobroniak's home town – has had to cope with a loss of population as people have moved abroad*

On the move again

Migration isn't static. Sometimes people move again, or go home, as the article below shows.

Changing migration patterns

Ania Rosiak is thrilled to be back home in Warsaw, Poland's capital. Tucking into waffles and ice-cream, she said 'I spent 14 months working for a bank in Glasgow. Like many Polish people, I was over-qualified, but now I'm hoping to set up a translation business'.

Ania is not the only one. By the end of 2008, almost half of the Eastern Europeans who had come to work in the UK had returned home. Why? Warsaw's unemployment rate is now lower than London's, and, while the UK's economy has been shrinking, Poland's has been growing slowly.

Adapted from 'Recession moves migration patterns' by Rob Broomby, BBC News, 8 September 2009.

Afghan refugees

In 2009, Afghanistan was constantly in the news. British and other NATO soldiers were often killed or injured there – but what about the Afghan people?

Afghanistan had an estimated population of over 28 million. But nearly 3 million were refugees. Life was hard, and a combination of conflict, drought, poverty, corruption, and lack of jobs led people to leave their homes.

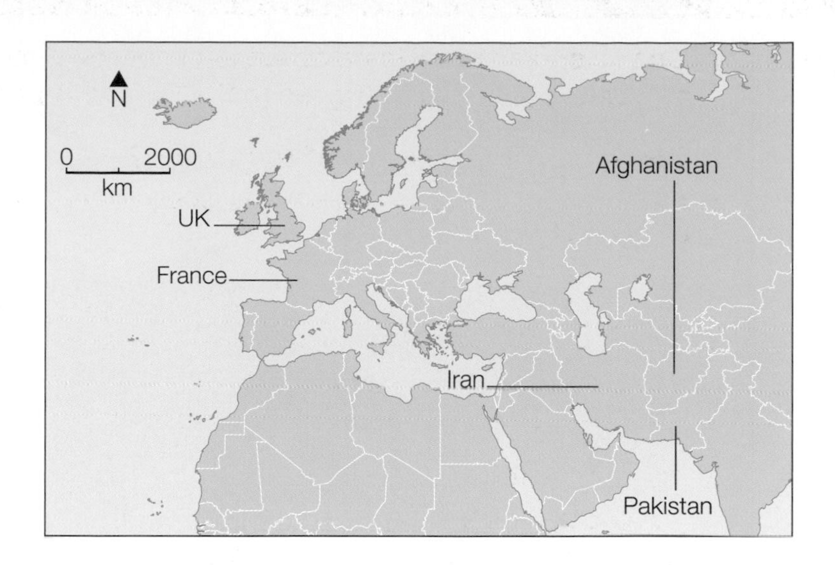

Where did they go?

Many refugees went to Pakistan and Iran. Others attempted the difficult journey to the UK – only to end up in 'The Jungle'.

The story of The Jungle

- The Jungle grew up on wasteland on the edge of Calais, in northern France. It was near the entrance to the Channel Tunnel.
- It was a makeshift, unsanitary settlement of homemade tents and shacks.
- In August 2009, between 700 and 800 Afghans were living there – waiting for a chance to get to the UK (either to be smuggled in, or to stow away on lorries or trains).
- Many of the Afghans said they didn't feel safe in France, but felt that England was a 'good' and safe' country for refugees.
- Early on 22 September 2009, the remaining migrants were arrested. Within 24 hours the settlement was bulldozed.
- Many people hoped that the migrants would be allowed to stay in France, or be sent home. However, most of them were just released, so they returned to northern France and kept trying to get into the UK.

◀ *Most of the migrants in The Jungle were single young men*

YOUR QUESTIONS

1 Write a newspaper article (no more than 300 words) about the Afghan refugees. Include a map (like the one on this page). Tell your readers why the people have become refugees, where they are trying to get to, and what has happened to them.

2 Complete your case study notes about Polish migration that you started on page 179.

Add a section about the impacts of the migration on the UK and Poland. You need to identify both positive and negative impacts.

Finish your case study with a section about how things seem to be changing.

Hint: 'Identify' means to select. So, here you need to select those impacts which are positive, and those which are negative.

9 » Changing urban environments

What do you have to know?

This chapter is from **Unit 2 Human Geography Section A** of the AQA A GCSE specification. It is about the issues facing cities in richer and poorer parts of the world, and how these can be managed, as well as sustainable urban living. The table shows how the pages in this chapter match the content in the specification.

Specification content	Pages in this chapter
Urbanisation has happened at different rates and at different times in the rich and poor world.	p184-185
Different parts of urban areas have different functions and land uses.	p186-187
There are many issues facing urban areas in the richer parts of the world including: housing, rundown inner cities and CBDs, traffic, and ethnic segregation.	p188-193 Case study of Birmingham
Rapid urbanisation in poorer parts of the world has led to the development of squatter settlements and an informal economy. Squatter settlements can be improved.	p194-195 p196-197 Case studies of improving squatter settlements: Dharavi (Mumbai, India) and Old Naledi (Gabarone, Botswana)
Rapid urbanisation in poorer parts of the world has caused environmental problems which need to be managed.	p198-199
Urban living can be made more sustainable – environmental, social and transport issues.	p200-202 p203 Case study BedZED. London

Your key words

- Urbanisation
- Rural-urban migration
- Megacities
- Concentric ring model
- Central Business District (CBD), inner city, suburbs, rural-urban fringe
- Ethnic segregation
- Rebranding
- Shanty towns, squatter, squatter settlements
- Informal economy
- Self help, site and service schemes
- Sustainability
- Integrated transport policy

Exam help ...

Advice See pages 297-299 for information on how to be successful in your exams.

Practice See page 308 for exam questions on this chapter.

What if ...

- everyone in the world lived in cities?
- cars weren't allowed in cities?
- cities could be sustainable?

On this spread you'll find out what urbanisation is, and how it differs between richer and poorer countries.

Sunita's story

I'm Sunita and this is my story. Two years ago, my parents brought my brother Rakesh and I to live in Mumbai – in an area called Dharavi. People say it's a slum. Maybe it is – we're all poor here – but my father says at least we have work. And one day maybe Rakesh or I will be rich.

Dharavi is very crowded and very noisy. Everyone is busy all the time. Just outside our house people wash laundry, sew clothes and bang the dents out of used oilcans, so they can be recycled. There are small workshops everywhere. Somebody told me there are 15 000 in Dharavi.

It's smelly, too, I suppose. There are lots of open sewers. I like to walk down to the biscuit factory where it smells nicer!

I go to school every morning. The lessons are literacy and maths. In the afternoon I help my mother clean our house. It only has two rooms, but we have electricity. Afterwards, I often go rag picking with my friends to earn a little money.

▼ *Dharavi, Mumbai*

Why go to Mumbai?

Sunita isn't alone. Every year, thousands of people move to Mumbai – in a process called **urbanisation**. Urbanisation is happening rapidly in India, as more and more rural people leave the countryside for a new life in India's expanding cities.

Many rural people in the world's poorest countries are really poor. And they know that in cities like Mumbai some people are very rich. So, it's tempting to think that moving to the city will make them better off as well. But many people who live in the cities are also very poor – and many of the migrants from the countryside end up in places like Dharavi.

Rural areas, especially in poorer countries, have few jobs – apart from working on the land. So there's little chance of breaking out of poverty. Factors like this help to 'push' people away from the countryside. In the city there are more jobs, better educational and health facilities, more entertainment options – and some people are better off. These factors help to 'pull' people to the cities. This movement of people from the countryside to the cities is called **rural-urban migration**.

Urbanisation is the rise in the percentage of people living in urban areas (towns and cities), in comparison with rural areas.

▲ *Why people move to cities*

Urbanisation in richer countries

Many richer countries in Europe and North America, including the UK, experienced urbanisation during the late eighteenth and nineteenth centuries – during the Industrial Revolution. As these countries industrialised and expanded their manufacturing, millions of people left the countryside to work in the new urban factories.

Today, up to 90% of the population of richer countries already live in towns and cities, so the rate of urbanisation is very slow – in the UK it's just 0.5% a year.

Urbanisation in poorer countries

Urbanisation in poorer countries has only really been happening since the 1950s (although some people did begin leaving the countryside to move to Mumbai in the mid-nineteenth century). As well as starting later, urbanisation in poorer countries is now happening much faster than in richer countries, as the poorer countries catch up:

- In India, 29% of the population now live in cities (about 340 million people). The rate of urbanisation is about 2.4% a year.
- In Botswana, in southern Africa, the percentage of people living in urban areas increased from 4% in 1966 to 60% in 2010.

The megacities

More than half of the world's population (over 3.3 billion people) now live in cities. As a result, the number of really big cities (with populations of over 5 million) is growing. And **megacities** – which have over 10 million people – are also growing. The graph and map on the right show where the world's megacities are, and how they're expected to grow.

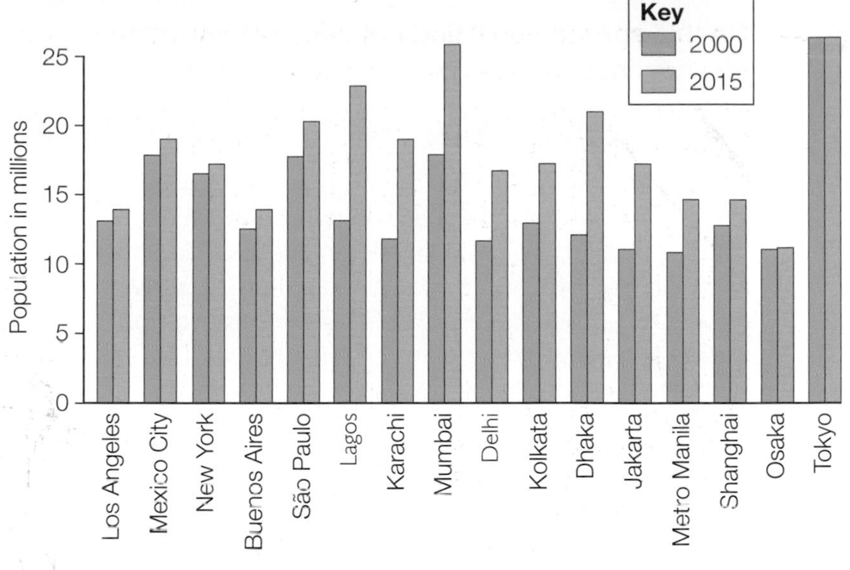

▲ *The expected growth of the world's megacities, 2000-2015*

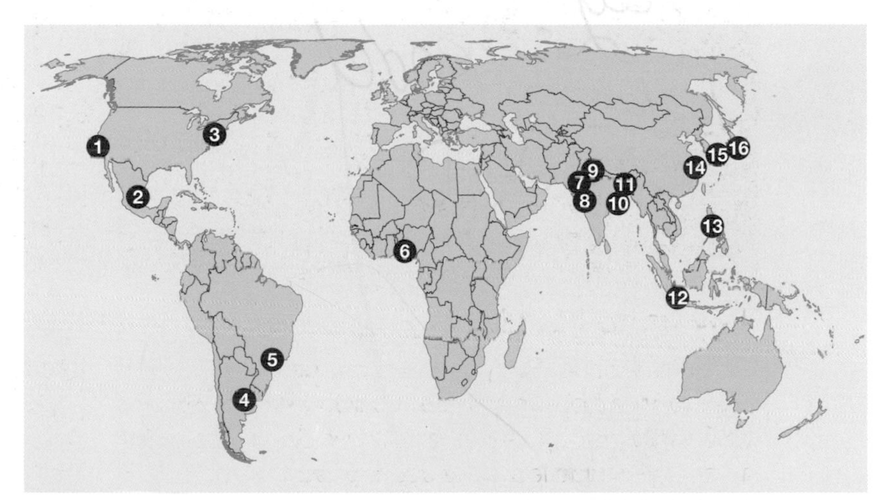

▲ *The locations of the world's megacities*

YOUR QUESTIONS

1. Write a definition of urbanisation and rural-urban migration, using your own words.
2. Why do people move to cities like Mumbai?
3. How does urbanisation differ between richer and poorer countries?
4. Use the map to describe the distribution of the world's megacities.

On this spread you'll find out about the structure of cities in richer and poorer countries.

Cities in richer countries

In the 1920s, an American called Ed Burgess studied land use in the city of Chicago. He noticed that the different parts of the city had different types of land use, and that they were arranged in a ring pattern. Burgess drew up an urban model – called the **concentric ring model** – to show this land use pattern (a model is a simplified version of what you see in the real world). Each ring in the model performs a job, or function, for the city. And these jobs or functions change with increasing distance from the city centre.

Key
- Central Business District (CBD)
- inner city
- inner suburbs
- outer suburbs
- rural-urban fringe

▲ *The Burgess concentric ring model of urban land use*

The CBD

The city centre is also called the **Central Business District (CBD)**. The CBD contains the city's main shops, together with leisure and entertainment facilities, e.g. cafés, bars, restaurants and theatres. It also includes buildings like town halls. Around its edge are the offices of professional people, like solicitors and accountants.

▲ *Birmingham's Central Business District*

The inner city

The inner city used to include a lot of manufacturing industry, but much of this has now closed down. Traditionally, it also provided cheap terraced housing for the factory workers. These houses were small, and the streets containing them were arranged in a grid pattern – with little or no green spaces.

Recently, however, inner cities have become more popular (especially with young adults), because they're close to office workplaces and the facilities of the city centre. They've been smartened-up and redeveloped, with expensive upmarket apartments for young professionals being built to replace the old terraced housing. Lozells and Aston in inner city Birmingham are being redeveloped like this.

The suburbs

The suburbs are more expensive places to live, and are popular with families. They can be divided into inner and outer suburbs. The further you go out from the city centre, the bigger the houses get – changing from semi-detached to detached. They also have bigger gardens, and there are more open spaces like parks.

▶ *A suburban street in Moseley, Birmingham*

The rural-urban fringe

The rural-urban fringe is on the edge of the city. It's almost in the countryside. New housing estates are often built here, as well as big retail and leisure parks. It has villages and fields, as well as recreational facilities like golf courses and riding stables.

Cities in poorer countries

Just like in richer countries, cities in poorer countries tend to have similar patterns of land use. The CBD is still in the centre, but then the pattern changes from the one seen in richer countries (as shown by the Burgess concentric ring model). The land use pattern in poorer cities is not as simple as that:

- The closest ring to the CBD isn't a zone of old industry and cheap housing. It's a high-class residential area, with large villas and modern expensive apartments.
- Cheaper housing is found further out of the city.
- Poor-quality housing can be found right next to the posh expensive areas, but shanty towns and slums are often found on the very edge of the city.
- Industry tends to be found along the main roads into the city.

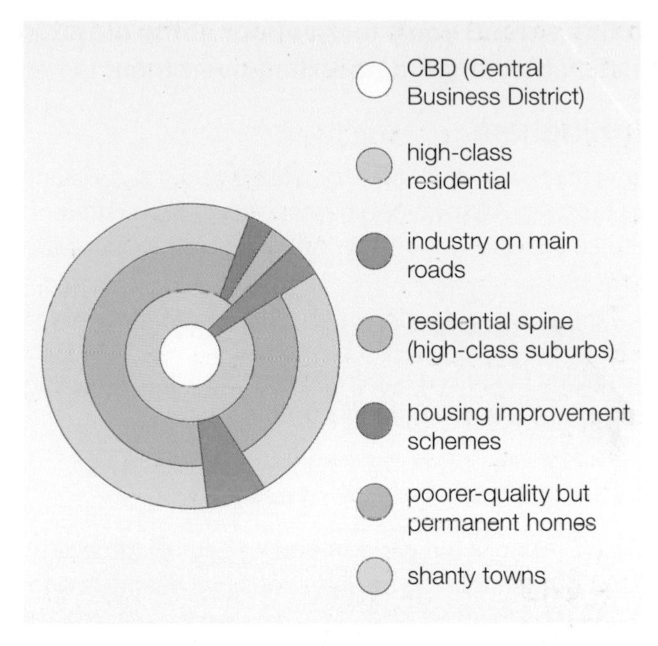

▲ *Urban land use in a city in a poorer country*

◀ *The world's most expensive home is in Mumbai – built right next to the CBD. This 27-storey tower block was built by India's richest man as a home for himself, his wife and their three children! It is worth $1 billion (£630 million) and has 600 servants, a cinema, three helicopter pads, several swimming pools – and a 160-car garage.*

▼ *In Mumbai, rich and poor live side by side*

YOUR QUESTIONS

1 a What is the CBD?

b What does the concentric ring model show?

2 Copy and complete the table on the right for a city in a richer country, like Birmingham.

3 Compare the patterns of land use in a city in a richer country, with those in a city in a poorer country.

Hint: 'Compare' means you need to identify the similarities and differences between the patterns of land use.

4 How well does the land use in a city you know fit the Burgess model?

Area of the city	What you find there
CBD	
Inner city	
Suburbs	
Rural-urban fringe	

On this spread you'll learn about some of the issues that face many cities in the UK, and how the government is helping to deal with them.

Birmingham

Birmingham is the UK's second-largest city. It's vibrant and buzzing. All sorts of different people from different cultures live and work there. And it's got something for everyone.

- The Balti Triangle is famous for its curry houses.
- Brindley Place is a super-modern development, with luxury hotels and cutting-edge restaurants.
- The nightlife of Broad Street (lined with bars and clubs) gives the city centre a real buzz.
- The Bullring is the glitzy shopping centre with over 140 shops stocking the latest designs, fashions and electronic gizmos.
- The City of Birmingham Symphony Orchestra and Birmingham Royal Ballet provide world-class culture.
- And St Paul's Square is the place to go for the latest in live popular music.

Issues facing cities

On the surface, a city can seem fast-paced, modern and exciting. But if you look at it more closely, you can often find a variety of problems. Birmingham faces a number of issues that can also be found in many other British cities:

- Not enough good-quality affordable housing.
- Too much traffic and pollution.
- A CBD with rundown or unused buildings.
- High unemployment in certain areas.
- A mixed culture with **ethnic segregation** (which means that people from different ethnic groups and religions live separately).

Over the next few pages, you'll see how Birmingham is trying to tackle some of these issues, in order to improve the city socially, environmentally and economically.

▲ *Birmingham by night*

▲ *Behind Birmingham's flashy exterior lie a lot of problems*

Help for inner cities

New Deals for Communities (NDC)

NDC was launched by the Labour government in 1999 as a way of helping struggling inner city areas. It identified 39 of the most-deprived inner city areas in the country. The local communities in those areas were then involved in helping to find solutions to the problems they were facing.

Aston, in inner city Birmingham, was identified as one of the 39 NDC areas. The scheme that was set up there is called Aston Pride. It covers three key areas:

- Health and regeneration – £400 000 has been spent setting up the Aston Pride Community Health Centre.
- Employment and business – local young people have been given help to enter the world of work, through a work-experience programme and a dedicated guidance team.
- Education and lifelong learning – broadband centres have been set up to give local people Internet access.

Rebranding

Some parts of cities can suffer from a bad 'image'. People don't want to live there – and companies won't move there – so the whole area suffers a downward spiral into deprivation. **Rebranding** an area involves giving it a new image, so it attracts development and employment and leads to an upward spiral into success.

Eastside, in Birmingham, is going through this sort of **revitalisation** (putting new life back into something). The area is being rebranded as a 'learning and technology quarter', because a lot of colleges and university sites are located there, e.g. Aston University and Matthew Boulton College. Eastside will also have new leisure and cultural attractions, such as a large new city park and a revitalised canalside area, including new housing. The hope is that a positive and modern environment can be created and advertised that will attract new businesses to the area and boost employment opportunities.

▼ *The revitalisation of Eastside*

YOUR QUESTIONS

1 Write a definition of rebranding.

2 a Look at the five issues bullets in the box opposite and decide whether each one is environmental, social or economic. You might decide that some of them fit in more than one category.

b Pick two of these issues and explain how solving them could help to rebrand an inner city area.

3 Discuss with a partner the benefits that Internet access can provide for a community.

4 Why does it matter what an area is called?

Hint: Think about Eastside as a 'learning and technology quarter'. How do you think this label might help its image?

On this spread you'll look at issues facing the CBD, as well as housing in cities.

Birmingham's CBD

Birmingham used to be a big manufacturing centre. Engineering and jewellery manufacture were particularly important industries for the city. But, by the 1980s, much of Birmingham's traditional industry had gone. Its factories couldn't compete with cheaper goods from abroad – and closed down. This had a big effect on the local economy and the CBD.

Today, the Bullring shopping centre (right in the middle of Birmingham's CBD) is one of the city's most famous landmarks. But it hasn't always been glitzy and modern. It was first built in the 1960s, but, by the 1980s, it was looking rundown and past its best. The decline of manufacturing in Birmingham, and its impact on the local economy, meant that people had less money to spend in the Bullring's shops.

▲ The Bullring shopping centre before redevelopment ...

What Birmingham did

It was decided that bringing the Bullring shopping centre back to life was vital for improving the physical environment of the CBD and getting the city's economy going again. So, the Bullring was completely redeveloped and reopened in 2003. It now has more than 140 major shops, including a flagship Selfridges store with a very unusual design! Special attention was paid to the redesign of the Bullring's buildings, so that they would help to make Birmingham's CBD look attractive and inviting to visitors from outside the city. The final designs had lots of comments – good and bad!

And elsewhere in the CBD:

- Brindley Place was an area of old warehouses by the canal. A £400 million project has now transformed it into a smart, modern pedestrian zone of upmarket apartments, offices, hotels, restaurants and cafes. Nearby visitor attractions include the National Indoor Arena (NIA) and the Sea Life Centre aquarium.
- Traditional shopping streets in the city centre have been pedestrianised, and buildings in the Jewellery Quarter have been restored – making it an attractive area for tourists to visit.

▲ ... and the Bullring after redevelopment – the Selfridges building on the right stands out!

How the redevelopments have helped

Birmingham's CBD has been brought back to life. The physical environment has been improved, so that people want to live there, businesses want to open there, and tourists want to visit. This all means that more money continues to be invested – which keeps the CBD alive! Big changes continue to be made, with a major (£550 million) government scheme planned to give New Street railway station (right next to the Bullring) a much-needed makeover.

your planet

More than 35 million people use the Bullring shopping centre every year – that's a lot of shopping!

Housing issues

Now that the inner cities are being regenerated, more people want to live there – for example, people with jobs in the shops, offices and leisure facilities of the CBD, as well as nurses and teachers who work in hospitals and schools in the inner city. Living near work saves money and time, and is less environmentally damaging than commuting long distances.

Universities and large colleges are often built in inner city areas. So, lots of students want to live in the inner city too.

▲ *New housing in Lozells replacing old Victorian terraces*

The right home

People who want to live in a city need a home there, but different people have different needs:

- Young professional people in their 20s/30s might be single, or live as a couple. A flat right in the city centre, near all the nightlife and shops, might suit them.
- Families with children are more likely to prefer a house with a garden in a quieter area, with good schools nearby.
- Students and lower-paid workers need cheaper housing – maybe a flat or a room in a large shared house in one of the older areas of the city.

Urban Living

Lozells and Birchfield are neighbourhoods in the north-west of Birmingham's inner city (see the map on page 188). Housing in these areas is being redeveloped as part of the government funded Urban Living scheme.

The Urban Living scheme in Birmingham is one of nine set up in the Midlands and the North of England in the first decade of the twenty-first century. It has tried to find ways to improve the quality of life for inner city residents.

Some buildings in south Lozells, particularly small unpopular Victorian terraced houses, have been demolished and replaced with new housing (some of it eco-developments). Other old houses there have been refurbished. Many large Victorian houses in north Lozells – that used to be split up into individual bedsit flats – have now been converted back into large family homes. The improved housing in Lozells has raised the image of the area and encouraged more families to live there.

▲ *In Birchfield, these three 1960s tower blocks have been demolished to create space for a new mixed residential development that includes affordable housing for families.*

YOUR QUESTIONS

1 a List three things that Birmingham has done to improve its CBD.

b How do you think these improvements have helped to make the city more attractive to visitors?

2 a What sort of accommodation are the people on the right likely to want?

b Which part of the city would each one probably want to live in (e.g. CBD, suburbs), and why?

i A 32-year-old single man who has just started a new office job in the CBD.

ii A married couple with three children, two dogs and a rabbit!

iii An elderly lady who lives on her own and can't drive.

3 Why was it better to improve the housing in Lozells, rather than build on an unused field on the edge of Birmingham? Try to think of three reasons – one social, one environmental, and one economic.

9.5 » Urban issues 3 – richer world

On this spread you'll find out about the multicultural nature of cities in the UK, and how the impact of traffic can be reduced.

Multicultural cities

Walk through most cities in the UK and you'll see signs of lots of different cultures – maybe a Chinese supermarket, or a Polish church, or a Halal butcher catering to the dietary needs of Muslims. But, quite often, people from a particular ethnic background will live in an area of the city separate from other groups. This called **ethnic segregation**. Why does it happen?

- People prefer to live near others who have the same background and speak the same language.
- They often live close to places that are important to their culture, like a place of worship.
- They may live where the housing is cheaper, and so end up in the same area.

The problem for cities is that some sections of their populations are being cut-off (isolated) from basic services. This can be because of difficulties with the language, or because of cultural differences. For example, in some cultures, women do not want – or are not allowed – to see a male doctor.

▲ *The Lozells area of Birmingham*

Supporting people

City councils in the UK have to try to find ways of supporting all members of the community. How?

- Many places now print information in a number of different languages (like the medical documents on the right). Places like hospitals employ interpreters to translate for people who can't speak English.
- Cultural leaders from different parts of the community can be involved in discussions and decisions that will affect the area.

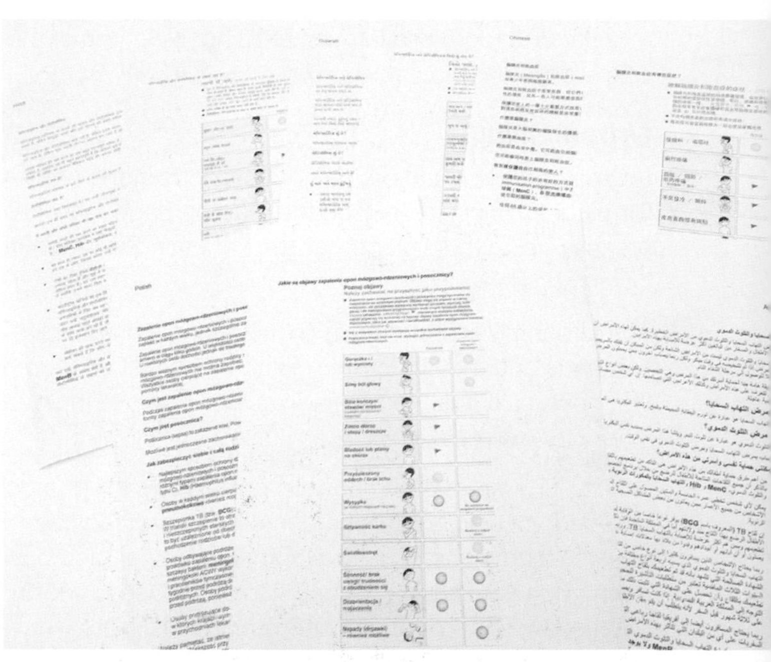

▲ *Medical documents translated into different languages, like Polish and Arabic, to help support those members of the community with poorer English language skills*

What's happening in Birmingham?

Remember the Aston Pride scheme from page 189? Part of this scheme is the COFSS. This stands for Community Outreach Family Support Service. It aims to help people make use of services such as healthcare. COFSS health workers are based in community centres, mosques and schools, so people can reach them easily. They aim to work together with local people to create services that meet specific needs.

Traffic in cities

There are now more cars on the UK's roads than ever before. British cities have big problems with traffic and congestion. People are always on the move – in and out of the CBD, to and from work. But more cars mean more air pollution, which affects people's health and causes damage to buildings.

So, what can be done to help tackle the joint problems of keeping people moving and limiting pollution? Different cities have tried different solutions, as the table shows.

What's happening in Birmingham?

Birmingham is trying to cope with its traffic in a number of ways. It:

- has a park-and-ride scheme
- has bus lanes (although it removed some of them, because they restricted the traffic flow)
- encourages car sharing
- has its own tramline, called the Birmingham Metro.

The Metro – Birmingham

Birmingham's Metro is a tramline that runs for 20 km from Wolverhampton into central Birmingham. The trams run frequently (every 7-10 minutes) and allow people to travel quickly and easily into the CBD.

The Metro opened in 1999 and cost £145 million to build. There are plans to add more lines in future, e.g. to extend the system to Birmingham International Airport and the National Exhibition Centre (NEC).

▲ *Gridlocked Birmingham. 25% of all carbon dioxide emissions in the UK come from car exhaust fumes.*

▼ *Some solutions to traffic problems*

Solution	How it works	Benefits
Park and ride	Drivers leave their cars in big car parks on the outskirts of the city. Regular buses then take them to the city centre.	There are fewer cars in the city, so there's less congestion and improved air quality.
Bus lanes	These are lanes that only buses (and sometimes taxis) are allowed to use.	• The buses are not held up by other traffic, which makes them more reliable and quicker. • Public transport is less polluting, because it reduces the number of cars on the road.
Trams	Urban tramways run on the roads, powered by overhead electricity lines.	There is less air pollution, and it's another travel option for people visiting the city.
Congestion charge	People have to pay to drive into certain zones of the city.	• The extra cost involved encourages people to look at different forms of transport, instead of taking their car. • The money raised by the charge can be used to help improve public transport options.

YOUR QUESTIONS

1 What does 'ethnic segregation' mean?

2 How do city councils help to support people from different ethnic groups?

3 Which of the solutions to traffic problems given in the table is the best, do you think? Rank the options from 1 to 4 (1 being best, 4 being worst) and explain your rankings.

4 Look back at pages 188-191. Use the information on those spreads, and this one, to draw up a mind map of the issues facing cities like Birmingham. Add as much detail as you can.

9.6 » Rapid urbanisation – 1

On this spread you'll learn why squatter settlements develop in poorer countries, and what they're like.

Modern Mumbai is a lively, busy Indian city. Its rapid growth has brought money and power for some people, who live in homes like the one on page 187. But 60% of Mumbai's population (like Sunita on page 184) live in poverty in places like Dharavi.

Dharavi is one of the world's biggest **shanty towns**. Over 1 million people live there – about the same number as the population of the entire city of Birmingham in the UK! But Dharavi's residents aren't alone – there are millions of people around the world living in similar places.

▲ *A squatter settlement in Mumbai*

A **shanty town** is a squatter settlement that springs up in an area that used to have no houses.

A **squatter** is someone who settles on land without the legal right to stay there.

Shanty towns are also called **spontaneous settlements** – because spontaneous means unplanned.

Why do squatter settlements develop?

They usually develop because of rapid **rural-urban migration** (see page 184). People move to cities for various reasons, but the rapid rate of urbanisation in many poorer countries means that their governments don't have the time (and sometimes don't have the money) to provide houses, drainage, clean water, schools, etc.

Where do squatter settlements develop?

Squatter settlements usually develop in the less-pleasant parts of a city, such as steep hillsides or swampy areas. They're found in many poor countries, such as Kibera in Kenya's capital Nairobi (on the right) and Roçinha, in Rio de Janeiro, Brazil.

What are they like?

You probably wouldn't choose to live in a place like Dharavi. Sunita on page 184 said it was smelly, but at least it provided her with a home.

- Shelters in squatter settlements are made from anything that's available, e.g. wood, cardboard, metal from oil drums, and plastic sheeting.
- They're usually overcrowded.
- There's little or no proper sanitation (drains and sewers).
- There are problems with pollution, disease and a lack of clean water.

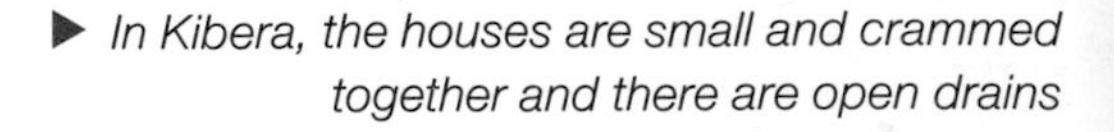

▶ *In Kibera, the houses are small and crammed together and there are open drains*

The informal economy

Many people in squatter settlements think that if you're old enough to walk and carry a bucket, you're old enough to work and earn your keep. Many people work for themselves in the **informal sector**.

People who work in the informal sector don't do a job that earns a regular wage. They make and sell goods and services unofficially – often on a 'cash-in-hand basis'. They don't have a contract, so there's no job security. They also don't have any health-and-safety protection, health insurance, or pension scheme to fall back on. If they can't work they don't earn anything. But, on the positive side, they don't pay any taxes either!

Living in Dharavi

Dharavi lies between two railway lines in Mumbai. A lot of the homes there are pretty solid – made from brick, wood and steel. And a lot of them have electricity (like Sunita's home). Although people live there illegally, Dharavi has well-established communities that provide self-help clinics, food halls and meeting places – as well as thousands of small workshops like the ones on the right.

But average incomes in Dharavi are low. Rakesh Pol, a leather worker, earns about £40 a month. He can rent a room for about £12 a month. Gradually, families can acquire extra building materials to improve their homes, but few of them can afford to move out of Dharavi, because the rest of Mumbai is far too expensive.

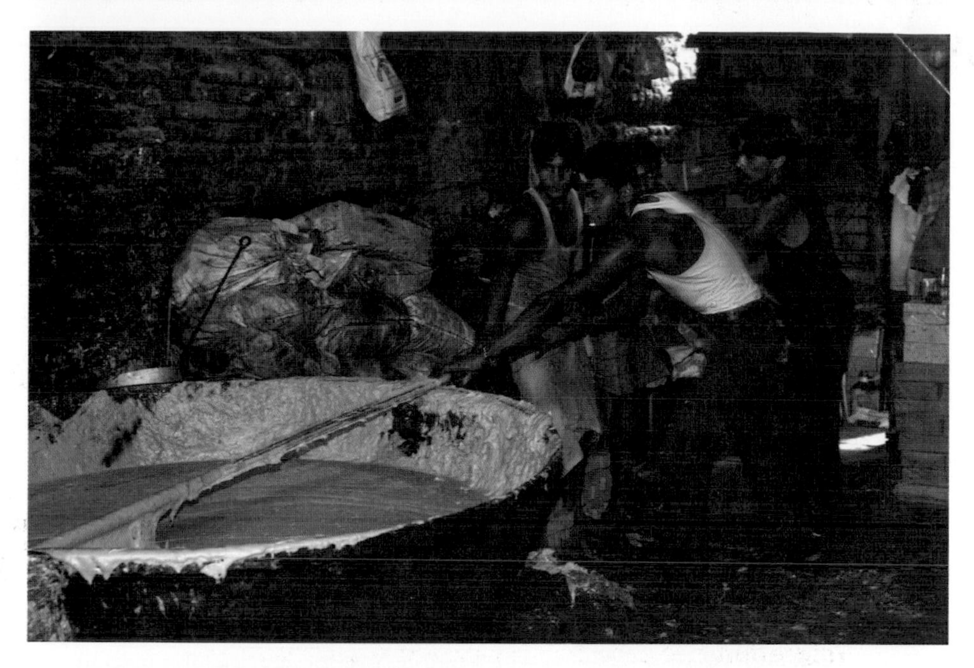

▲ *Recycling soap (top) and making pots – just two of the thousands of small businesses in Dharavi*

YOUR QUESTIONS

1. What do these terms mean: shanty town, squatter settlement, informal sector?
2. Why do squatter settlements develop?
3. Outline the advantages and disadvantages of working in the informal sector.
4. Put yourself in the shoes of someone who lives in Dharavi. Explain to a visitor why you live there, what it's like, and how you make a living.

your planet
Economic activity in Dharavi generates nearly $1 billion for India's economy.

On this spread you'll find out how squatter settlements in poorer countries can be improved.

Improving squatter settlements

Around the world, people are trying to improve their quality of life. If you live in a squatter settlement, improving your home is a good place to start. Ways of doing this might involve the local authorities, or the residents themselves – or both.

Vision Mumbai

The city authorities in Mumbai have a big plan, called *Vision Mumbai*. Part of this plan is to try to tackle the poor quality of life of many Mumbai residents. Over the years, Mumbai's slums have multiplied and grown out of control, and pollution and water problems have rocketed.

Dharavi's buildings might be poor quality, but the land they're built on is worth a fortune – US$10 billion! As part of *Vision Mumbai*, the plan is to demolish Dharavi's existing buildings and sell the land to property developers. As part of the deal, these developers will have to use some of the cleared land to build better homes for Dharavi's current residents. 1.1 million low-cost, but higher-quality, homes could be built (many of them in high-rise tower blocks to fit in more homes in a smaller land area). This should cut the number of Mumbai residents living in slum housing by 90%. The water supply, sanitation, education and healthcare would all be improved too.

But what's in it for the property developers? Well, *Vision Mumbai* has encouraged the developers to get involved by offering them the land for less money than it's worth. Plus, as well as building high-rise tower blocks for Dharavi's existing residents, the developers will be able to use the land area saved by building upwards to build profitable shopping malls, office blocks and upmarket apartments for sale and rent to Mumbai's richer residents.

So, everyone's a winner?

Not quite. Remember Sunita on page 184? She says, '*Vision Mumbai* has a problem. What happens to the people who live in Dharavi now, while their new homes are being built? Where do they go? Some people have already been forced to leave their homes, so that they can be demolished to make way for the new buildings.' And Dharavi doesn't just provide homes – it provides jobs too. Where will all the small workshops and businesses, like those on page 195, go when the area has been redeveloped?

▲ Vision Mumbai *plans to replace Dharavi with buildings more like those in the background*

Global slowdown

In 2009, it was announced that the plans to 'makeover' Dharavi would be delayed, because of the global economic crisis. Some of the organisations that had signed up for the Dharavi project had also dropped out. So, perhaps Sunita will keep her home, but life won't get much better.

Gabarone and Old Naledi

Gabarone, Botswana's capital, is another city with a problem. Most of Gabarone's rural migrants have ended up in run-down areas like Old Naledi. It started life in 1964, as a temporary settlement for builders – squeezed between the main road and the railway line. By 1971, it was dirty and dangerous:

- 6000 people lived there illegally, on a 24-hectare site.
- It lacked water, waste disposal, drainage and electricity.

In 1973, Gabarone's city council wanted to bulldoze the site and resettle the residents elsewhere. But it was too expensive. Instead, the Self Help Housing Association was set up to improve the settlement. Starting in 1975, Old Naledi was made legal and upgraded under this scheme. The text box below shows how the scheme works.

The Self Help Housing Association scheme (SHHA)

Phase 1: Basic site and service provision

- The SHHA marked out the land area and provided the basic services – unsurfaced roads, basic water supply and main drains.
- Building plots were allocated to families, so they could build their own homes.
- The SHHA provided cheap building materials and loans.

Phase 2: Improved site and service provision

- The water supply, drains and pathways were improved.
- Water standpipes and rubbish collection points were put in for every five homes.
- Schools, shops and community facilities were developed.

Phase 3: Improvements up to 2009

- Standpipes and proper sanitation were provided for every home.
- Electricity was provided to every home for a few hours a day.
- Local small business enterprises were set up.

Self-help – residents improve their homes themselves. They might get some help from the local authorities, or from non governmental organisations (NGOs), e.g. by providing things like cheap building materials.

Site and service schemes – local authorities provide a site for new building (i.e. the land), and set up basic services (i.e. water, drainage and electricity supplies), but the rest is down to the new residents.

▲ *Old Naledi today*

YOUR QUESTIONS

1. Explain what these terms mean: self-help, site and service schemes.
2. Write a letter to Sunita explaining either why you think Dharavi should be redeveloped, or why you think it should stay as it is.
3. Choose either Dharavi or Old Naledi and produce a PowerPoint presentation about why, and how, the squatter settlements are being improved.

   Hint: For questions 2 and 3, use the information on pages 194-195 as well as here to help you.

9.8 » Rapid urbanisation – 3

On this spread you'll find out about environmental problems caused by rapid urbanisation in poorer countries.

Water pollution

Rapid urbanisation and industrialisation in poorer parts of the world has created big problems for the environment there. For example, getting rid of waste (of all kinds) has led to serious pollution in the Mithi River, which flows through Mumbai. For a long time, this river has been treated as a watery waste disposal unit – leading to pollution from a number of different sources:

- Big industries in Mumbai dump their untreated industrial waste straight into the river.
- The airport uses it to dump untreated oil.
- 800 million litres of untreated sewage go straight into the river – every day!
- And it's also used for dumping other waste, including food and cattle slurry, metals and old batteries – some of which is very toxic.

And in Dharavi, which sits right next to the river – apart from dumping human waste – the river is also used for things like washing out used oil drums.

▲ *Mumbai and the Mithi River*

Flood risk

The solid waste dumped in the Mithi River (the metals and plastics) clogs it up and blocks the drains. Plants then grow on some of this waste, which helps to increase the risk of flooding.

In July 2005, the Mithi River flooded after a metre of rain fell in just 24 hours. Nearly a quarter of Mumbai was flooded. Roads and railway lines were under water for more than 24 hours. The airport was closed and many areas had no electricity for several days. People had to wade through water that was sometimes neck deep. 406 people died and the floods cost the city US$100 million.

▼ *The Mithi River is heavily polluted*

What's being done?

After the 2005 flood, the Mithi River Project was set up to try to prevent such a serious flood happening again:

- The river channel was dredged to make it deeper and allow it to hold more water. It was also widened and obstacles were removed – and the banks were smoothed near bends in the river. All of this was designed to allow the water to flow more easily down to the sea.
- But none of those actions made the river any cleaner, so waste discharges from factories are now checked. And more public toilets have also been built, to reduce the amount of raw sewage being dumped in the river.

Part of *Vision Mumbai* (see page 196) involves rebuilding homes in Dharavi and improving the area's water supply, sanitation and drains. That should mean that less untreated sewage ends up in the river.

Dharavi's workshops are also a source of pollution. But many are recycling materials that would otherwise be thrown away – adding to Mumbai's waste disposal problem. So, keeping the workshops going – but in a more environmentally friendly way – will help to reduce the overall amount of waste. Education projects are also needed to help people understand why they shouldn't dump rubbish straight in the river.

▲ *Dredging the Mithi River helps the water to flow faster to the sea (along with all of its waste and pollution)*

Air pollution

Air pollution is a major problem in Mumbai. Exhaust gases from vehicles, and smoke from burning rubbish and factory chimneys, pollute the air. And, as the Indian economy grows, more and more electricity is needed – most of which is generated by burning fossil fuels like coal. As a result, large amounts of greenhouse gases, including carbon dioxide, are being released into the air.

Mumbai's residents, especially those who live in squatter settlements like Dharavi, suffer from very high rates of breathing problems. And illnesses like bronchitis are common.

What's being done?

Mumbai has concentrated its efforts to cut air pollution on transport. Vehicle exhausts are the biggest single source of air pollution there.

- A new metro system in the city aims to encourage people to use more public transport. Its trains are due to start running in 2011. By 2021, the planned metro system should have nine lines. 32.5 of its 146.5 kilometres of track will be underground.
- The city has also banned diesel as a fuel in all of its taxis. Many of Mumbai's 58 000 taxis now use compressed natural gas instead, which reduces greenhouse gas emissions.
- The main roads in and out of the city have been upgraded with 55 new flyovers. Smoother-flowing traffic should mean less congestion and less pollution.

▲ *Serious air pollution in Mumbai*

YOUR QUESTIONS

1 Draw a spider diagram to show the sources of the pollution in the Mithi River.

2 Why do cities such as Mumbai experience water and air pollution?

3 Design a poster to raise environmental awareness in Mumbai.

4 a Explain how the Mithi River is being improved.

b Do you think that these plans will make the river cleaner? Explain your answer.

9.9 » Making urban living sustainable – 1

On this spread you'll find out how cities can be made more sustainable.

Unsustainable cities

What images do you picture when you think of cities? Parks and open spaces? Preserved old buildings? The chances are that you think of traffic jams, hordes of people, and buildings crammed together. Cities are pretty unsustainable for two main reasons:

- They 'suck in' and consume enormous quantities of resources, e.g. energy, water, food and raw materials.
- They produce enormous amounts of waste. This is usually got rid of in the surrounding land, rivers, sea and air.

Cities have problems with:

- the way they use their buildings and land
- waste disposal
- traffic congestion.

But there are ways in which cities can be made more sustainable, as the spider diagram shows. This spread looks at how we can sort out some of the environmental issues that cities have. Spread 9.10 looks at social and transport issues, as well as an example of a sustainable community.

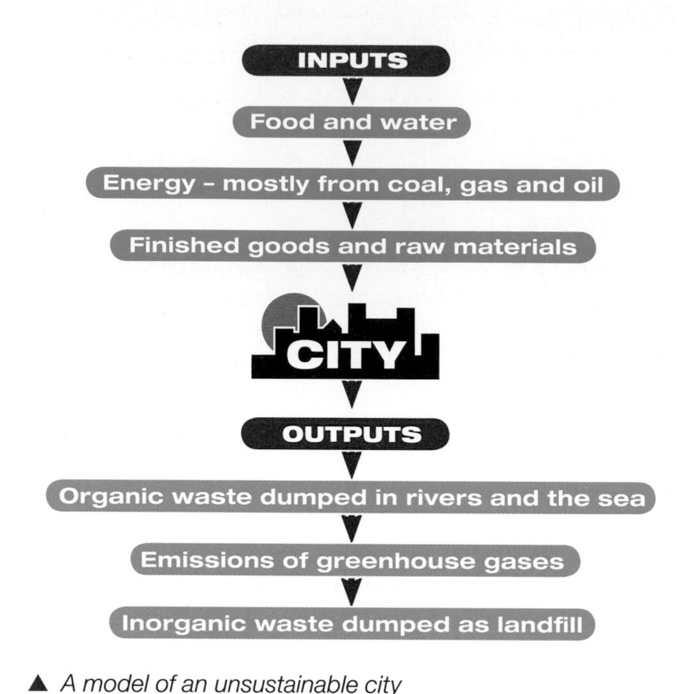

▲ *A model of an unsustainable city*

> **Sustainability** means meeting the needs of people today, while not putting the needs of future generations at risk.

Sustainable urban strategies

- Recycling water to conserve supplies
- Providing green spaces
- Reducing the reliance on fossil fuels – and rethinking transport options
- Keeping city wastes within the capacity of local rivers and oceans to absorb them, and making 'sinks' for the disposal of toxic chemicals
- Involving local communities and providing a range of employment
- Conserving cultural, historical and environmental sites and buildings
- Minimising the use of greenfield sites by using brownfield sites instead

Environmental issues

One of the main ways of making cities more sustainable is by reducing the amount of waste being produced in the first place, and by reusing and recycling as much as possible of the waste that is produced (see page 264)

Using brownfield sites

Brownfield sites are disused and derelict plots of land in urban areas that used to have buildings on them. Using them for new building is a way of recycling land. This avoids using greenfield sites (new land on the edge of the city), and stops the city growing in size at the expense of the surrounding countryside.

Conserving the environment

In order to be more sustainable, cities need to conserve their historic environment (e.g. the buildings they already have), as well as the natural environment.

The historic environment can be conserved in many different ways. For example:

- old industrial buildings, like warehouses, can be turned into apartments
- rundown old houses can be redeveloped to provide housing that will last into the future
- canals in cities can be rebranded and regenerated as leisure facilities.

The natural environment can be conserved by cities:

- using more electricity generated renewably, e.g. by solar and wind power
- collecting and recycling water, instead of piping it in from reservoirs in the countryside
- running fuel-efficient public transport systems that cause less pollution.

Open spaces

Open spaces, or green spaces, are vital to make cities more sustainable. They can act like 'green lungs' for the city, reducing its impact on the environment and recycling carbon. Plants take in carbon dioxide and return oxygen to the atmosphere, and so help to reduce pollution.

Epping Forest (see pages 80-82) is Greater London's largest open space. It stretches over 12 miles from east London to just north of Epping in Essex. And it's almost 6000 acres in size.

Epping Forest brings a range of benefits:

- It supports a wide variety of wildlife.
- It provides a range of recreational opportunities for the millions of people who visit every year – from cycling to sailing.
- It provides opportunities for conservation and sustainable management.

▼ *Epping Forest*

YOUR QUESTIONS

1. Define the term 'sustainability' in your own words.
2. Find a photo of a city and stick it in the middle of a large piece of paper. Add notes around the photo to show how cities can be made more sustainable.
3. Look at the model of an unsustainable city opposite. Redraw the model to show a more sustainable city.

On this spread you'll learn more about how urban living can be made sustainable, and investigate the example of BedZED in London.

Social issues

Involving local people in decisions about their communities means that they are more likely to accept the decisions that are made, as they feel they have been consulted and that their views are of value – leading to a more sustainable community. Local councils consult people about a wide range of things – from where new homes should be built, to how local health and social care services are planned and run.

An important part of the Aston Pride scheme in Birmingham (see page 189) was involving the local community. Residents, community groups and faith groups have been involved in the decision-making process to help regeneration in the area, and to make the Aston Pride NDC area a better place to live.

Sustainable living

Living sustainably is another matter. It depends on individual people doing things like recycling their waste and using resources like water and energy as little as possible. And perhaps even generating their own electricity, using solar panels or mini wind turbines.

Transport issues

There are a variety of ways to manage transport, and you're probably aware of the types of schemes in the checklist below.

Each of the schemes in the checklist could be applied successfully on its own. But the aim is to link them together into an **integrated transport policy** – providing a 'door to door' service that rivals the use of the car. However, truly sustainable transport systems involve:

- reducing the number of heavy goods vehicles
- encouraging low-emission vehicles (electric cars and LPG buses)
- reducing the need for long journeys.

Land use planning that puts homes close to places of work, shops and services, or around major public transport hubs is needed.

Scheme	
• One way systems that manage traffic flow	✓
• Restricted parking that prevents streets from becoming clogged	✓
• Traffic calming measures and speed restrictions that aim to keep traffic moving at a steady pace and reduce accidents (pictured)	
• Red routes that restrict roadside stopping	
• Park and ride and strategic multi-storey car parks to divert traffic to manageable locations and free town and city centres from traffic	
• Ring roads that keep traffic out of urban centres	
• Cycle lanes and footpaths	
• Cheap public transport and dedicated bus lanes	
• Clean transport using alternative fuels	

◀ *Transport options checklist*

Sustainable urban living – BedZED, London

The Beddington Zero Energy Development (BedZED), near Croydon in Greater London, is the largest carbon-neutral (it doesn't add any extra carbon to the atmosphere) eco-community in the UK. It was built on reclaimed land and focuses on social and environmental sustainability, while promoting energy conservation. There are nearly 100 apartments and houses, as well as offices/workplaces there.

BedZED's homes use 81% less energy for heating, 45% less electricity, and 58% less water than an average British home. They also recycle 60% of their waste. Energy consumption at BedZED has been reduced by:

- using building materials that store heat in warm weather and release it at cooler times
- using natural, recycled or reclaimed building materials
- building the homes facing south – to maximize 'passive solar gain'
- backing the offices onto the homes – facing north – so there's less solar gain and no need for air-conditioning
- using 300 mm insulation jackets on all buildings
- producing at least as much renewable energy as that consumed
- using heat from cooking and everyday activities for space heating
- using low-energy lighting and appliances throughout
- using energy tracking meters in kitchens
- providing a combined heat and energy power plant, using urban tree waste/off-cuts
- providing homes with roof gardens, rainwater harvesting and wastewater recycling
- providing a green transport plan – the community layout promotes walking, cycling and public transport with bus, rail and tram links
- introducing the ZEDcars car sharing club and local free electric charging points

▶ *BedZED – a carbon neutral community that opened in 2002*

YOUR QUESTIONS

1 What does the term 'integrated transport policy' mean?

2 Describe the ways in which transport is managed in a city that you know.

3 Explain why involving local people can make sustainable living more likely.

4 How are the homes in BedZED designed to encourage sustainable living?

5 Design a poster to advertise the benefits of living in a sustainable city development, such as BedZED.

10 » Changing rural environments

What do you have to know?

This chapter is from **Unit 2 Human Geography Section A** of the AQA A GCSE specification. It is about changes in rural areas and the rural-urban fringe; making rural living sustainable; farming in the UK; and change and conflict in rural areas in sub-tropical and tropical countries. The table shows how the pages in this chapter match the content in the specification.

Specification content	Pages in this chapter
The rural-urban fringe is under pressure.	p206-209
Changes in remote rural areas – depopulation, declining services, declining villages and second homes.	p210-213 Case study of East Anglia
Making rural living sustainable. Supporting people in rural areas, the economy and the environment.	p214-217 Case study of making rural living in East Anglia sustainable
A commercial farming area in the UK: farming and the environment; supermarket chains and food processing firms; the global market; reducing the environmental effects of farming.	p218-221 Case study of farming in East Anglia
Sub-tropical and tropical rural areas: cash-cultivation, forestry and mining and impacts on subsistence food production; soil erosion; changes to farming caused by irrigation; the impact of rural-urban migration.	p222-225

Your key words

- Greenfield sites
- Rural-urban fringe
- Urban sprawl
- Green belts
- Suburbanised villages
- Commuters, commuter villages
- Regional shopping centres, retail parks
- Depopulation
- Rural services
- Second homes
- Sustainability
- Diversification
- Agribusiness
- Eutrophication
- Food supply chain
- Food processing
- Environmental Stewardship
- Single Payment System
- Cash-crop production
- Slash and burn
- Irrigation

Exam help ...

Advice See pages 297-299 for information on how to be successful in your exams.

Practice See page 309 for exam questions on this chapter.

What if ...

- the countryside was turned into a giant playground?
- all the countryside was built on?
- all our country pubs and shops closed?
- everyone left the countryside?

On this spread you'll find out why the rural-urban fringe in Gloucestershire is under pressure.

Fox's Field homes get the go-ahead

Banner-waving protestors were out in force. It was the start of a public inquiry into plans for 105 new homes on the last green space – Fox's Field – between Stroud and Stonehouse, in Gloucestershire. The protestors' banners read 'Foxes not boxes'. The spokeswoman for the group said 'We're worried about the traffic, drainage, and effect on the environment in this Area of Outstanding Natural Beauty.'

But the protest was in vain. Plans to build the new homes were given the go-ahead by the Planning Inspector. Stroud District Council had already refused permission twice for Barratt Homes to build on Fox's Field, but they were over-ruled. Councillor Barbara Tait said: 'This decision could seriously affect the Council's ability to protect **greenfield sites**.' The Council was concerned about the effect of the development on the character of the area. But the Planning Inspector disagreed. He thought the development would be 'sympathetic to the location on the edge of an existing urban area'.

The rural-urban fringe

The **rural-urban fringe** – where the town or city meets the countryside – is under pressure. The growth of towns and cities into the countryside is often called **urban sprawl**. The word 'sprawl' tells us that this really isn't a good thing.

Places at the rural-urban fringe that haven't been built on are known as greenfield sites, like Fox's Field. But many of them are under pressure. Not just from house-building and expanding towns and villages, but also from:

- leisure activities, like golf courses and country parks
- out-of-town shopping centres, like Bluewater in Kent
- transport developments, like high-speed rail links.

The rural-urban fringe has a number of advantages for developers:

- There's more space to expand.
- Land on the fringe is cheaper than in towns and cities.
- Motorway access is easier.
- It's a more pleasant environment.

your planet
It's estimated that, by 2026, about 5 million new homes will be needed across the UK – putting more pressure on the rural-urban fringe.

Protecting the rural-urban fringe

Green belts are areas of protected rural or undeveloped land surrounding urban areas. There are 14 in England – covering roughly 13% of the land area. Their purpose is to:

- protect the countryside from urban sprawl, and stop towns from merging into each other
- encourage the regeneration of towns and cities by forcing developers to use brownfield sites instead, such as abandoned factories and warehouses
- protect the countryside location of old towns and cities.

However, building on green-belt land isn't completely banned. Even though it's hard to get planning permission for new developments, the green belts are gradually being nibbled away.

Increasing demand for housing and jobs

Fox's Field isn't the only green space in the Stroud area that's likely to be concreted over. In 2010, Stroud District Council consulted local people to decide where the best place or places would be to build 2000 new homes that are going to be needed in the area between 2010 and 2026. But, for every house built, two jobs also need to be created. And that means more building – of workplaces to provide the jobs. The challenge for District Councils is to try to manage growth in a sustainable way – to provide homes, jobs and facilities for communities, while at the same time protecting and improving the environment.

Transport developments

Building bypasses, motorway links and their service stations, and Park and Ride car parks, also increases the pressure on the rural-urban fringe.

- The Gloucestershire Gateway Trust wants to build a new service station on the M5, near Gloucester. It would cost £35 million and employ 300 people. But, as the map shows, more countryside would be lost as a result.
- Building more roads also increases dependence on cars.
- Much of the M25 was built right through London's green belt.

YOUR QUESTIONS

1. Start a dictionary of key terms for this chapter. Begin with greenfield site, rural-urban fringe, urban sprawl, and green belts. Add a definition for each term.
2. Write a letter to Stroud's MP explaining why you are either in favour of or against the housing development at Fox's Field.
3. The proposed new service station on the M5 would create 300 jobs. Hold a class discussion about which is more important – protecting the countryside or providing jobs.

▼ *Stroud District Council faced a choice. It could plan for the building of a single development of 2000 houses at Cam, Eastington or Stonehouse (plus business premises), or a small number of houses on each of 40 different sites across the whole district.*

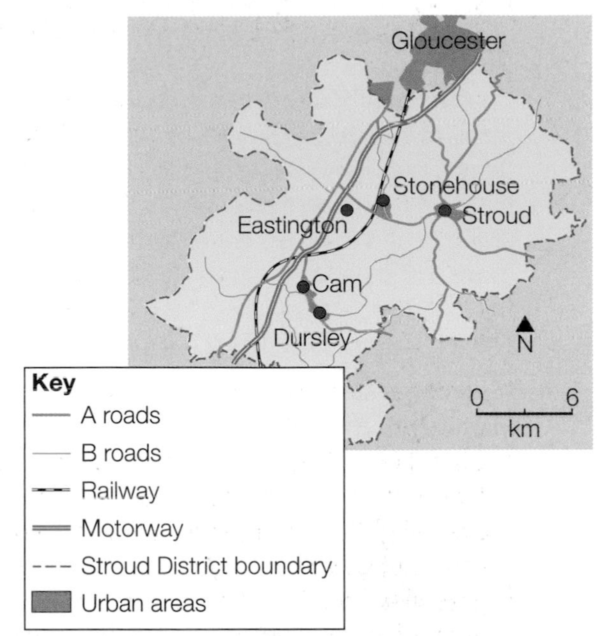

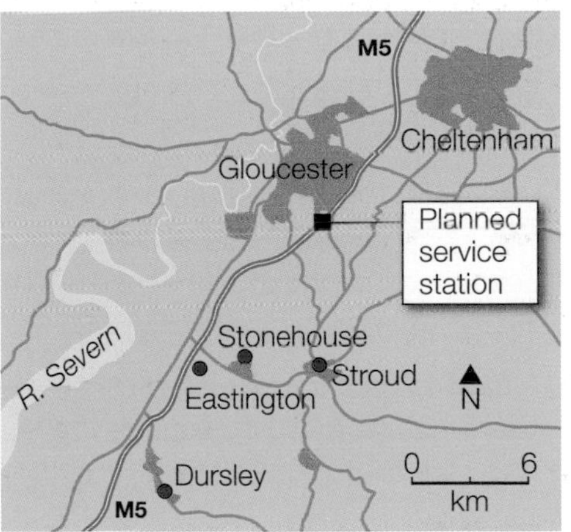

▲ ▼ *The planned new service station on the M5 will use land on both sides of the motorway*

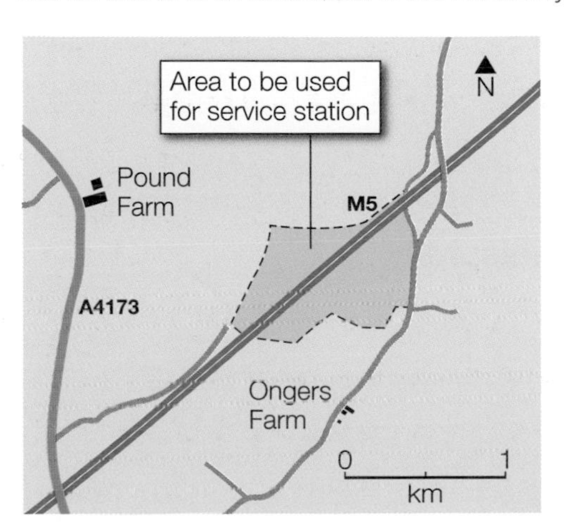

On this spread you'll learn more about pressures on the rural-urban fringe.

Suburbanised villages

An increasing overall population, growing demand for more homes (especially as more people are now living on their own) – and people moving out of towns and cities into the countryside – is leading to the creation of **suburbanised villages**. As villages grow in size, they start to look more like the suburbs of towns or cities. That's where the name 'suburbanised' comes from. Suburbanised villages are found in the rural-urban fringe, because they're close to the countryside for living in and to urban areas for work. They grow and develop over a period of time, as the diagram shows.

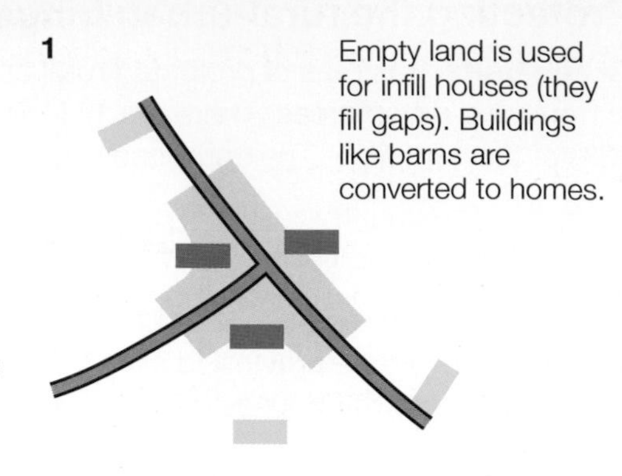

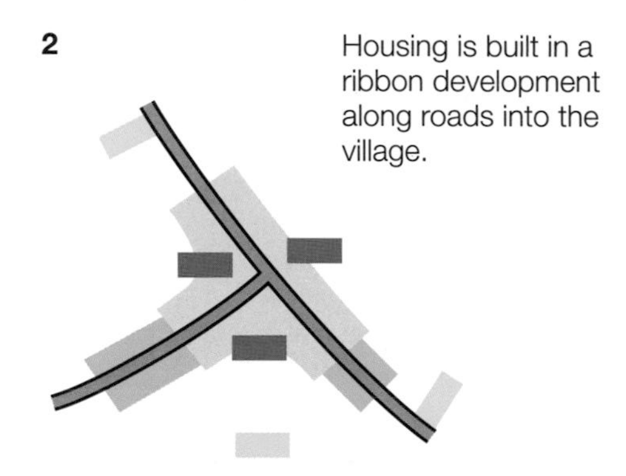

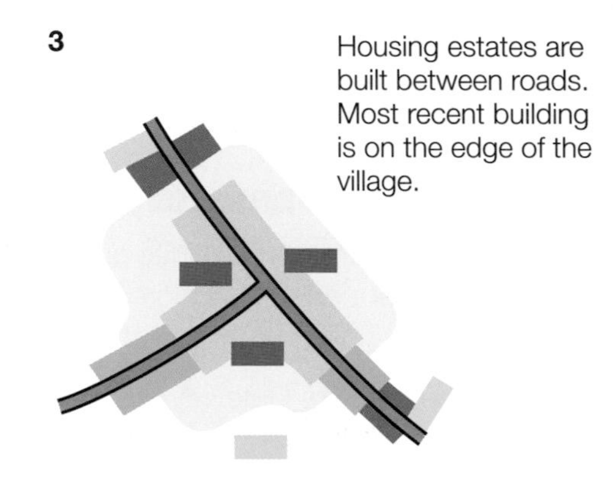

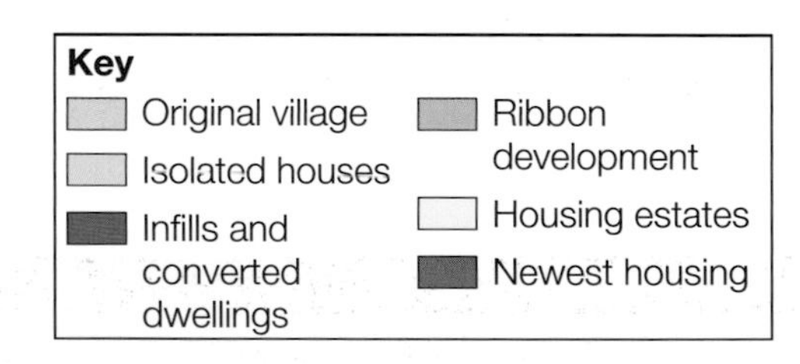

▲ *How suburbanised villages develop*

How suburbanised villages change character

Over time, suburbanised villages begin to look and feel very different. For a start, there are more modern houses than old traditional buildings. But they also change in other ways:

- The outsiders moving into the village are likely to work elsewhere (probably in urban areas), or be retired.
- The new residents also tend to push up house prices. This means that many young local people can't afford to buy homes where they grew up, and they have to move away.
- Many small local shops and post offices also close, because the new **commuters** shop nearer to where they work.
- Most of the people moving into the village own their own cars, so they don't use local bus services. As a result, these services can be cut – creating problems for those who rely on them, such as the elderly and people with disabilities.
- Attractive countryside is lost and the environment changes. There's more traffic on country roads, as people travel to work.

Villages with a large number of commuters are called **commuter villages**. Many of the people who live in them are out all day working somewhere else. Although these villages might lose their shops and schools, they may still have pubs and restaurants for returning commuters in the evening. Commuting is really common. In 2008:

- 50% of workers living in South Gloucestershire commuted to Bristol.
- 11% of workers living in Stroud commuted to Gloucester.

Leisure

The rural-urban fringe also provides space for golf courses, country parks and other leisure pursuits. At Great Blakenham, in Suffolk, a winter sports facility is planned! In 1999, the Blue Circle cement works and quarry near the village was closed, causing the loss of hundreds of jobs. Now there are plans to build Europe's biggest indoor winter sports facility there – called SnOasis.

Commuters are people who live in one place and travel to work somewhere else, usually a town or city. They might commute to work by rail or by car.

SnOasis should open in 2013, and it plans to provide 2000 permanent jobs. Not only that, the scheme will:

- double the size of the village
- provide affordable homes
- improve the road system and construct a new railway station
- bring £50 million a year into the region.

Some people are in favour of SnOasis. They can see the benefits that it will bring. But others think that there will be big negative impacts. They claim that:

- about a third of the site is currently arable land, lakes and grassland – and home to a variety of wildlife
- traffic congestion will increase
- many of the jobs created will be low-skilled and low-paid.

▲ *It may be on the site of an old quarry, but SnOasis will have a big impact on this bit of rural-urban fringe*

Retail outlets

The rural-urban fringe is also under pressure from **regional shopping centres, retail parks** and large stores. Developers build there because:

- the land is cheaper than in town and city centres
- there's space for expansion and for big car parks
- better transport links mean that people can get there more easily.

The Bluewater regional shopping centre in Kent is close to the M25 – and 10 million people live within 1 hour's drive of it. But the bad news is that the buildings and new transport network have put the rural-urban fringe under yet more pressure, and the area is more built up and crowded than it used to be.

In Norfolk, things are different. In 2007, North Norfolk District Council rejected Tesco's bid to build a store on the edge of Sheringham town centre. The view was that the supermarket would damage the town. The small independent shops would suffer, as well as local suppliers.

▲ *The Bluewater regional shopping centre has encouraged further development of the rural-urban fringe, both through housing and new transport links*

Retail parks have large shops and are found on the edges of towns and cities.

Regional shopping centres are indoor shopping centres with a large variety of shops, and masses of car parking. They're usually close to large urban areas and near to motorway junctions.

YOUR QUESTIONS

1. Add these to your dictionary of key terms: suburbanised villages, commuters, commuter villages, regional shopping centre, and retail parks.
2. Use all the information on this spread, and Spread 10.1, to draw a spider diagram. It should show how the rural-urban fringe is under pressure, e.g. from housing, and the impact that building on the rural-urban fringe has.
3. Work in small groups. Hold a role-play about SnOasis. Each member of the group should choose one of the following roles:
   - A retired person living in Great Blakenham
   - A teenager living in Great Blakenham
   - Someone working in nearby Ipswich
   - A local farmer.

   Work in role to think of questions you want to ask the developers about the impact that SnOasis will have on the rural-urban fringe.

On this spread you'll find out why people leave rural areas like East Anglia.

Endangered species

When talking about endangered species, you'd expect to be discussing leopards and tigers – not pubs and shops. But, as the article shows, if we're not careful country pubs and village shops might soon become as rare as the Amur leopard.

▼ *Country pubs are under threat*

1000 country pubs and village shops to close in the next year

More than 1000 pubs and village shops could close in the next year, it has been warned. According to the Rural Shops Alliance and British Beer and Pub Association, around 1200 shops have closed in the countryside in the last two years, and 600 pubs have closed in the last twelve months. These mass closures are ripping the heart out of community life in hundreds of villages across the country.

Adapted from Mail Online, June 2009

East Anglia

East Anglia's counties are in crisis – particularly Norfolk, Suffolk and north Cambridgeshire. But the problems they face can be seen in other remote rural areas of the UK as well – from Cornwall to Ceredigion.

- Norfolk, Suffolk and north Cambridgeshire are sparsely populated and relatively remote from the rest of the UK.
- Older people dominate the populations of these areas. Look at the age breakdowns for Southwold and Brancaster in the table.
- The birth rate is declining in the rural communities of East Anglia.

The north Norfolk coast Area of Outstanding Natural Beauty (AONB) is home to 40 000 people in the most remote and rural part of East Anglia.

However, large numbers of young people have been leaving this area. The number of 16-29 year olds has fallen by 22% since 1991.

At the same time, the number of 45-69 year olds has increased by 24%.

	Southwold (Suffolk)	Brancaster (Norfolk)	England
% aged 0-4	3.3	2.4	6.0
% aged 5-15	10.8	7.3	14.2
% aged 16-19	4.0	3.1	4.9
% aged 20-44	19.5	20.3	35.3
% aged 45-64	24.3	32.0	23.8
% aged over 65	38.1	34.9	15.8
Total population (2001)	4025	1484	49 000 000

▲ *The age breakdown in Southwold and Brancaster, compared with England as a whole*

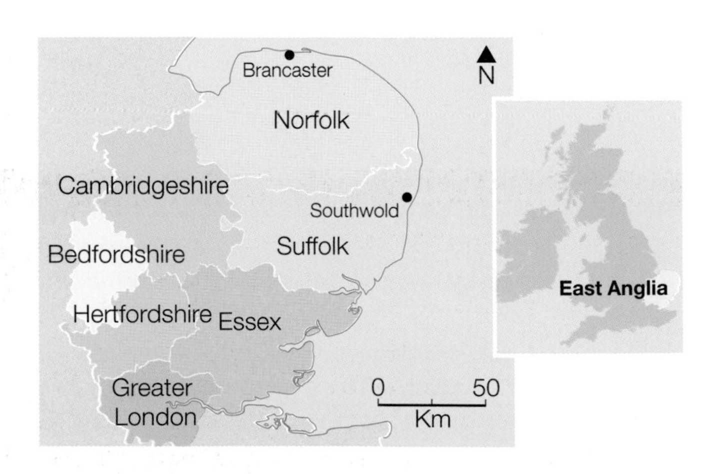

Why are people leaving East Anglia?

There are a number of reasons why people leave rural areas (known as **depopulation**). Poverty and a lack of good jobs and affordable housing are usually high up on the list.

Low wages and poverty

- 8.3% of the working population in north Norfolk, and 3.9% in Suffolk, work in farming. That compares with an average of just 1.4% across the UK as a whole. And farm incomes fell by 75% between 1991 and 2005.
- As well as those working in farming, many of the people in north Norfolk and Suffolk depend on badly paid food-processing jobs for their living.
- The average rural wage was £20 289 in 2008, compared with £27 487 in towns and cities.
- From 2006 to 2008, the proportion of the rural population living below the poverty line went up by 3%. In urban areas, it went up by just 1%.

A lack of jobs

- The traditional rural industries are disappearing in East Anglia.
- Young people are leaving farming and moving to the cities because of low pay and long working hours in agriculture.

The housing crisis

- Norfolk has three times the UK's average number of holiday or second homes. As a result, house prices there trebled between 2005 and 2010.
- In 2004, 84% of local households in the north Norfolk AONB couldn't afford an average terraced house there.

▲ *As well as being low-paid, a lot of farm work is seasonal*

▲ *There's a lack of affordable housing in East Anglia. In north Norfolk, more than 300 affordable homes were built between 1998 and 2005, but there are still 30 000 people on waiting lists for affordable homes in Norfolk alone.*

YOUR QUESTIONS

1. Add 'depopulation' to your dictionary of key terms for this chapter.
2. Annotate a blank map of East Anglia with reasons why people are leaving the countryside.
3. Draw a table like the one on the right and use this spread to add information to it. You are going to complete this table on the next spread, so leave space to add more things.

   Hint: 'Social' means to do with people. 'Economic' means to do with money and the economy.

Social changes in rural areas	Economic changes in rural areas
	High house prices

4. Why does the closure of country pubs and village shops 'rip the heart out of community life' in villages? Discuss this as a class.

On this spread you'll explore what happens when people leave rural areas like East Anglia, and about the problems of second homes.

What happens when people leave?

A number of different things happen when people start to move out of rural communities:

- Because there are fewer people to spend money, many pubs and local shops close, and bus services are reduced or removed completely.
- People who don't have cars, or who can't drive, become more isolated.
- Village primary schools and post offices close.
- The population becomes unbalanced and older, because it's usually the younger people who leave.

It's a bit like a downward spiral, as the diagram shows.

The table shows how **rural services** (shops, schools, pubs, etc.) in Suffolk and Norfolk compare with rural services in the rest of the UK.

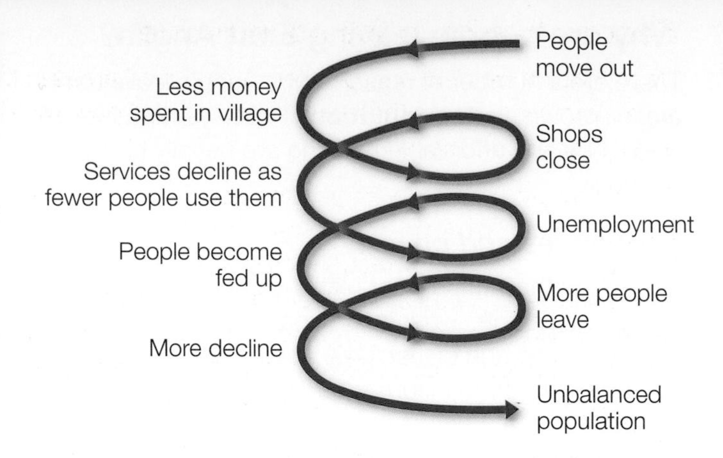

▼ *In the UK, 99% of the areas suffering from a lack of services are rural*

The percentage of rural parishes without a ...	Suffolk	Norfolk	UK
doctor	88	86	83
chemist	83	82	79
general shop	79	72	70
post office	48	40	43
daily bus service	80	84	75
pub	33	38	29
village hall	23	26	28
primary school	64	54	50

Tibenham – a village in decline?

Tibenham, in south Norfolk, is very old. It's mentioned in the Domesday Book of 1086, and its name comes from Tibb's ham (in Old Saxon that means Tibb's home).

In 1845, there were 13 shops and businesses in Tibenham – including a grocer and general store, a wheelwright and two blacksmiths. In 1845, it had a population of 749.

By 1982, the population had fallen, and shops and businesses had declined dramatically. All that was left was:

- the Post Office Stores, which sold basic food, groceries and newspapers
- the Greyhound Pub
- and a primary school with 2 teachers and 25 pupils.

And it got worse. By 2001, the population had fallen to just 453:

- The Post Office Stores was closed, and so was the school – which had been replaced by new housing.
- The Greyhound Pub was still open.
- A village hall had been built to provide a focus for the village.

▲ *The Greyhound Pub, Tibbenham*

Second homes in the countryside

Depopulation isn't the only problem facing rural areas in the UK. What sometimes happens is that – as the young and poor move out – older and richer people move in.

Southwold and Walberswick, on the Suffolk coast, have been given the nickname 'Hampstead-on-Sea'. This is because so many people from north London have bought properties there. Southwold's total population is about 4000, but only 1500 people live there permanently. It has 1250 properties, but 450 are **second homes** or holiday cottages – that's 37%! Second homes are owned by people who have their main homes somewhere else.

Second homes are a big problem in rural areas across the UK. In the Yorkshire Dales National Park, 15% of homes are second homes or holiday cottages. But, in 2005, people who didn't live there were banned from building new homes in the National Park. This was an attempt to provide more homes for local people.

There's a smell of money in Southwold, with its Regency homes, neat greens, delis, and posh clothes shops. But it shares the same problems as other parts of East Anglia – a lack of jobs, young people leaving, and high house prices.

Local estate agents say: 'Southwold looks wealthy, because of the type of people living here. But it has so many second homes that aren't being lived in year round. You need people to live here and spend money locally. We've got an ageing population, and there just aren't the people to support the elderly in the community. This isn't real life.'

Adapted from an article in *The Guardian* in 2007

More problems in Southwold

Suffolk District Council has found that in places with a lot of second homes, like Southwold:

- traditional businesses decline, but there's a demand for more decorators
- local shops close because the owners of second homes aren't always there to use them
- there's a decline in the use of local schools
- the demand for waste collections and transport varies.

▶ *Local shops are what people need, but when the owners of second homes arrive, restaurants, art galleries and bookshops are what they get, like these in Holt, Norfolk*

YOUR QUESTIONS

1. Explain what 'rural services' and 'second homes' are, and then add them to your dictionary of key terms.
2. Annotate a blank map of East Anglia to show what happens when people leave the countryside.
3. Complete the table you started on page 211 of the social and economic changes in East Anglia. Add information from this spread.
4. Southwold might look wealthy, but it's all a front. Work in small groups to create a mind map of Southwold's problems.

On this spread you'll explore how people in rural areas, such as East Anglia, can be supported.

Stopping the decline

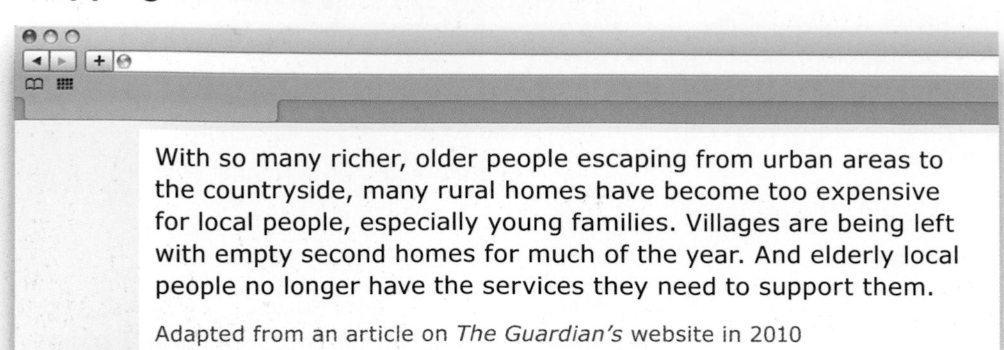

With so many richer, older people escaping from urban areas to the countryside, many rural homes have become too expensive for local people, especially young families. Villages are being left with empty second homes for much of the year. And elderly local people no longer have the services they need to support them.

Adapted from an article on *The Guardian's* website in 2010

The above extract sums up some of the problems facing rural areas. The article's headline was 'A slow decline in sustainability'. **Sustainability** is about meeting the needs of the present generation while allowing future generations to meet their own needs. This spread, and Spread 10.6, looks at how rural living can be made more sustainable.

Supporting rural areas

Rural areas need help. Many of the people who live there need better-paid jobs, affordable housing, and improved services. Not only that, but the economy needs supporting and the environment needs protecting. Some big rural issues need the help of Government agencies. For example:

- DEFRA (Department for Environment, Food and Rural Affairs). A lot of DEFRA's work is rural in one form or another. It's responsible for access to the countryside and for overseeing our National Parks and AONBs.
- Natural England is the Government's advisor on the natural environment in England. Its responsibility is to see that England's environment can adapt and survive for future generations to enjoy.
- Regional Development Agencies (RDAs) were responsible for creating sustainable economic growth across England. The East of England Development Agency (EEDA) was responsible for East Anglia. In 2010, the Government decided to replace the RDAs with Local Enterprise Partnerships.

▲ *The environment in rural areas like the Broads in East Anglia (pictured) needs protecting*

your planet

Floating pennywort is a plant that's invading the Broads in East Anglia. It can grow up to 20 cm a day – and needs to be controlled.

Supporting people

Providing jobs

In rural areas like East Anglia – where many people have low-paid jobs working in farming and food processing – there's a need to widen the range of jobs available, and also broaden people's skills. This is known as **diversification**. There are all sorts of examples of farm diversification:

- 27 farms in East Anglia have become vineyards, producing wine.
- In rural-urban fringe areas, farmers are now opening their own farm shops, and changing to Pick-Your-Own.
- Some working farms are now opening to the public and offering extra educational and recreational facilities, such as school visits, clay-pigeon shooting and golf ranges.

▲ *Farm shops, vineyards, rare-breed farms – and even owl sanctuaries – all help to widen the range of jobs and skills available in rural areas of Suffolk*

Tourism provides 250 000 jobs in East Anglia. It can be all-year round, e.g. Center Parcs in Thetford Forest, or it can depend on the weather, e.g. coastal resorts like Southwold or Southend. Tourism provides a lot of jobs and income that many people rely on. But often it simply means swapping low-paid seasonal work in farming, for low-paid seasonal work in tourism!

Providing homes

David Orr was in charge of the National Housing Federation (NHF) in 2009. He said: 'Affordable housing lies at the centre of the battle to save traditional village life'. The NHF estimates that around 100 000 affordable homes need to be built in rural areas of England by 2020 in order to meet local people's needs.

Across East Anglia, local schemes involve building on land where it wouldn't normally be allowed, e.g. in rural villages where housing is needed. Normally, the people who live in the new affordable homes need to:

- be connected to the village by birth
- have lived there for a number of years
- work in the village.

YOUR QUESTIONS

1 Write your own definitions of 'sustainability' and 'diversification', and add them to your dictionary of key terms for this chapter.

2 Begin a spider diagram to show how rural living could be made more sustainable. Copy the diagram on the right as a start. You will add more to your diagram after working on Spread 10.6.

Hint: This question is asking you for ideas that could apply in any rural area.

3 How can widening the range of jobs available help people in rural areas?

4 Was David Orr right when he said 'Affordable housing lies at the centre of the battle to save traditional village life'? Explain your answer.

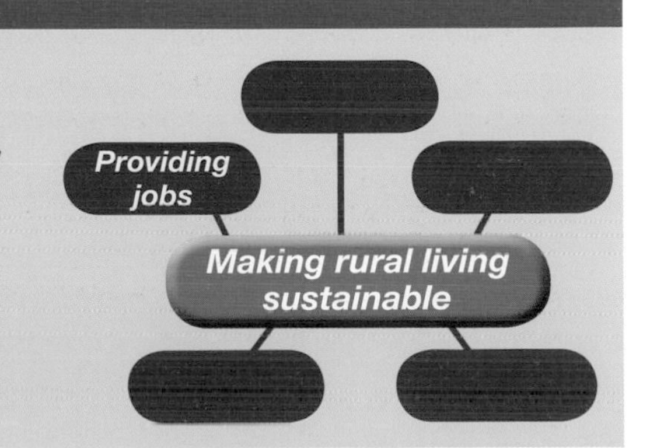

10.6 » Making rural living sustainable – 2

On this spread you'll find out how services are provided in rural areas, and how the economy and environment can be supported.

Supporting people

Providing services

Community transport Having a car is important in rural areas – to get to work, to go to the shops, and to meet friends. But 47% of adults in rural Suffolk don't have access to a car – and 30% of adults in north Norfolk don't even have a driving licence!

- The Bittern Line railway in north Norfolk serves local people and tourists, and is supported by the Bittern Line Community Rail Partnership. It runs between Sheringham and Norwich, and links together facilities for cyclists, buses and car users. Some of the money to run it comes from Government grants and local funding.
- The Vital Villages Programme manages local dial-a-bus services that can be booked in advance.

Combined services In East Anglia, some communities have grouped together to save their services:

- The Brancaster Village Shop (and at least 23 other community shops across East Anglia) combines the services of several retail outlets that have closed. Ventures like this are co-ordinated by the Village Retail Services Association.
- Eight rural post offices in north Norfolk now include a police desk as well, which makes reporting and dealing with rural crime much easier.
- Some very small primary schools have merged, to stop them from closing completely.

▲ *The Bittern Line railway – a Community Rail Partnership*

▲ *The Brancaster Village Shop remains the hub of the village*

Supporting the rural economy

EEDA (see page 214) helped businesses, supported people through training, and breathed new life into places – all with the aim of improving the economy.

Between 1999 and 2009, EEDA invested over £111 million into the Suffolk economy. In this period:

- nearly 4000 jobs were created or safeguarded
- 16 000 people were helped to improve their skills
- 700 businesses were helped to start and grow – and over 4000 saw their businesses improve.

Two of the ways in which EEDA helped in Suffolk are shown on the right.

North Suffolk Skills Centre

A vocational training centre aiming to help young people still at school gain Level 1 skills in careers such as vehicle maintenance, hairdressing, and catering. The overall aim was to increase skill levels in the workforce.

Waveney Sunrise Scheme

A £14.7 million scheme to regenerate Lowestoft. It has improved buildings, roads and public areas in the town centre, and created a more attractive environment for residents, traders and visitors.

Supporting the environment

Protection

The East Anglian countryside is a working landscape, but it also includes areas protected as a National Park (the Broads), AONBs, and Nature Reserves (see the map).

The Broads Authority aims to:

- conserve and enhance the area's natural beauty
- promote understanding and enjoyment of the area
- protect the interests of navigation (boats)
- consider the needs of agriculture, forestry, and the economic and social interests of people who live or work in the Broads.

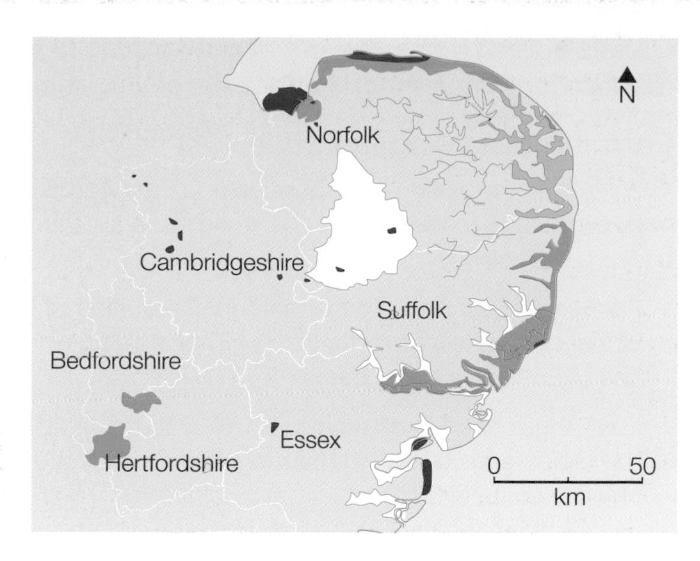

▲ *Protecting the environment in East Anglia*

Conservation

Natural England's job is to conserve and enhance England's natural environment. It works with a range of different people and organisations, and provides funding for farmers and other land managers to manage the land in ways that improve the environment. It also makes sure that there are rules to stop damaging activities.

Natural England's 'Future of Farming Awards' recognise those farmers who have made the greatest contribution to conserving England's special wildlife and landscape, and who have helped people gain access to the natural environment.

In 2008, White House Farm in Saxmundham, Suffolk, was Highly Commended in the Future of Farming Awards. It has 120 acres of land and supports livestock. Projects at the farm include enriching the wildflowers, thinning woodland (shown here) and keeping the dead wood. Each year, the farm runs open events for six weeks. Lamb reared on the farm is sold at the local butchers, and served at the local pub.

YOUR QUESTIONS

1 Finish off the spider diagram that you started on page 215 about how rural living can be made more sustainable.

2 Create a poster with the title 'Making rural living sustainable in East Anglia'. Use information from this spread and from pages 214-215.

Hint: For this question, just pick out information relevant to East Anglia.

3 In your opinion, how successful have the attempts been to make rural living sustainable in East Anglia? Justify your answer.

4 Use the information on pages 214-217 to explain how the Government supports the rural economy and the environment.

10.7 » Farming in East Anglia – 1

On this spread you'll find out about farming in East Anglia, and how modern farming methods affect the environment.

Farming East Anglian style

According to the National Farmers Union, 'East Anglia is one of the most productive landscapes in the world'. And here are some facts to back up their claim:

- Farmers in East Anglia grow more than a quarter of England's wheat and barley, almost a third of England's potatoes, and over half of England's sugar beet.
- East Anglia has 1.1 million pigs on 1900 farms, and its hens lay 2.2 million eggs *every day*. Farmers there supply a quarter of the chicken eaten in England.
- The sheep, beef and dairy herds are small compared with other regions of the UK, but they're important for looking after the landscape.
- East Anglia is a major centre for horticulture – growing everything from peas and beans, to apples, strawberries, salad crops and flowers.
- About 50 000 people work directly in agriculture and horticulture, but the farming industry supports many other jobs, ranging from animal-feed manufacturers to vets.
- About three-quarters of the land in East Anglia is used for agriculture, so farmers there play an important role in looking after the countryside.

Farming facts

Farming can be classified in different ways:

1 By what is grown or produced

- **Arable** farms grow crops.
- **Pastoral** farms rear animals.
- **Mixed** farms grow crops and rear animals.

2 By how much input there is

- **Intensive** farms have large inputs of labour, money or technology.
- **Extensive** farms have smaller inputs and are usually larger.

3 By output

- **Commercial** farmers sell all of their output to make a profit, e.g. arable farms in East Anglia.
- **Subsistence** farmers produce food to feed themselves and their families, possibly with a little left over to sell.

Physical and human factors

A farmer's choice about which type of farming to go for depends on a range of physical and human factors:

- Physical factors include things like climate, soils and relief. East Anglia tends to have warm sunny summers and cold winters. There's also a lot of flat land, and the soils are fertile, so farmers tend to go for crops rather than animals.
- Human factors include things like the use of machinery, fertilisers and pesticides, the demands of supermarkets and food-processing firms, and Government policies. These all have an impact on farms in East Anglia.

How does farming affect the environment?

Modern farming methods, like those used by big **agribusinesses**, can have a harmful effect on the environment.

Soil erosion

Repeated ploughing, especially in autumn and winter, means that the soil is exposed at the wettest, windiest time of the year. The Fens in East Anglia are very flat – with little to slow down gales – so the soil is easily eroded.

Removal of hedges

Hedges take up valuable farmland, so a lot have been removed to create larger fields – and to make it easier for farmers to use big machinery.

But removing the hedges means that important habitats for insects, birds, mammals, and reptiles are lost.

Hedges also provide windbreaks – and their roots help to bind the soil together. Removing them means that the soil is more easily washed away by the rain, or blown away by the wind.

Use of chemicals

Many farmers use a range of different chemicals on their land, including pesticides and fertilisers:

- Pesticides control pests, diseases, and weeds in crops.
- Fertilisers replace nutrients removed from the soil by plant growth.

Pesticides, fertilisers, and slurry (animal manure) all end up in our water supply – and are a major cause of pollution. Fertilisers that find their way into rivers can cause rapid algal growth, because of the increased nutrients. This algal growth on the surface restricts photosynthesis for underwater plants, and uses up oxygen in the water. This reduced oxygen level then kills many fish and other aquatic species. This process is called **eutrophication**.

Agribusiness

This type of large-scale farming is run like a big business. A lot of farms are owned and operated by one large company. Agribusiness farms are likely to use a lot of modern machinery and chemicals, and be very capital-intensive – that means they need a lot of money to operate them.

In East Anglia, many of the small family farms have gone – swallowed up by agribusinesses. As one old farmer said: 'It's just one big tractor now and a thousand acres. There's nobody on the land today'.

▲ *Wind erosion in East Anglia*

YOUR QUESTIONS

1. What do these farming terms mean – intensive, extensive, commercial, subsistence, agribusiness, eutrophication? Add them to your dictionary of key terms, along with a definition.
2. Produce a PowerPoint presentation about farming in East Anglia, including text and photos or diagrams. Include information about what is grown or produced there, and the factors that affect farming.
3. Write a 60-second News broadcast to tell people how modern farming methods affect the environment.

10.8 » Farming in East Anglia – 2

On this spread you'll learn about some of the problems facing farmers, and how the environmental effects of farming can be reduced.

The problems facing farmers

The power of the supermarkets

Many of today's supermarket chains started life as market stalls. Tesco started in 1919, when Jack Cohen opened a stall in the East End of London. Over the years, the **food supply chain** that connects producers (farmers) with consumers (you and me) has got longer and more complicated.

- As they've got bigger and bigger, the supermarkets' control over the food supply chain has become stronger.
- The supermarkets decide how much of the retail or shop price they will pay to the farmers who supplied the food. When the supermarkets decide on a price, they don't always take into account how much it actually costs farmers to produce the food.
- Supermarkets also set very high quality standards, and reject food that doesn't meet them. This means that a lot of perfectly edible food that farmers have spent money producing can then be thrown away and wasted.
- Farmers don't like to speak out against the supermarkets, for fear of losing their business.

The power of food-processing firms

Food processing is when raw ingredients, like sugar beet, are changed into something we can eat, like sugar. The food and food-processing industry is worth £3 billion to East Anglia alone.

British Sugar is the only British producer of sugar from sugar beet. So, the company buys up the whole UK sugar beet crop and uses it to produce over 1 million tonnes of sugar each year. British Sugar agrees contracts in advance with the farmers who grow the sugar beet. These contracts include a guaranteed price for the crop, which takes into account the costs to the farmers of producing the beet and adds a sustainable profit on top.

But sometimes the relationship between farmers and food-processing firms can go wrong – as the article on the right shows.

Competing in the global marketplace

In early September, fruit and vegetables grown in the UK (like apples and green beans) are in season. But, despite this, British supermarket shelves will be filled with apples from South Africa and beans from Kenya – alongside, or instead of, the British produce. Why?

The food market has now become **global**, and the size and power of the supermarkets means that they can pick and choose the cheapest suppliers – whether they're at home or abroad. So British farmers have to compete with those who can produce food more cheaply elsewhere.

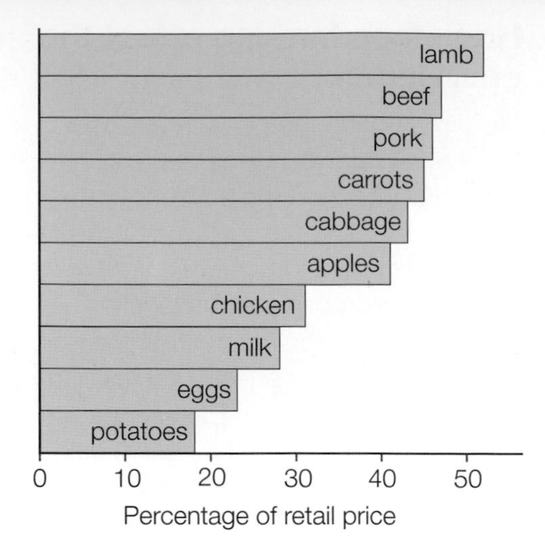

▲ *The amount paid to farmers as a percentage of the retail or shop price*

▲ *A sugar beet processing factory in East Anglia, operated by British Sugar*

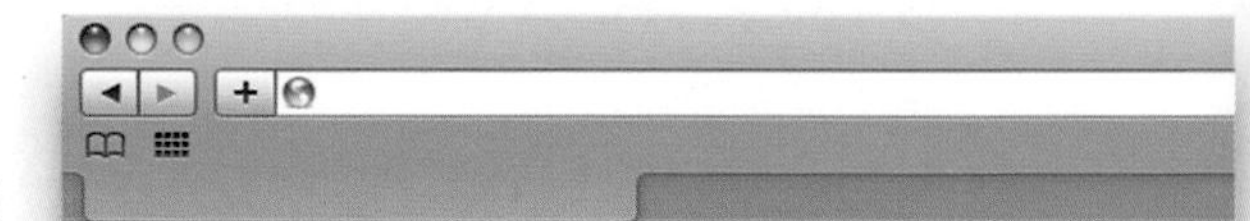

Birds Eye cancels 180 pea farm contracts

Birds Eye cancelled the contracts of nearly half of its UK pea growers in Suffolk and Norfolk when an Italian export deal collapsed. This decision will cost growers an estimated £5.5 million in lost income, and many workers are likely to lose their jobs. Some of the affected farms have been supplying Birds Eye for 65 years. Farmers will now have to grow other, less-profitable, crops.

Adapted from articles on the BBC News website in February 2010

Reducing the environmental effects of farming

Organic farming

Organic farming aims to protect both the environment and wildlife. In organic farming:

- pesticide use is restricted. Instead, farmers encourage wildlife to help control pests and diseases naturally, e.g. ladybirds eat greenfly.
- artificial chemical fertilisers are banned. Farmers keep the soil healthy and fertile by rotating their crops each year, and using clover to fix nitrogen in the soil.
- the regular use of drugs, antibiotics and wormers isn't allowed. Farmers try to avoid the need for them by moving animals to fresh grazing and keeping smaller herds.
- Genetically modified (GM) crops are banned.

Land farmed organically in the UK accounts for 4% of all agricultural land. Not a huge amount, but overall the demand for organic food is increasing. About 75% of organic food is sold in supermarkets, but over the last few years the amount sold through box schemes, farmers' markets and independent shops has been growing.

▲ *Organic food might look the same as any other food, but it's better for the environment – and many people believe that it's healthier too*

Government policies

Environmental Stewardship is a government scheme that pays farmers and land managers to look after their land and manage it to improve the environment. It helps people to:

- look after wildlife and their habitats
- make sure that the land keeps its traditional character
- protect historic features and natural resources
- ensure traditional livestock and crops are conserved
- provide opportunities for people to visit and learn about the countryside.

The **Single Payment System** is a European Union (EU) agricultural subsidy paid to farmers. In order to receive the subsidy, the farmers need to meet animal welfare standards and keep land in good environmental condition.

▲ *Lapwings, like this one, are in decline. But at Frostenden Hall (a farm in Suffolk), they are breeding in increasing numbers – thanks to the Environmental Stewardship scheme.*

your planet

Environmental Stewardship has funded over 200 000 km of hedgerow planting, restoration and environment-friendly management.

YOUR QUESTIONS

1. Add these terms, plus a definition, to your dictionary of key terms: food processing, organic farming, Environmental Stewardship, Single Payment System.
2. Find a photo of farming in East Anglia. Annotate it to show how supermarkets, food-processing firms and the global market can cause problems for farmers.
3. Write a 45-second News broadcast to explain how the environmental effects of farming could be reduced.
4. Find out where the food eaten in your home over a weekend is grown. List as many items as you can – particularly fresh food.
   - **a** Describe what you find.
   - **b** What problems could this cause for farmers in the UK?

On this spread you'll find out how rural areas in Thailand and the Amazon Rainforest are changing.

Thailand – rice and shrimps

Dulah Kwankha used to be a rice farmer in Thailand. He earned $400 a year, but now he earns six times that. After taking out a $12 000 bank loan, he converted his rice padi into a shrimp pond, which produces three crops of shrimp a year. Dulah now has money to spend for the first time in his life. But he still has to pay off his original bank loan – and borrow more money each year to run the shrimp farm.

In Thailand, thousands of poor rice farmers – who hardly made enough money to live on – have become rich by changing to shrimp farming.

Shrimp farming in Thailand

Shrimps have now been farmed in Thailand for over 60 years. It's a type of **cash-crop production** – where crops are produced and sold (often overseas) for profit. Between the 1970s and 1990s, Thailand's coastal shrimp industry expanded dramatically, and it became much more intensive (as the middle photo shows). By 2007, Thailand was the second largest shrimp producer in Asia (which itself produces 75% of all farmed shrimp).

The impacts of shrimp farming

Although shrimp farming has had some positive impacts – such as increased wealth – many of the impacts have been negative:

- The waste from shrimp farms has to be removed before the next crop cycle begins (see the bottom photo). The sludge is full of decaying food, shells and chemicals.
- Conflicts arise over where the sludge should be dumped.
- There's a constant threat of disease and the spread of infection.
- Rice fields and canals experience salinisation.
- The number of plant and animal species found in and around shrimp farms declines.

And, as Dulah Kwankha says, 'With rice farming, I only made enough money to feed my family, but I didn't worry much. Shrimp farming is easier, but I'm worrying a lot more – especially about debt'.

▲ *Growing rice is hard work – a lot of effort is needed to build irrigation channels, prepare the fields and plant, weed and harvest the rice*

▲ *The coastal area of southern Thailand was once covered in rice paddies, but there are now shrimp ponds as far as you can see.*

▲ *Fancy a shrimp? This is what's left after harvesting – a toxic sludge*

Traditional farming in the Amazon Rainforest

Traditionally, the Amerindians who live in the rainforest adopt a **slash-and-burn** approach to farming:

- A small area of land is cleared of trees and other vegetation. This vegetation is then burned, which adds nutrients to the soil.
- The small patch of cleared land is then farmed for a few years. The Amerindians also hunt and fish for food in the surrounding forest.
- After a few years, the Amerindians move on to clear and farm a different area.
- This type of farming is seen as sustainable, because only a small area of forest is cleared at a time. And afterwards the forest is left to regenerate and recover for future use.

▲ *A small patch of forest that has been cleared for slash-and-burn farming, before being allowed to regenerate*

Threats to traditional farming

It's claimed that one hectare of trees (about the size of a football pitch) is cleared in the Amazon Rainforest *every second*. Much of this massive destruction is done by loggers and miners – who move into the forest from outside.

Forestry

Loggers clear the rainforest for timber (such as mahogany) and timber products (such as pulp and paper).

Mining

The Amazon Rainforest developed above a rich base of minerals. The Carajas Mine in Brazil is the world's biggest iron ore mine (see page 86). Manganese, copper, tin, aluminium and gold are also found in the area. Vast areas of rainforest have been cleared to create mines like Carajas and the roads to get to them.

Soil erosion

Removing trees in the rainforest means that there's nothing to protect the soil from the heavy rain. As the rain falls, the soil is washed away and the land can no longer be used.

▲ *Soil erosion in the Amazon Rainforest*

The impacts on traditional farming

Clearing the forest means the Amerindians lose out:

- The area of land available for them to farm is reduced.
- They can't use the cleared forest for hunting, or as a source of traditional medicines.
- Rivers carry more sediment, because of the increased soil erosion, and they flood more easily. The extra sediment can seriously affect fish populations, which removes a source of food.
- Mining also pollutes the rivers with toxic waste.
- Conflicts arise over land rights between the Amerindians and those who want to use the forest for mining and logging.

YOUR QUESTIONS

1. Explain to your partner what the terms 'cash-crop production' and 'slash-and-burn farming' mean, and then add them to your dictionary.
2. Draw a diagram to show the impact of mining and logging on the traditional Amerindian way of life.
3. Explain why conflicts might occur between different groups of people over shrimp farming.
4. Put yourself in Dulah Kwankha's shoes. Write a letter to a rice farmer whose fields are being ruined because of shrimp farming. Your letter needs to explain why you chose to farm shrimp, and how your life has changed.

10.10 » Irrigation and migration

On this spread you'll find out about changes to rural areas in poorer countries, due to irrigation and migration.

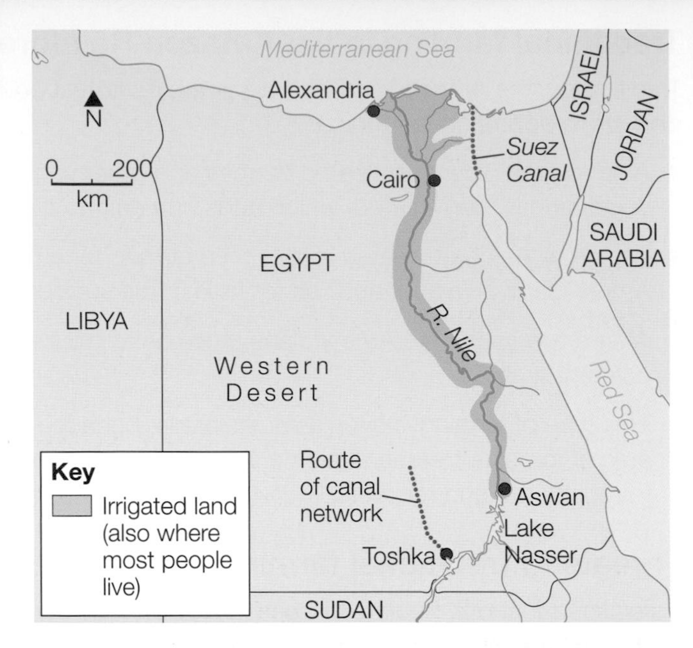

Farming and irrigation

In many places where rainfall is unreliable, or it falls at the wrong time, farmers need to use **irrigation** (the artificial watering of land). Sometimes this just means people being able to feed themselves and their families. But elsewhere it allows cash crops to be grown for export.

Egypt is hot and dry, and has a growing population of over 80 million. Most Egyptians live in the Nile Valley. Without irrigation, Egypt wouldn't be able to grow enough food to feed everyone. Egypt relies on farming – not just to feed its people, but also to provide 13% of its GDP, by selling cash crops.

Toshka

Egypt has now begun work on a scheme to irrigate more land away from the densely populated Nile Valley. The Toshka Project will cost $70 billion and will use pumps and canals to transfer water from Lake Nasser into the Western Desert. It will:

- increase Egypt's irrigated land area by 30%
- allow high-value cash crops, like olives, citrus fruits and vegetables, to be grown.

But there are problems ahead:

- Lake Nasser is silting up, so long-term water supplies may not be reliable.
- The canals which transfer the water might fill up with desert-blown sand, and water in them will also evaporate in the desert heat.
- The irrigation water could lead to salinisation (see page 93)
- The Toshka Project could fail if countries upstream – like Ethiopia and Sudan – take more water from the Nile, causing water levels to fall.

▲ *Irrigation at work in the Nile Valley*

▶ *The open canals will transfer water from Lake Nasser into the Western Desert*

Ethiopia

In the past, Egypt has had the most benefit from the Nile's water. But 85% of it actually comes from the Blue Nile in Ethiopia. While Egypt supported its growing population and developed its irrigated farms, Ethiopia's economy lagged behind. But now Ethiopia also wants to develop irrigated farming. So it will probably want to keep more of the water that Egypt has come to rely on.

High-tech solution – the Tekeze Dam

Northern Ethiopia produces a lot of coffee, which needs large-scale irrigation to grow it successfully. Ethiopia has a lot of water, but also has frequent droughts. So the Tekeze Dam has been built on the Tekeze River (which eventually joins the Nile). It is the highest dam in Africa and the water it stores will be used to irrigate 60 000 hectares of land, as well as generate electricity.

Appropriate technology – small dams

The Relief Society for Tigray believes that instead of a massive dam, it would be better to build smaller dams. A small dam (about 15 metres high and 300 metres long) was built near the village of Adis Nifas. The reservoir it created is close to the village's fields. Each family in the village has been given a quarter hectare of irrigated land, as well as fruit tree seedlings. The irrigated areas are full of crops – providing a permanent food supply for the villagers.

▲ *Here it comes! Opening the irrigation valves on a small local dam*

Rural-urban migration

The UN estimates that, sometime in 2007, someone migrating from their rural home to begin a new life in the city tipped the global rural-urban balance. From that point on, more of the world's population were living in towns and cities than in rural areas. But who migrates and why?

- In poorer countries, like Ethiopia, it's often the young men who leave. They move to the cities to find paid work.
- In Ethiopia, farms are often small (60% of households have less than one hectare of land). These are unsustainable and cannot support large families.
- Frequent droughts and variable rainfall lead to crop failures, which force people to move or starve.
- Natural disasters and conflicts also force people to leave their homes and head for the cities.

When the young men leave, the women and children might be left behind to continue farming the family plot. But with fewer people working the land, output can fall. The men might send money back to their families (called a **remittance**). In most sub-Saharan countries, remittances amount to around 15% of GDP, and contribute to the rural economy.

YOUR QUESTIONS

1. Add the terms 'irrigation' and 'remittance' to your dictionary for this chapter.
2. Why does Egypt need to use irrigation?
3. Write a newspaper article of about 300 words on: 'How irrigation affects farming in Ethiopia and Egypt'. Explain what irrigation is, where the irrigation water comes from, what crops can be grown, and also some information about the problems that irrigation water can cause.
4. Draw a mind map to show why people in poorer countries migrate, and the impact that this can have on rural areas.

11 » The development gap

What do you have to know?

This chapter is from **Unit 2 Human Geography Section B** of the AQA A GCSE specification. It is about contrasts in development, the factors affecting development, and how the development gap can be closed. The table shows how the pages in this chapter match the content in the specification.

Specification content	Pages in this chapter
The world can be divided up in different ways.	p228-231
Using development measures, and the links between them.	p228-231
Quality of life and standard of living, and improving quality of life.	p232-233
Global inequalities are made worse by physical and human factors, including environmental, economic, social and political factors.	p234-235 p236-237 Case study of the impact of Hurricane Mitch on Honduras
Reducing the development gap needs international efforts.	
• World trade – reducing the imbalance, Fair Trade and Trading groups.	p238-241
• Reducing debt	p242-243
• Aid and development – different types of aid, and sustainable development.	p244-245 (Including case study of FARM Africa and the Moyale Pastoralist project)
Different levels of development in the EU, and attempts to reduce the differences.	p246-247 (Comparing the UK and Poland)

Your key words

- Development, development indicators
- GDP per capita ppp, GNI per capita ppp
- Correlation
- Human Development Index (HDI)
- Standard of living, quality of life
- Informal settlements
- Global inequalities
- Trade, exports, imports, trade balance, trade surplus, trade deficit
- Newly Industrialised Countries (NICs)
- Transnational Corporations (TNCs)
- Tariffs, quotas, free trade
- Trading groups, World Trade Organisation, Fair Trade
- Highly Indebted Poor Countries
- Conservation Swaps (debt-for-nature swaps)
- Aid, short-term, long-term, tied.
- Top-down, bottom-up aid projects
- Donor, recipient
- Non-governmental organisation (NGO)

Exam help ...

Advice See pages 297-299 for information on how to be successful in your exams.

Practice See page 310 for exam questions on this chapter.

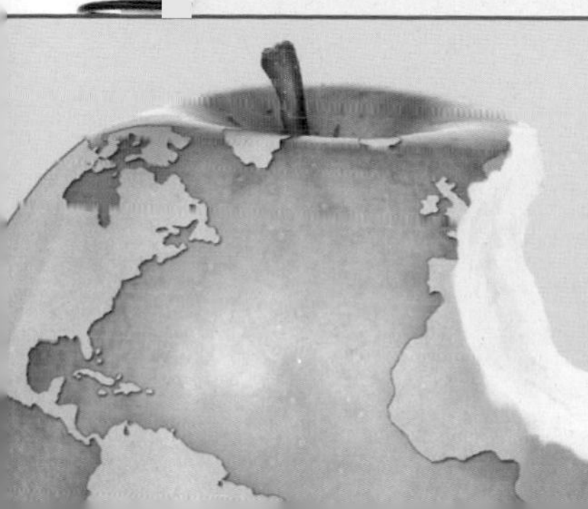

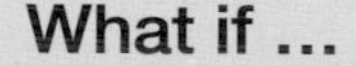

What if ...

- everyone had clean water to drink?
- all countries were equal?
- no-one had to live in a slum?
- trade was fair?

11.1 » Differences in development

On this spread you'll find out what development is, how we measure it, and about the gap between the richest and poorest countries.

Meet Blessing Kithuku. A few weeks before this photo was taken, she was so ill with typhoid that her family thought she would die.

Blessing lives in the Matopeni slum in Nairobi, Kenya's capital. Life in Matopeni (which means 'In the mud' in Swahili – the local language) is desperate. There are no proper roads or houses, and no clean piped water or proper toilets either.

This chapter is about development, and people like Blessing.

What is development?

Development means change. When we're talking about development in Geography, we're usually talking about change for the better. For example, changing the situation for Blessing and her family, so that they do have access to clean water and proper toilets, and they can improve their home.

Different factors affect a country's level and speed of development:

- ***Environmental factors***, like natural hazards (e.g. earthquakes).
- ***Economic factors***, like trade and debt.
- ***Social factors***, like access to safe water and proper education.
- ***Political factors***, like government corruption or internal conflict.

You'll learn more about these on Spreads 11.4-11.6.

Can you measure development?

Yes, in all sorts of ways. A popular one is to use **indicators** of wealth to measure a country's economic development. The two most common indicators of wealth are:

- **GDP** (Gross Domestic Product), which is the total value of the goods and services produced by a country in a year
- **GNI** (Gross National Income), which is like GDP – but also includes money earned from overseas.

The World Bank and the United Nations use four levels of income, or wealth, to group countries together (see the map opposite). This information is then used to describe countries as richer or poorer.

your planet

Angola, in Africa, has the lowest life expectancy in the world – just 38.2 years. Macau, in southeast Asia, has the highest – 84.4 years.

- GDP and GNI are always measured in US dollars. This allows direct comparisons to be made between countries.
- If the total figure for GDP or GNI is divided by the total population, it gives a figure **per capita** (or per person).
- To improve the accuracy of comparisons between countries, GDP and GNI figures also tend to be given in terms of **ppp** (**purchasing power parity**). The figures are adjusted, because a dollar buys more in some countries than in others. Generally speaking, goods and services are cheaper in less-developed or poorer countries than they are in more-developed or richer countries, like the UK or USA.

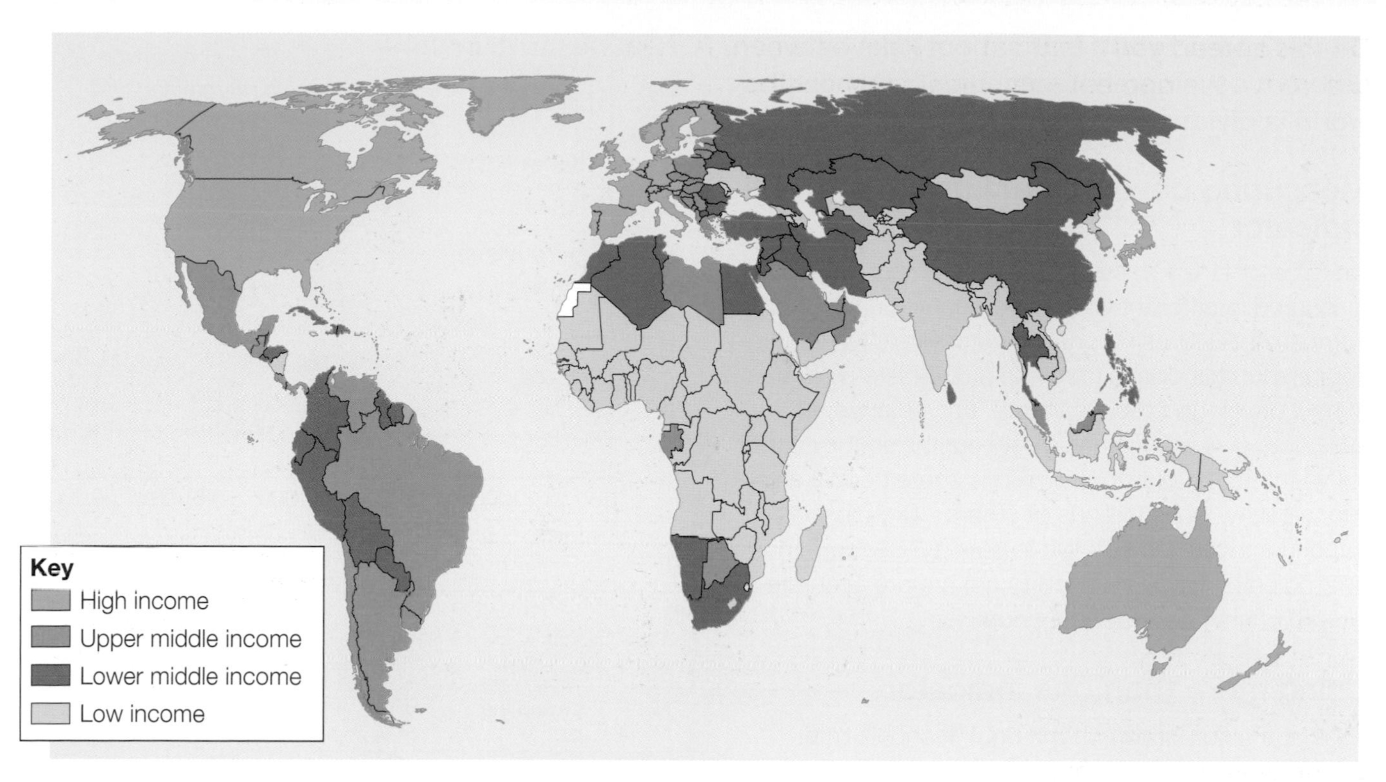

▲ *Different groups of countries in 2006, based on GNI per capita*

How else can you measure development?

GDP and GNI are economic indicators, but social indicators (e.g. covering population change, health and education) can also be used to measure development.

The table below shows some social indicators for a range of different countries. If all those numbers look a bit confusing, try looking at them one column at a time – comparing the figures for different countries. You should begin to see that there's a gap between the world's richest countries (those with the highest GNI figures) and the poorest.

Country	GNI per capita ppp	Birth rate (per 1000 population)	Infant mortality (per 1000 live births)	% adult literacy rate (over 15s who can read and write)
USA	46 970	14	6	99
UK	36 130	11	5	99
Japan	35 220	8	3	99
France	34 400	13	3	99
Poland	17 310	10	7	99
Mexico	14 270	20	18	91
China	6020	14	20	91
India	2960	22	51	61
Kenya	1580	37	55	85
Afghanistan	n/a	38	153	20

YOUR QUESTIONS

1 Start a dictionary of key terms for this chapter. Begin with the terms: development, indicators, GDP per capita ppp, GNI per capita ppp.

2 Look at the map above. Describe the distribution of income, or wealth, that it shows.

3 a Go to www.worldmapper.org and search for the infant mortality map.

b Describe what the map shows.

Hint: This map displays the size of countries in proportion to their infant mortality rate – the more babies that die in a given country, the bigger that country appears to be.

4 Look at the table on the left. Choose one high-income country and one low-income country (use the map above to help you). Describe the main differences between them.

On this spread you'll learn about links between different development measures, and how the world is divided up.

Measuring development using only one indicator

On Spread 11.1, you saw that development can be measured in different ways. Two countries can be compared easily using one indicator (like GNI per capita), but this doesn't provide a complete picture of how developed they really are. For instance, all indicators give an average for a country, so they can't tell you that, for example, in Kenya some people are very wealthy, while millions of others – like Blessing Kithuku and her family – live in poverty. So more than one indicator is usually used to get a more accurate picture of how developed a country is.

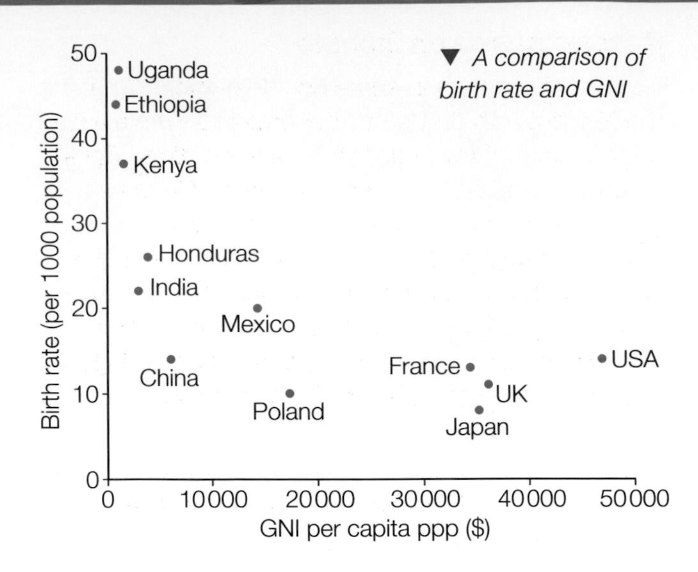

▼ *A comparison of birth rate and GNI*

Using more than one indicator

Look at the first graph on the right. It shows both the birth rate (per 1000 population) and the GNI (per capita ppp). The graph shows that there is a good link, or **correlation**, between birth rate and GNI. The wealthier a country is (i.e. the higher the GNI), the lower the birth rate – look at France, the UK and Japan. The opposite is also true. The poorer the country is, the higher the birth rate.

Now look at the second graph, which shows life expectancy and income (this time measured as GDP per capita). The coloured dots represent individual countries in the different regions. Again, you can see a clear correlation between the two development indicators.

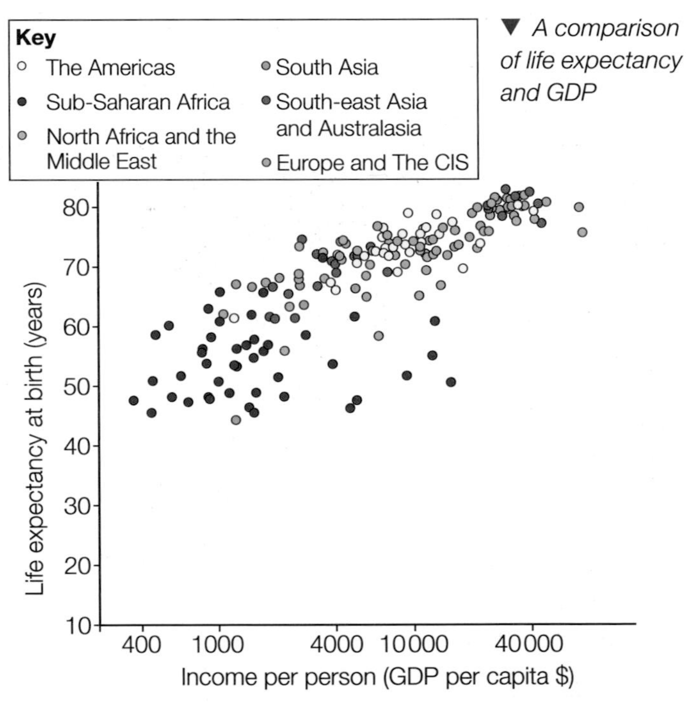

▼ *A comparison of life expectancy and GDP*

Dividing up the world

The North-South divide

In 1981, a report was published about global development. It was called the Brandt Report and it showed a divided world. A wealthy 'North' controlled 80% of the world's wealth, and a poorer 'South' only 20%. The map on the right shows the division of the world, according to the Brandt Report.

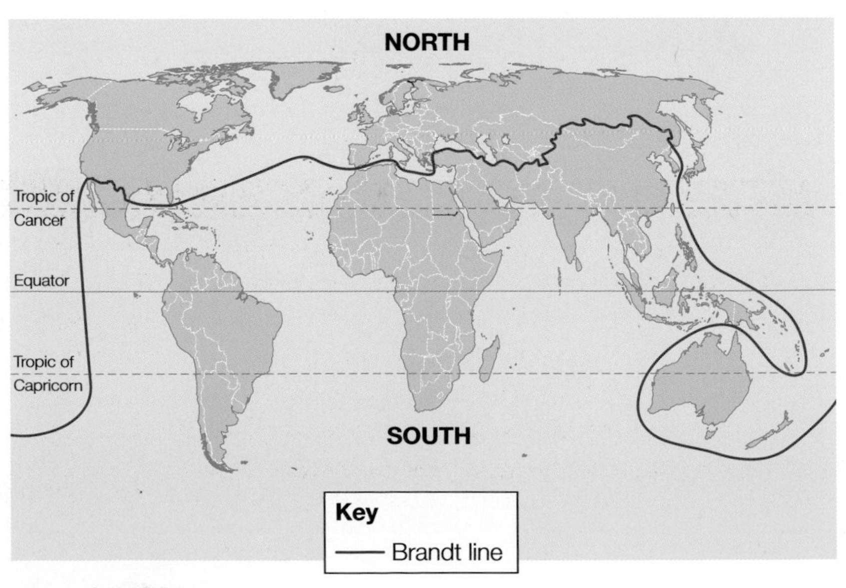

▶ *The North-South divide, according to the Brandt Report*

A world based on income

Rapid development in some countries means that the North-South divide has become too simple a way of dividing up the world. That's why the World Bank and the United Nations use four levels of income (high, upper middle, lower middle, and low) to group countries together, as the map on page 229 shows. However, there are still problems with this, because – for example – the group of low-income countries includes India (which has a large, rapidly growing economy), and Ethiopia (which has a small, slow-growing economy).

HDI rankings – the top 3 countries in 2010 ...	
1	Norway
2	Australia
3	New Zealand

... and the bottom 3	
167	Niger
168	D. R. Congo
169	Zimbabwe

The Human Development Index

The UN has devised another way of measuring development. It's called the **Human Development Index**, or HDI for short, and it uses four indicators:

- Life expectancy
- Education – both the literacy rate and the average number of years spent at school
- GDP per capita ppp

Each indicator is given a score, and the HDI is the average of the four scores. 1.000 is the best score, and 0.000 is the worst. Countries are ranked (put in order) from 1 to 169, according to their overall scores.

The UN uses the HDI because – by linking GDP to health and education – it shows how far people are benefiting from a country's economic growth.

▼ *The HDI around the world in 2007*

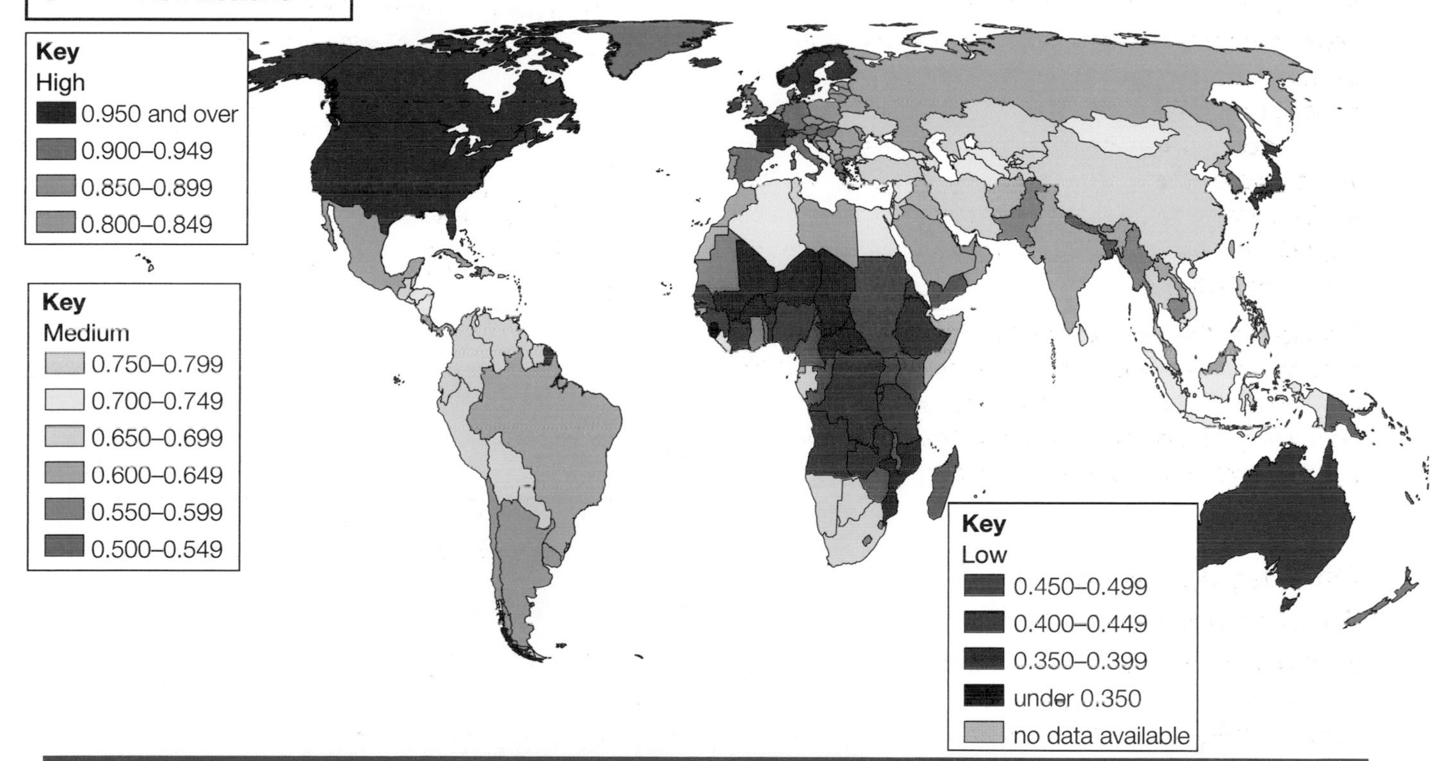

YOUR QUESTIONS

1 Add these terms to your dictionary of key terms for this chapter: North-South divide, Human Development Index. Now explain what they mean.

2 a Look at the graph opposite comparing life expectancy and GDP. Describe the correlation between these two indicators.

b Why do you think there is a correlation between them?

Hint: You won't find the answer on this spread. Think about it!

3 Use the information in the table on page 229 to help explain why Afghanistan has one of the lowest HDI scores in the world.

4 Do you think the HDI is a better way of measuring development than just income? Explain your answer.

On this spread you'll learn about the difference between standard of living and quality of life, and about how people in poorer countries improve their quality of life.

your planet
Almost half of the population of sub-Saharan Africa live on less than a dollar a day – the highest rate of extreme poverty in the world.

What's important?

What's really important to you? Money? Family? Friends? Being healthy and happy? And what about having enough to eat, clean water straight from the tap – all the things we take for granted – how important are they? And why does it matter?

It matters because things like money, being healthy, and having clean water, are all to do with development.

Standard of living

In its simplest terms, **standard of living** refers to how much money people have – so it's measured as GDP per capita. Do they have enough money to pay for the basics of food and housing? Is there any left for anything else? Or, are people surviving on a dollar a day – or less? Hundreds of millions of people across the world are, as the map on the right shows.

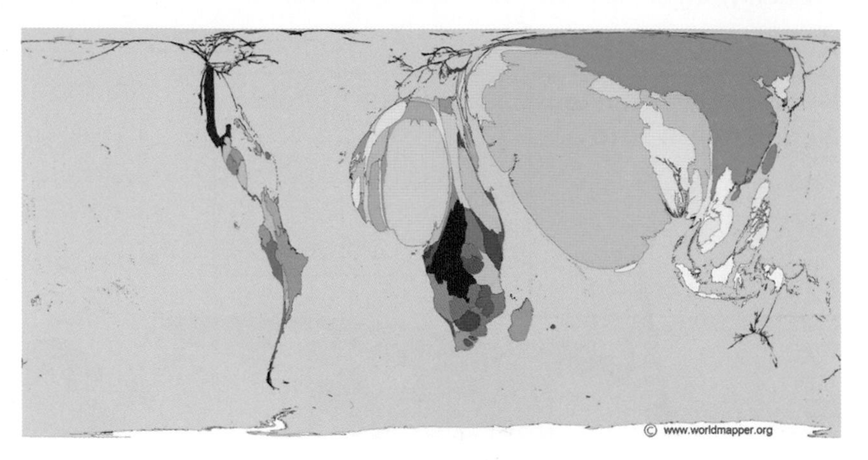

▲ *The distribution of people across the world living on a dollar a day – or less (once ppp has been taken into consideration, see page 228). The size of each country on the map is in direct proportion to the percentage of people living there on a dollar a day or less. So, the bigger a country is on the map, the more people it has surviving on very little, e.g. India.*

Quality of life

Quality of life is different to standard of living. The UN's HDI (see page 231) measures whether people have a long and healthy life (life expectancy), knowledge (literacy and the number of years spent at school) and standard of living (GDP per capita). So, the HDI is one measure of quality of life.

Other aspects of quality of life can often be harder to measure. For example, a safe, clean environment, the ability to vote without interference, the right to privacy – but they can all improve quality of life.

It must also be remembered that what is considered to be important for a good quality of life in a country like the UK, might not be considered important to someone living a very different lifestyle somewhere else in the world, like Kenya – or Bhutan!

▲ *Bhutan uses the idea of Gross National Happiness to measure quality of life. It is based on sustainable development; looking after Bhutan's culture and environment, and good government.*

Quality of life in Kenya's informal settlements

Blessing Kithuku (see page 228) lives in the Matopeni slum in Kenya's capital, Nairobi. She's not alone in living in overcrowded and unsanitary conditions. 60% of people in Nairobi live in places like Matopeni, without access to basic things like clean drinking water, proper toilets, education and healthcare. In 2009, 4 million people living in Kenya's urban areas were short of food – and malnutrition was increasing, especially in children.

Informal settlements, like Matopeni, are often unplanned. They're also known as slums, shanty towns, or favelas in different parts of the world.

Improving quality of life in Nairobi

In Kiambiu ...

Kiambiu is in northeast Nairobi. It's a slum like Matopeni, but things are starting to change there. The charity Christian Aid has a Kenyan partner called Maji na Ufanisi (MNU) – it means 'Water and Development'. With MNU's help, the residents of Kiambiu have built five toilet and shower blocks, and have employed local people to clean and maintain them. They charge people a small fee to use them, and then use that money to improve life further for the community, e.g. by building more toilet blocks and providing emergency healthcare for families. They've also got clean drinking water now. It has made a huge difference to people's lives, their health and their children.

In Matopeni ...

Blessing's mother, Catherine Kithuku, has formed a group that organises rubbish collections and educates people about health in Matopeni. Catherine dreams of an environment fit for people to live in. If MNU can find the money, it plans to begin work in Matopeni – and Catherine could see her environment change for the better. The photo below shows living conditions in Matopeni today.

(Adapted from 'Let's end poverty' Christian Aid, January 2010)

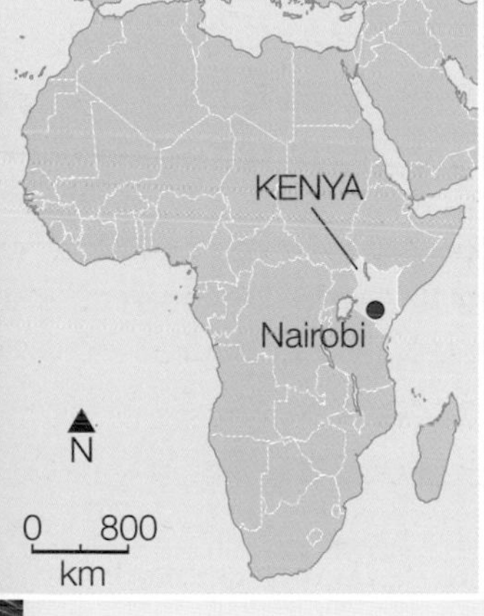

YOUR QUESTIONS

1 Explain what these terms mean: standard of living and quality of life. Add them to your dictionary of key terms for this chapter.

2 Describe what the map shows about people living on up to a dollar a day.

3 Hold a class discussion about quality of life.

a What's important to your quality of life in the UK?

b What do you think is important to Blessing Kithuku's quality of life in Matopeni, Nairobi?

c How have people in Kiambiu improved their quality of life?

11.4 » Widening the gap

On this spread you'll explore how physical and human factors can make the gap between richer and poorer countries wider.

The term **global inequalities** is another way of describing the fact that our world is unequal – with a big development gap between the richest and poorest countries. Different sorts of factors can worsen the problem and widen the gap.

Environmental factors

These include natural hazards like earthquakes, volcanic eruptions and hurricanes. Poorer countries tend to suffer most from natural hazards, because they lack the money to prepare for – and recover from – them. Honduras was devastated in 1998 by Hurricane Mitch – taking years to recover (see pages 236-237). Some countries also face several (or multiple) hazards, as the map shows.

Honduras

GNI per capita ppp $3750.1 (2008 est.)

HDI 0.604 (2010)

HDI rank 106 out of 169 countries (2010)

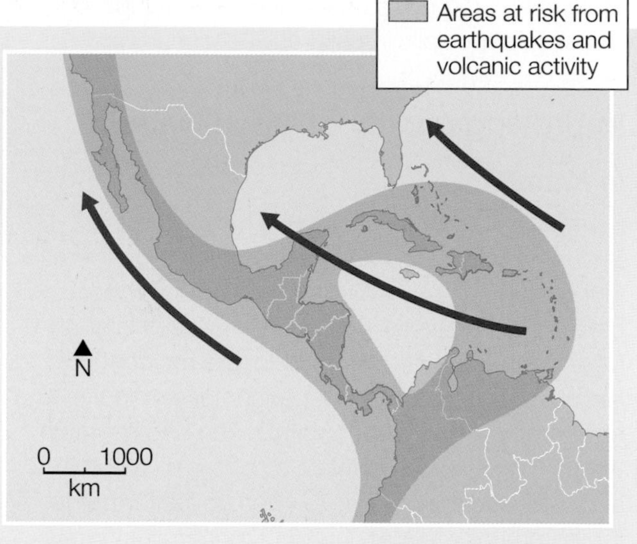

▲ *The natural hazards facing Central America and the Caribbean*

Economic factors

These include things like **trade** – when countries buy and sell goods and services. Countries tend to buy in goods and services that they don't have enough of, or that they can buy more cheaply from somewhere else. But global trade can cause problems and inequalities:

- Most of the world's trade is between the richer countries.
- Richer countries tend to sell 'expensive' manufactured goods and buy 'cheap' primary goods (e.g. coffee, metal ores, fish).
- Poorer countries, like Kenya, tend to sell the cheap primary goods and buy the expensive manufactured goods they can't make. So, as a result, they spend more than they earn from the products they sell. This trade imbalance helps to widen the development gap.

Kenya

GNI per capita ppp $1627.7 (2008 est.)

HDI 0.470 (2010)

HDI rank 128 (2010)

Political factors

In Zimbabwe, the government's increasingly violent redistribution of land from white farmers to the majority black population aimed to close the development gap there. But it had the opposite effect, because much of the land was given to government supporters with little experience of farming. As a result, the production of crops – both to eat and to trade – collapsed, the economy failed, and many Zimbabweans were left dependent on international food aid.

Zimbabwe

GNI per capita ppp $176.2 (2008 est.)

HDI 0.140 (2010)

HDI rank 169 (bottom)

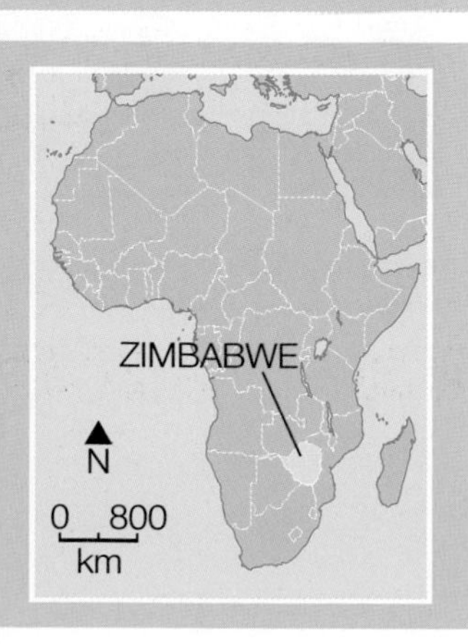

Social factors

Water availability is one of the social factors that impacts directly on the development gap. Water isn't used evenly around the world:

- 12% of the world's population uses 85% of its water.
- On average, domestic daily use is 47 litres per person in Africa, 334 litres in the UK, and 578 litres in the USA.

And the quality of water – whether it's clean or not – varies too. The map shows the percentage of people around the world with access to safe water. And there's a direct link between access to safe water and standard of living (as measured by GDP per capita), as the graph shows.

Not only that, but water really is a matter of life and death:

- A child dies from a water-borne disease every 15 seconds.
- In Ethiopia, nearly 74 000 children die from diarrhoea every year.

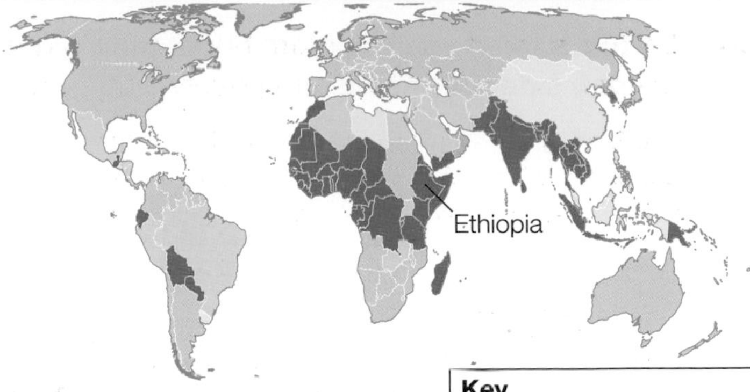

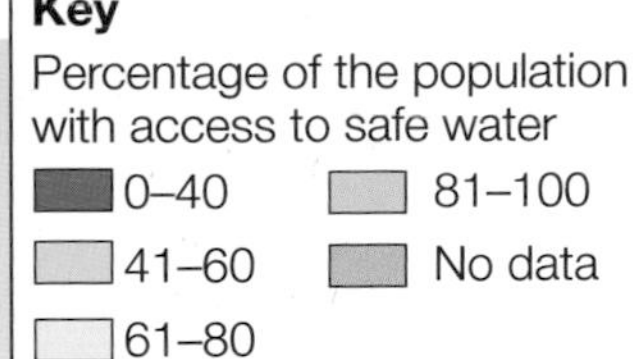

▲ *Access to safe water around the world. Out of a global population of about 6.6 billion, 1.4 billion don't have clean drinking water and 0.5 billion face water shortages every day.*

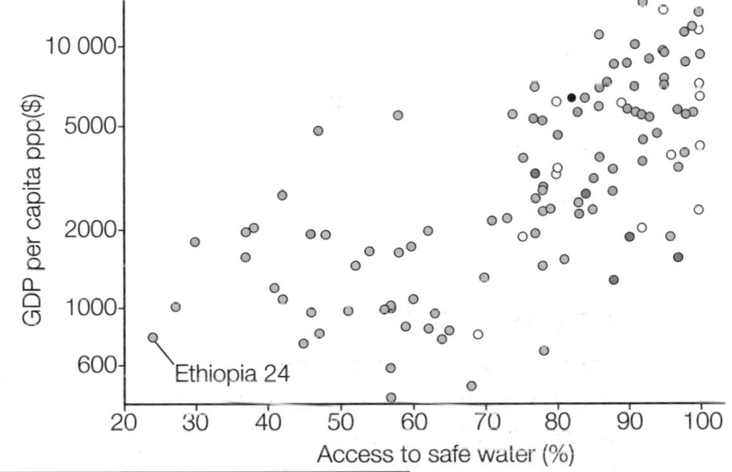

Key

- Arab States
- Central & Eastern Europe & CIS
- East Asia and Pacific
- High income OECD
- Latin America and the Caribbean
- South Asia
- Sub-Saharan Africa

▲ *The link between standard of living and access to safe water*

Reducing the gap

WaterAid is a charity which aims to overcome poverty and improve people's lives by helping them to get safe water and sanitation. One of the people they have helped is Mehari Abraha in Ethiopia (pictured below right). WaterAid has provided spring-fed wells to ensure clean water supplies for his family and others in the area.

Ethiopia	
GNI per capita ppp $992 (2008 est.)	
HDI 0.328 (2010)	HDI rank 157 (2010)

Before, we only had unprotected sources of water. My family suffered badly, and my 3-year-old daughter died. There were parasites which made us ill and we had to go to the health clinics all the time. I used to spend hours walking there and queuing to be seen. But now this has changed and I can save time and money. I have bought 20 chickens and 1 goat with the money I've saved. And with my spare time I can work on my maize and pepper crop.

YOUR QUESTIONS

1. Define the terms global inequalities and trade, and add them to your dictionary of key terms for this chapter.
2. Draw a spider diagram with 'Development gap' in the middle, and with four legs labelled 'Environmental factors', 'Economic factors', 'Political factors' and 'Social factors'. Add information to the spider diagram to show how these factors affect the development gap.
3. Explain how access to safe water can improve people's standard of living.

On this spread you'll find out about the impact of a natural hazard on a country's development.

your planet

Hurricanes are engines of destruction that can cause wind speeds of over 160 miles an hour, and unleash over 9 trillion litres of rain a day.

Honduras

'Military rule, corruption, a huge wealth gap, crime and natural disasters have made Honduras one of the least developed and least secure countries in Central America.' So starts the BBC's country profile of Honduras. But let's look more closely.

The indicators in the table below are the same as those given for a range of countries on page 229, plus some extra ones.

GNI per capita ppp ($)	3870
Birth rate (per 1000 population)	26
Death rate (per 1000 population)	5
Infant mortality (per 1000 live births)	22
Doctors (per 1000 population)	0.6
Life expectancy (years)	70
Adult literacy (%)	80
Employed in agriculture (%)	39
Employed in industry (%)	21
Employed in services (%)	40
Below poverty line (%)	59

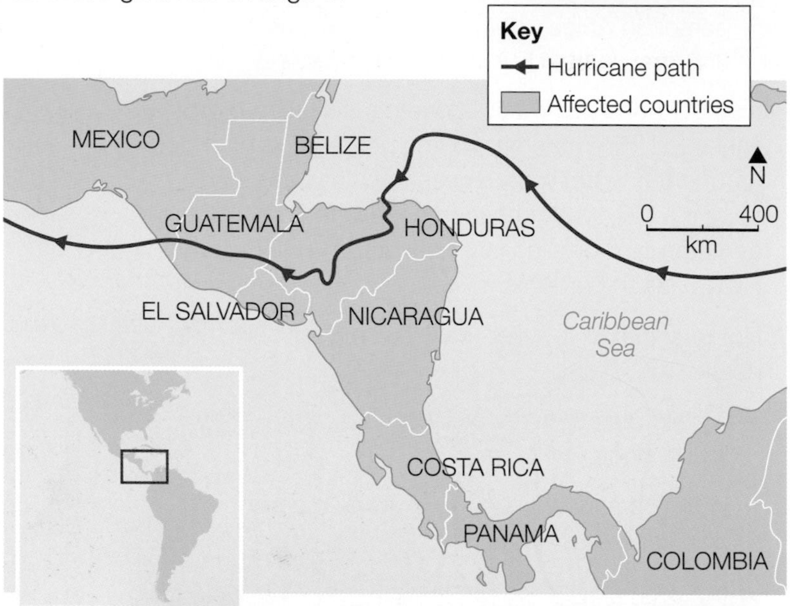

The percentage breakdown of people employed in agriculture (farming), industry and services is important, because it's a good indicator of a country's level of development. Richer countries tend to have fewer people employed in agriculture (in the UK the figure is just 1.4%) and more in service industries (in the UK the figure is 80.4%). In contrast, poorer countries like Honduras still have a lot of people employed in agriculture.

Another reason why the percentage of people employed in agriculture is important is because the Honduran economy depends on selling crops like bananas and coffee to bring in foreign income. So when a natural disaster like Hurricane Mitch occurs, the economic results can be devastating for the whole country, not just individuals.

▲ *Hurricane Mitch hits Honduras, 1998*

Hurricane Mitch

Hurricane Mitch roared into Central America in 1998 – devastating Honduras and putting its development back by decades. Ramon Espinal was the mayor of Morolica at the time. It used to be one of the southern region's most beautiful towns, but not one house was left standing. With all roads cut and phone lines down, Ramon walked for 24 hours to raise the alarm. Today, he wipes away a tear as he remembers the hurricane: 'Nearly 180 years of history gone – in one night. It was like a nightmare.'

Hurricane Mitch – the facts

- At least 5000 people were killed.
- 70% of Honduras's crops were destroyed.
- More than half of all homes were destroyed or damaged, and whole villages and neighbourhoods disappeared.
- 160 bridges were swept away.
- More than 300 km of roads were wrecked.
- Heavy rain caused flooding and landslides.
- The cost of repairing the damage was estimated at $2-3 billion.
- The Honduran Ambassador to London said 'Honduras has no money to pay for reconstruction. Its economy was wiped out by the hurricane.'
- Hurricane Mitch had such a big impact on the Honduran economy, that smaller disasters since 1998 have had a bigger impact than they should have done.

▲ *Hundreds of thousands of people were left homeless as a result of Hurricane Mitch*

Honduras – ten years on

Marco Burgos is the Honduran government minister in charge of dealing with natural disasters. He's standing on a bridge across the river that divides the capital city in two – and pointing to another bridge. 'You see that? It was a temporary bridge put up ten years ago, but it's still being used. It's fallen down three times, because people have stolen the screws!'

Marco is very critical of the Honduran government's response to Hurricane Mitch. 'The National Congress has never taken this seriously. We aren't properly prepared, and there still isn't a national emergency plan.'

$1 million was spent in 2007 to deal with the impact of flooding and mudslides, but around 4 million people (that's about half the Honduran population) still live in vulnerable places.

Hector Espinal, from UNICEF, said: 'If Honduras suffers another Mitch, more people will definitely die. Honduras hasn't learned its lessons'.

▲ *Many people in Honduras still face risks from flooded rivers or landslides*

(Adapted from 'Honduras struggles 10 years after Mitch' on the BBC News website, October 2008)

YOUR QUESTIONS

1. Use the information in the table opposite to show that Honduras is a poor, or less-developed, country.

   Hint: Compare the figures in the table opposite with those on page 229 and use the comparisons in your answer.

2. Write about 250-300 words about the impact of Hurricane Mitch on Honduras's development.

3. In a group of five, take on the following roles: Hector Espinal (from UNICEF), a Honduran farmer, Marco Burgos (Minister for Natural Disasters), the Honduran Ambassador to London, another government minister (whose view is that Honduras is preparing for hurricanes). In your roles, discuss the topic: 'Honduras hasn't learned its lessons'.

11.6 » Trade and the development gap

your planet

In the UK we drink 165 million cups of tea a day, or 62 billion a year. And half of our tea comes from East Africa.

On this spread you'll explore the link between trade and the development gap.

Trouble brewing

As you made your cup of tea this morning, Elizabeth Miheso will have been at work for hours in Kenya, picking tea on a small tea estate – for 30p a day.

And as you stirred in the sugar, Ibrahim Shikanda will have been helping his parents harvest the sugar cane.

You're unlikely to meet Elizabeth or Ibrahim. But your life and theirs are linked – by trade.

The world of trade

You looked briefly at trade on page 234. Now you need to look at it in more detail, because the patterns of trade – and the problems with trade – go a long way towards explaining why there is a development gap between richer and poorer countries.

Take Kenya and Japan as examples. The diagrams on the right show their **imports** – the goods and services they buy from elsewhere, and their **exports** – the things they sell to other countries.

The difference between imports and exports is called the **trade balance**. What most countries want is a **trade surplus**, which should mean that the country becomes richer and people's standard of living improves. If it's the other way round, there'll be a **trade deficit**. A country with a trade deficit will stay poor, get into debt and people's lives won't improve.

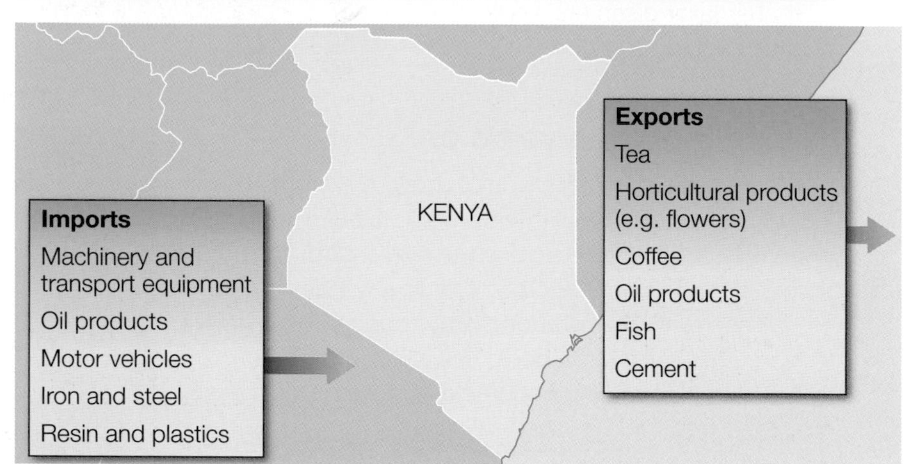

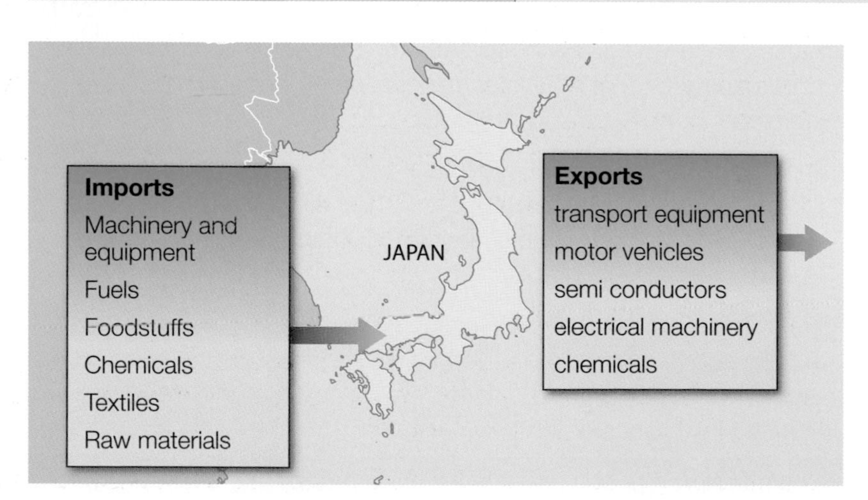

Patterns of trade

Look at Kenya and Japan again.

Kenya 2009
Exports = $4.48 billion   Imports = $9.03 billion.
You don't need to be a genius at maths to see that the sums don't add up. Perhaps that's one of the reasons why Kenya is ranked number 128 in the HDI rankings?

Japan 2009
Exports = $ 516.3 billion   Imports = $490.6 billion.
So, Japan has a trade surplus, and that helps to explain why Japan is ranked at number 10 in the HDI rankings.

Kenya is a poorer country and Japan is a richer one:

- Japan exports expensive manufactured goods, like vehicles, and imports cheaper primary products, like raw materials.
- Kenya exports cheap primary products, like tea and coffee, but then imports more expensive manufactured goods.

And there are other patterns:

- Most countries trade with their nearest neighbours, so Japan's biggest trading partners are China and the USA.
- Most world trade is between richer countries; there has been relatively little trade between poorer countries.

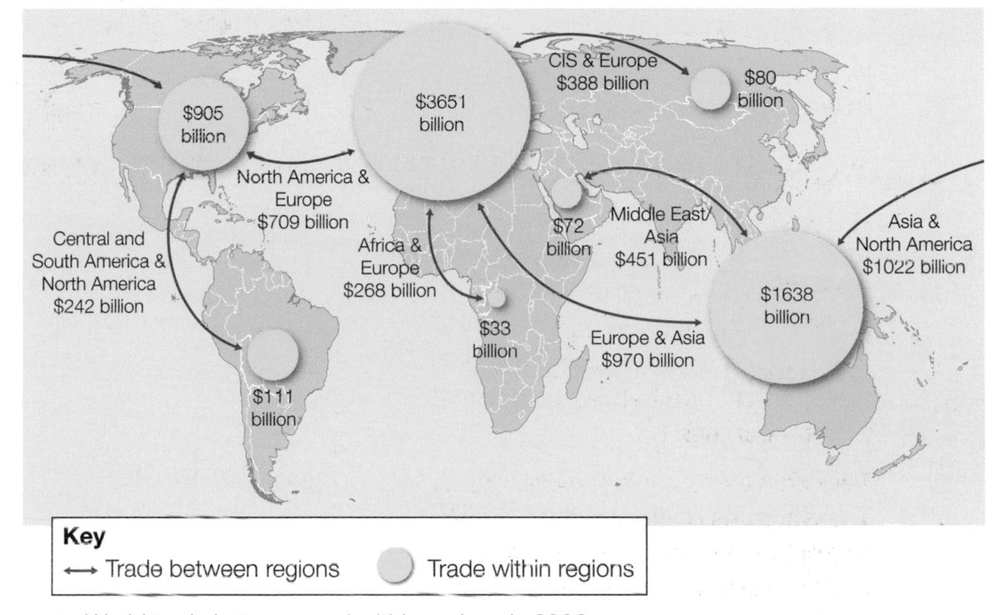

▲ *World trade between and within regions in 2006*

- Countries like Singapore, South Korea and Taiwan (the **Newly Industrialised Countries**, or **NICs**) – plus Mexico and Brazil – have become more important in world trade. China is also rapidly becoming a very powerful trading nation.
- **Transnational corporations**, or **TNCs**, also have a powerful role to play in world trade, and many have their headquarters in the world's richer countries.

What's the problem?

- Primary products – like tea, sugar and coffee – are sold very cheaply, and the price paid to the growers often changes dramatically. When the price drops, people like Elizabeth and Ibrahim suffer.
- Manufactured goods are sold for more money in the first place, and their price usually stays steady.

Because of the above situation, there's a trade imbalance. This results in a:

- trade deficit for many poorer countries, making it harder for them to develop
- trade surplus for most richer countries
- continuation of the development gap.

YOUR QUESTIONS

1. Define these terms: imports, exports, trade balance, trade surplus, trade deficit. Add them to your dictionary of key terms for this chapter.
2. Describe the differences between Kenya and Japan's exports.
3. Annotate a blank world map to show some of the patterns and problems arising from trade.
   Hint: Annotate means to add labels that explain things.
4. Explain the link between trade and the development gap.

11.7 » Reducing the gap 1 – trade

On this spread you'll find out about attempts to reduce the development gap by making trade fairer.

Trade – fair or not?

Most trade is in the hands of the world's richer countries. They try to control it by creating barriers to protect their own jobs and industries. They do this by using **tariffs** and **quotas**.

- Tariffs are taxes or customs duties paid on imports. They're used to make imported goods more expensive and less attractive to buyers than home-produced goods.
- Quotas are precise limits on the quantity of goods that can be imported. They're usually restricted to primary goods, so they work against poorer countries.
- **Free trade** is when countries don't discourage, or restrict, the movement of goods with tariffs and quotas.

Tariffs can hit poorer countries hard, as the example below from Ghana shows.

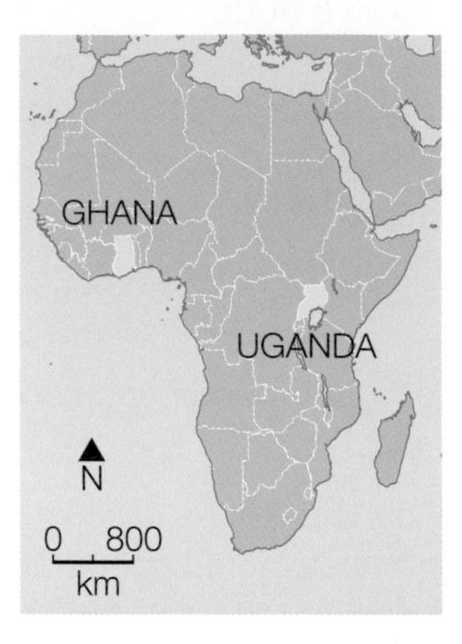

Ghana's cocoa

Ghana exports raw cocoa beans. Most of the processing and packaging of the cocoa is then done in Europe.

EU import tariffs are much higher for processed cocoa than for raw cocoa beans. For example, in 2007, the EU charged a 7.7% import tariff on cocoa powder and 15% on chocolate containing cocoa butter – but no tariff at all on raw cocoa beans. So, Ghana is forced to export the raw beans and lose out on any extra money they might have made by processing them. As a result, people like William Korampong stay poor.

> Cocoa is seasonal, so we have long periods of poverty. After paying my debts, there's no money left to send the children to school or pay for food. Allow the government to process the cocoa here, and not abroad – then there will be more money in our pockets.
>
> Cocoa farmer, William Korampong

Trading groups

These are countries that have grouped together to increase the amount they trade between them, and the value of their trade. Two of the biggest **trading groups** are the European Union (EU – which the UK belongs to) and the North American Free Trade Agreement (NAFTA). Creating formal trading groups means that member countries can cut the tariffs in place between them – making goods cheaper. But, as trading groups like the EU try to increase trade within the group, poorer countries from outside lose out and the development gap widens.

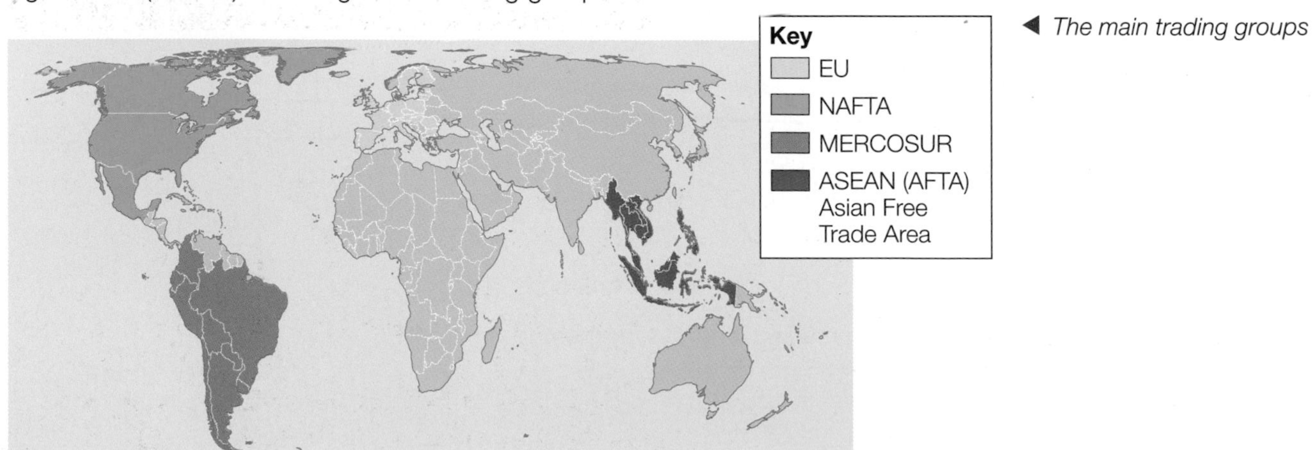

◀ *The main trading groups*

The World Trade Organization (WTO)

The **WTO** deals with the rules of global trade. It aims to make trade easier and get rid of anything hindering it. It negotiates new trade agreements and settles trade disputes.

The WTO is focussing on helping poorer countries by reducing and removing farm subsidies – grants paid to farmers to encourage them to produce more.

In theory, the WTO should be against subsidies, because:

- it believes in free trade
- only richer countries can afford to give big subsidies to their farmers – poorer countries can't.

But the WTO's Agreement on Agriculture still allows the EU and the USA to spend massive amounts of money on subsidies. As a result, large farmers in those countries carry on producing huge volumes of food, which the EU and US governments then buy and 'dump' on poorer countries in the form of food aid. This then reduces the price of food actually produced in those countries, and puts the livelihoods of local farmers at risk.

Fair Trade

One way of making trade fairer is for producers in poorer countries to join together to produce goods for the **Fair Trade** market. Fair Trade in the UK started as a way of helping farmers in poorer countries to improve their lives. Goods produced through Fair Trade have a logo that many people now recognise. Through Fair Trade, producers are paid an agreed minimum price – plus a premium to be invested in projects that benefit them and their communities.

Gumutindo Coffee Co-operative, Uganda

Uganda's biggest export crop is coffee. It was worth US$350 million in 2007. Most of the coffee beans are grown by small farmers, and an increasing number of them now grow coffee for the Fair Trade market. One example is the Gumutindo Coffee Co-operative in eastern Uganda. Here, 3000 farmers – 91% of whom depend on coffee for their main income – produce coffee beans, which undergo primary processing on the farms themselves. This involves processing the ripe red coffee cherries by removing the skin and pulp and drying the bean. Secondary processing (milling) is done at a nearby warehouse. The beans are then packed for export, leaving the final roasting to be done at their destination.

The money from the Fair Trade premium last year helped me to pay for my daughter's school fees, which are very expensive. I tell my children and neighbours to spend time producing good-quality coffee. Since the other farmers have seen us receive the Fair Trade premium, they have tried to copy what we are doing and the quality is getting better.

Mr Difasi Namisi, a Gumutindo farmer

YOUR QUESTIONS

1 Explain these terms and add them to your dictionary of key terms for this chapter: tariffs, quotas, free trade, trading groups, World Trade Organization, Fair Trade.

2 a In pairs, discuss the advantages for a country like the UK of being a member of the EU.

b Are there any disadvantages?

3 Still in your pairs:

a Draw a spider diagram to show how trade helps to keep countries poor.

b Draw a second spider diagram to show how trade can be made fairer to allow countries to close the development gap.

On this spread you'll find out how cancelling the debts of poorer countries helps to reduce the development gap.

Debt

Many poorer countries are in debt, and the sums of money they owe are huge. But how did they get into debt, and what has it got to do with development? The box on the right will help to explain.

How poorer countries got into debt

A In the 1970s, international banks lent large amounts of money to poorer countries to build expensive infrastructure projects, like dams and power stations.

B By the 1980s, interest rates had more than doubled. This increased the amount of money which poorer countries had to find to 'service' their loans (or debts) – in other words, to pay back some of the original loan amount, plus interest.

C Many countries couldn't afford to pay their debts, so the unpaid interest was added to the original loan amounts – and their debts began to grow.

D In return for reorganising the debts to make them more manageable, the banks expected the governments of the indebted countries to cut spending. In Uganda, for example, the biggest cuts were to healthcare and education – crucial for a country's development.

Debt and trade

Central America once had about 500 000 km^2 of rainforest. By the late 1980s, more than 80% of it had gone. Every year, more than 800 km^2 of rainforest are cut down in Honduras – to be replaced by ranches, banana plantations and small farms. The connection between the rainforest and debt is this:

- The trees that are cut down provide hardwood. This is sold overseas to earn foreign money to repay Honduras's debt.
- The new fruit plantations also earn money to repay the debt.

Every year, 20% of the money Honduras earns from exports is spent on debt repayments.

Cancelling debt

In July 2005, ten Live 8 concerts were held around the world at the same time, and the photo shows just one of them. They were part of a campaign to Make Poverty History and cancel world debt. A few days later, at a meeting of the G8 (the world's eight richest countries), an agreement was made to cancel all debts (worth $40 billion) owed by 18 **Highly Indebted Poor Countries**.

The **Highly Indebted Poor Countries (HIPC)** are a group of 38 of the poorest countries with the greatest poverty and debt. They include Uganda and Honduras.

But, two conditions had to be met before the debts were cancelled:

- Each government had to show that it could manage its finances, and that it wasn't corrupt.
- It also had to agree to spend the saved debt money on education, healthcare and reducing poverty.

By 2008, 27 of the 38 HIPC countries had met these conditions, and had had $85 billion of debt cancelled. The bad news is that African countries still owe $300 billion, and there's little chance that they'll be able to repay it.

Cancelling debt – the good news

Cancelling $1.5 billion of Uganda's debt has had a major positive impact on the country's development, as the table shows.

	Before cancelling debt	During and after cancelling debt
Population using an improved water source (%)	44	60
Population undernourished (% of total population)	24	19
Spending on education (% of GDP)	1.5	5.2
Adult literacy rate (% aged 15 and older)	56.1	66.8
Percentage of income from exports (plus net income from abroad) spent on debt repayments	81.4	9.2

◀ *Progress in Uganda*

Conservation swaps

Conservation swaps, or **debt-for-nature swaps**, are a way of tackling debt, and benefiting nature and conservation at the same time. The most common type of debt-for-nature swaps work like this. A country (the creditor) which is owed money by another country (the debtor), cancels part of the debt in exchange for the debtor country's agreement to pay for conservation activities. NGOs like the WWF often help to arrange the swaps.

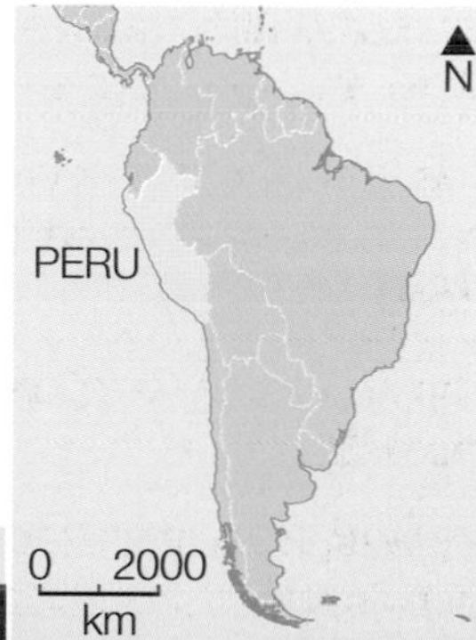

Debt-for-nature swaps protect Peru's endangered forests

In 2002 and 2008, Peru and the USA agreed to a debt swap worth almost $40 million. Peru agreed to conservation activities to preserve more than 27.5 million acres of endangered rainforest. The rainforest provides a habitat for many rare species, including jaguars and pink river dolphins.

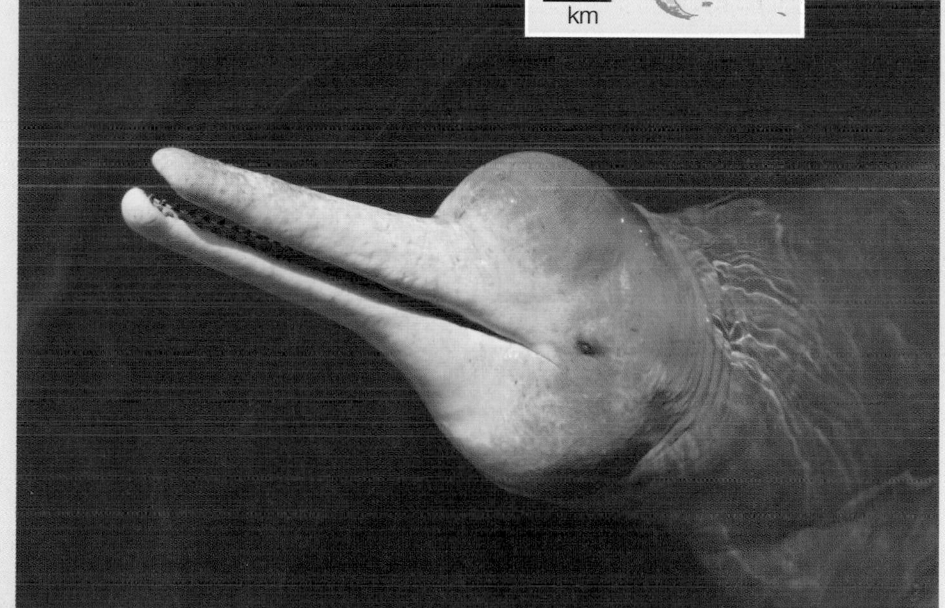

▶ *The pink river dolphin, one of the rare species whose habitat has been preserved due to debt-for-nature swaps*

YOUR QUESTIONS

1. Add these to your dictionary of key terms: Highly Indebted Poor Countries, conservation swaps (debt-for-nature swaps).
2. Draw a flow chart to show how poorer countries got into debt.
3. Describe two ways in which debt can affect the environment.
4. Explain how Uganda's development has benefited from debt cancellation.

   Hint: Use the information in the table, and give figures in your answer.
5. Work with a partner to produce a PowerPoint presentation on the link between debt and the development gap.

11.9 » Reducing the gap 3 – aid

On this spread you'll investigate aid, and explore whether trade or aid can close the development gap.

What is aid?

Aid is when a country receives help from another country, or an organisation such as an **NGO** (e.g. Oxfam or UNICEF), to help it to develop and improve people's lives.

The aid could be:

- money in the form of gifts/grants, or loans which have to be paid back
- emergency supplies (such as tents and food), machinery or technology
- people with special skills, such as engineers, doctors or teachers.

It can be:

- **short-term**, e.g. coping with the immediate problems caused by natural disasters, such as the Haiti earthquake (see pages 22-25). This type of aid often depends on people responding to particular emergencies and donating money to organisations like Oxfam.
- **long-term**. Sustainable long-term aid can prevent emergencies from happening in the first place, e.g. by providing deep wells to reduce the effects of drought in the future, or immunising people against diseases.
- **tied**. For example, a condition of the aid might be that the recipient has to spend the aid money on the donor country's products.

The **donor** is the country or organisation giving the aid. The **recipient** is the country receiving the aid.

Top-down or bottom-up?

Aid projects can be organised in either a **top-down** or a **bottom-up** way. Both approaches have advantages and disadvantages for donors and recipients.

Top-down development

- All decisions about development projects are made by national governments or large organisations.
- The development projects are often large-scale across the whole country, e.g. improving healthcare or education.
- The projects often involve huge sums of money.
- Outside experts help to plan the development.
- Local people have no direct involvement in decision-making.

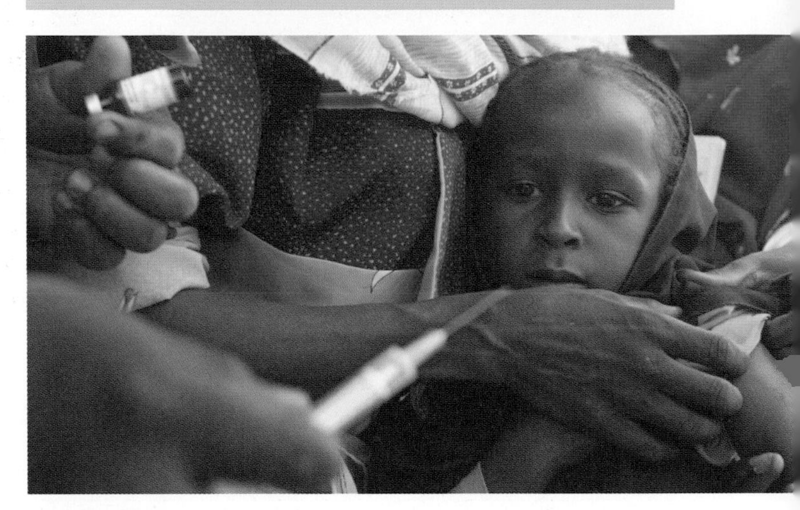

▲ *UNICEF's programme of immunising children against diseases is an example of top-down development*

Bottom-up development

- Experts work with local communities to identify their needs.
- Much of the work is done by NGOs, such as FARM-Africa (see opposite).
- Local people have control over improving their lives.
- Technical experts help with projects.

▶ *Equatorial College School in Uganda is funded by a school in North London. It is an example of bottom-up development, and provides education in a remote rural area. All of the decisions about the school are made by local people in Uganda, but the school is in regular contact with its London partner.*

Aid and sustainable development

The official United Nations' definition of **sustainable** development is that it 'meets the needs of the present without compromising the ability of future generations to meet their own needs.' Organisations like FARM-Africa encourage sustainable development through the aid they provide.

FARM-Africa and the Moyale Pastoralist Project

Life is tough in northern Kenya, but for some people things are getting better. FARM-Africa has set up the Moyale Pastoralist Project. The Project is helping communities in northern Kenya to survive, by reducing their dependence on their animals for all of their income. The Project is:

- helping communities to form Local Development Committees, which identify problems and find solutions, e.g. improving access to markets, so that farmers are able to sell their animals and the milk they produce more easily
- helping families to adapt the way they manage their crops, animals and forests to improve their sustainability, e.g. avoiding deforestation, which worsens drought conditions by allowing soil erosion
- providing small-scale loans to set up alternative businesses, such as beekeeping or small shops to sell local produce
- training local people to identify clean, safe water sources, and to dig and protect wells.

▲ *Northern Kenya often suffers from prolonged drought – people lack food and suffer extreme poverty*

Saku's story

As a member of a local women's group, Saku applied to her Local Development Committee for a small loan. She used this to set up a shop selling local produce, like honey, milk and eggs. Her shop now provides most of her family's income, and they can now afford food, healthcare and education – life's essentials.

Saku has now repaid her loan, so the money can be lent to someone else to start a new business. Saku says 'I feel I have contributed to the development of my family and the whole community.'

Trade or aid?

For many years, development has been linked to giving aid. Individuals give generously after major disasters, but governments are less generous. The UN set a target that every year the richer countries should give 0.7% of their GDP to poorer countries as aid. By 2008, only five countries had managed to do this – Norway, Sweden, Denmark, Luxembourg, and the Netherlands.

But, in the long term, trade is much better than aid for helping poorer countries to develop. Trade creates jobs, which provide wages that people can spend on improving their standard of living and quality of life.

YOUR QUESTIONS

1 Write a definition of sustainable development in your own words. Add it to your dictionary of key terms for this chapter, along with definitions for aid, donor, recipient, top-down and bottom-up development.

2 a Complete a table listing the advantages and disadvantages of top-down and bottom-up aid.

b Highlight the advantages for donor countries in one colour, and those for recipients in another colour.

c Now highlight the disadvantages for the donor countries in a third colour and those for the recipients in a fourth colour.

3 Debate this as a class – 'Trade not aid is what's needed to help poorer countries to develop.'

On this spread you'll learn about the different levels of development in the European Union (EU), and attempts to reduce the differences.

EU funding to decline

What do Gateshead's Millennium Bridge, a children's centre in Newcastle and a shop worker in Gateshead all have in common?

The answer is that they wouldn't be there if money from the EU hadn't helped to fund them. The bridge and the children's centre were partly funded by the EU. The training which led to the shop worker getting her job was also paid for by the EU.

But the rules have now changed. The EU's politicians have decided that the richer member countries, like Germany, France and the UK, should get less money, while the poorer member countries, like Poland, should get more.

(Adapted from 'North-East and Cumbria: EU funding' on the BBC News website, May 2006)

▲ *Gateshead's Millennium Bridge, partly funded by the EU*

The EU

The EU was set up in 1957 to achieve economic and political cooperation after the Second World War. Twenty-seven countries now belong to the EU, and they account for 31% of global GDP.

Differences in development

The EU may be one of the richest parts of the world, but there's still a big gap between its richest and poorest countries and regions.

- The richest regions (in terms of GDP per capita) are all cities – London, Brussels and Hamburg.
- The richest country – Luxembourg – is more than seven times richer than Bulgaria or Romania, the poorest members of the EU.

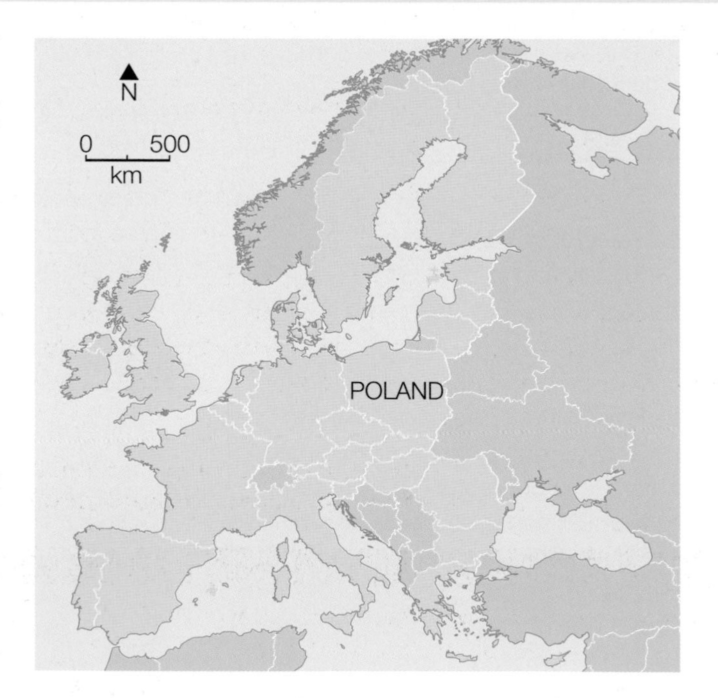

▲ *The 27 EU member states in 2010*

Comparing the UK and Poland

UK	Poland
Europe's core The UK is part of Europe's 'core', or centre. Core regions tend to have larger, wealthier populations – that create more wealth. They produce and consume more goods and services than other regions, and have the best communications.	**Europe's periphery** Poland is on the periphery – or edge – of the EU. Countries on the periphery tend to have poorer populations and poorer communications. Belonging to the EU has helped some countries on the periphery to develop, such as Ireland.
Social and economic factors • The UK joined the EU in 1973. • It's one of the EU's richest members, but still has 14% of its population living in poverty. • The UK's economy was in recession in 2010, so it had a trade deficit (see import and export figures below).	**Social and economic factors** • Poland used to be a communist country. It joined the EU in 2004. • 17% of its population live in poverty. • Hundreds of thousands of Polish migrants came to the UK to find work after Poland joined the EU (see pages 178-180).
GNI per capita ppp: $36 130	**GNI per capita ppp:** $17 310
HDI rank: 21	**HDI rank:** 41
Percentage employed in different sectors Agriculture 1.4% Industry 18.2% Services 80.4%	**Percentage employed in different sectors** Agriculture 17.4% Industry 29.2% Services 53.4%
Exports: $351.3 billion	**Exports:** $134.7 billion
Imports: $473.6 billion	**Imports:** $141.7 billion

The EU's regional policy

The EU's regional policy transfers resources from richer to poorer areas, so the poorer areas can catch up with the rest of the EU. From 2007-2013, regional spending will use up 36% of the EU's budget – a staggering €350 billion. The focus is on countries in Central and Eastern Europe – like Poland. The money comes from three different sources:

- The European Regional Development Fund, which pays for things like general infrastructure.
- The European Social Fund, which pays for things like training and job creation programmes.
- The Cohesion Fund, which covers environmental and transport infrastructure projects, as well as the development of renewable energy. This fund is reserved for countries with living standards that are less than 90% of the EU's average, so it includes Poland, Portugal and Greece.

▶ *Funding from the EU's regional policy helped to transform Ireland's roads from pot-holed lanes (top) to modern transport routes (bottom)*

YOUR QUESTIONS

1 a Draw two pie charts to show the percentage of people employed in different sectors in the UK and Poland. Round the figures up or down to help you.

b Describe the differences between the pie charts.

2 a Describe Poland's location in relation to the rest of the EU.

b Why is this important in terms of Poland's development?

3 Use the information comparing the UK and Poland to explain why there's a development gap between these two countries.

4 Do you think that the EU's regional policy will help to bridge the development gap between the UK and Poland? Explain your answer.

12 » Globalisation

What do you have to know?

This chapter is from **Unit 2 Human Geography Section B** of the AQA A GCSE specification. It is about the impacts of globalisation on the world in the twenty first century – looking at the changing importance of manufacturing, and increasing demands for energy and food. The table shows how the pages in this chapter match the content in the specification.

Specification content	Pages in this chapter
Understanding what globalisation is, and interdependence (the connections and links between countries).	p250-251
Globalisation has meant that manufacturing and services have developed worldwide. • Developments in ICT. • Development of call centres. • Transnational corporations (TNCs).	p252-254, p255 Case study of Nokia
Manufacturing is becoming more important in some parts of the world, and less important in others.	p256-257
China – the new economic giant.	p258-259 Case study of China
The global demand for energy is rising; the impacts of increased energy use.	p260-261
Sustainable development, and protecting the environment.	p262-265
• Using renewable energy to achieve sustainable development.	p262-263 Case study of renewable energy in India
• Reducing the costs of globalisation from local to global	p264-265
The global demand for food can have positive and negative impacts (environmental, political, social and economic). Encouraging the use of locally produced food.	p266-269

Your key words

- Globalisation
- Interdependent
- Information and communications technology (ICT)
- Call centre
- Transnational Corporation (TNC)
- Deindustrialisation, industrialisation
- Newly Industrialised Countries (NICs)
- Special Economic Zone (SEZ), Export Processing Zones
- Non-renewable, renewable, sustainable energy
- Biogas
- Sustainable development
- Marginal land
- Food miles
- Carbon footprint

Exam help ...

Advice See pages 297-299 for information on how to be successful in your exams.

Practice See page 311 for exam questions on this chapter.

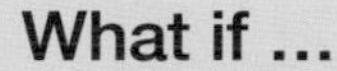

What if ...

- everything was made in China?
- you worked in the factory opposite?
- we ran out of oil next year?
- countries went to war over water?

12.1 » Twenty-first century world

On this spread you'll find out what globalisation is, and how places around the world are connected.

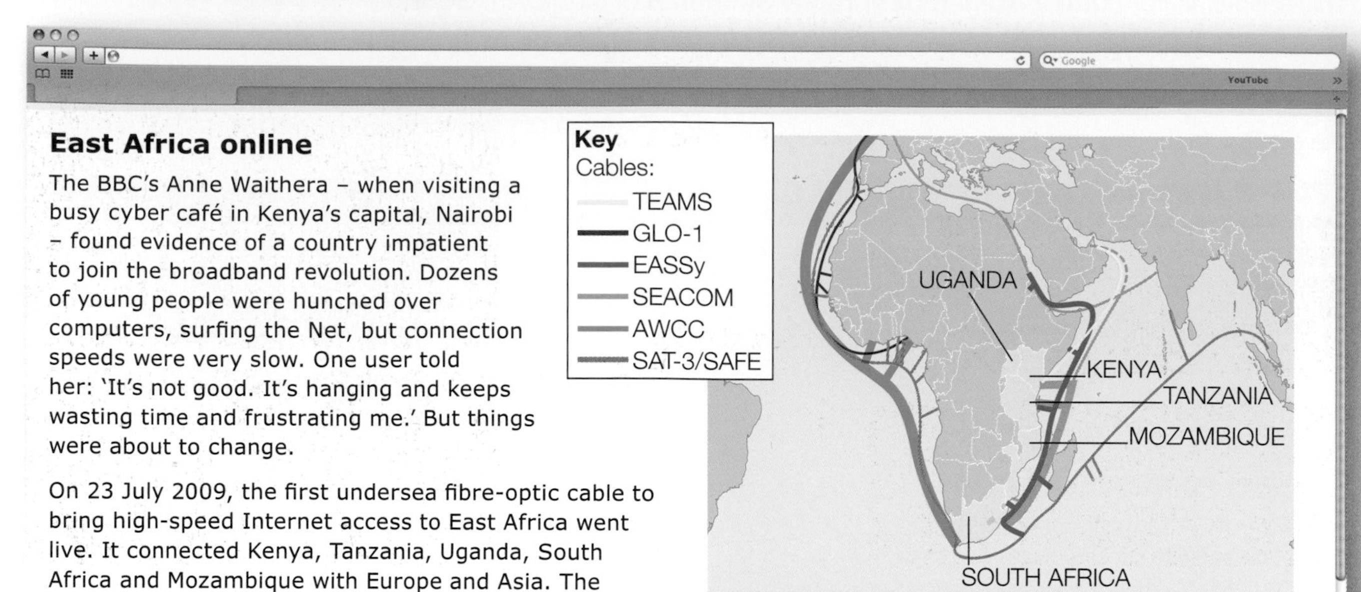

East Africa online

The BBC's Anne Waithera – when visiting a busy cyber café in Kenya's capital, Nairobi – found evidence of a country impatient to join the broadband revolution. Dozens of young people were hunched over computers, surfing the Net, but connection speeds were very slow. One user told her: 'It's not good. It's hanging and keeps wasting time and frustrating me.' But things were about to change.

On 23 July 2009, the first undersea fibre-optic cable to bring high-speed Internet access to East Africa went live. It connected Kenya, Tanzania, Uganda, South Africa and Mozambique with Europe and Asia. The African company SEACOM is operating the cable, and says that it will help to boost East Africa's industry and business.

▲ *Africa's existing and proposed fibre-optic cable connections*

Some reactions to the SEACOM cable going live were:

> *Hopefully, this means that the new website I want to create for my business can now happen. Before this, some customers would give up halfway through the online ordering process, because the web pages were loading so slowly.*

Greg, Cape Town, South Africa

> *Upgrading to fast broadband will boost our online communication with the rest of the world, and especially our exchange programme with schools in the UK.*

Simon, from a school in Nairobi, Kenya

▲ *Internet cafés, like this one in Nairobi, serve many purposes – including helping Kenyans to get jobs abroad*

> *How will this cable benefit people in rural areas of Tanzania who don't even have electricity?*

Johannes, Dar es Salaam, Tanzania

Globalisation

Globalisation is a word we hear a lot these days. The simplest explanation is that it represents the growth and spread of ideas around the world. This relies on connections and links between different peoples, countries and regions. And that's why the East Africa fibre-optic cable is important, because it brings the newly connected East African countries closer to the rest of the world.

The term **globalisation** has changed – from meaning the ways in which companies do business around the world, to the ways in which people, cultures, money, goods and information 'move' between countries.

The connections and links between different countries and regions mean that they've become **interdependent**. In terms of trade, for example, they depend on each other for goods and services. So, while Kenya – in East Africa – exports goods like cut flowers and vegetables to Europe, it also imports oil from the Middle East.

Several big factors have helped the process of globalisation, as this diagram shows:

Factors helping globalisation

Improvements to transport – In 1840, the fastest ships sailed at 10 mph. By the 1960s, jet aircraft flew at 500-700 mph.

The growth of computer and Internet technology – Information can now be sent around the world in seconds.

Expanding markets in developing countries – Thanks to the expansion of its economy, India's middle class is growing in size. This is increasing the demand for overseas goods and technology, e.g. the latest iPhone.

The growth of international organisations – For example, banks and transnational corporations (TNCs), like HSBC and Nike.

The impacts of globalisation

◀ *This is Egypt, but you can still get a McDonalds*

On people

- Skilled people now move around the world more frequently – global companies need global workers.
- Migrants move to countries where they'll have a better standard of living.

On finance

- Trillions of dollars are exchanged electronically every day around the world.
- Some TNCs make more money than entire countries!

On culture

- People around the world can now share common interests in sport, music and films, e.g. live Premiership football can now be watched simultaneously in places as far apart as the UK and Borneo!
- Cheaper telephone and Internet connections bring information by e-mail and global media, e.g. BBC World.

On politics

- Global companies can influence the ways in which people vote and think about issues, e.g. News Corporation owns The Sun and The Times newspapers in the UK (among other things), and also many other media outlets around the world, such as Fox News in the USA.
- International political organisations – like the EU – have expanded, leading to a reduction in national identity.

YOUR QUESTIONS

1. Define these terms using your own words: globalisation, interdependent.
2. List the factors that have helped globalisation.
3. What evidence does the information about East Africa provide for globalisation?
4. Work in a small group.
   - **a** Where were the clothes and shoes that you're all wearing made?
   - **b** Discuss how they are examples of globalisation.
5. Hold a class debate. The topic is 'Globalisation is making the world a better place'.

12.2 » Global connections

On this spread you'll learn about the importance of developments in ICT.

One clumsy ship cut off the Web

In January 2008, a ship trying to anchor off the coast of Egypt in bad weather cut through an undersea fibre-optic cable. This caused an Internet blackout that left tens of millions of people in the Middle East and Asia struggling to communicate with the rest of the world.

In the end, it only took a few days to fix the cable and restore the Internet, but the impact was felt by many businesses. British Airways was one of the worst-affected British companies, because its telephone call centres in India couldn't function properly.

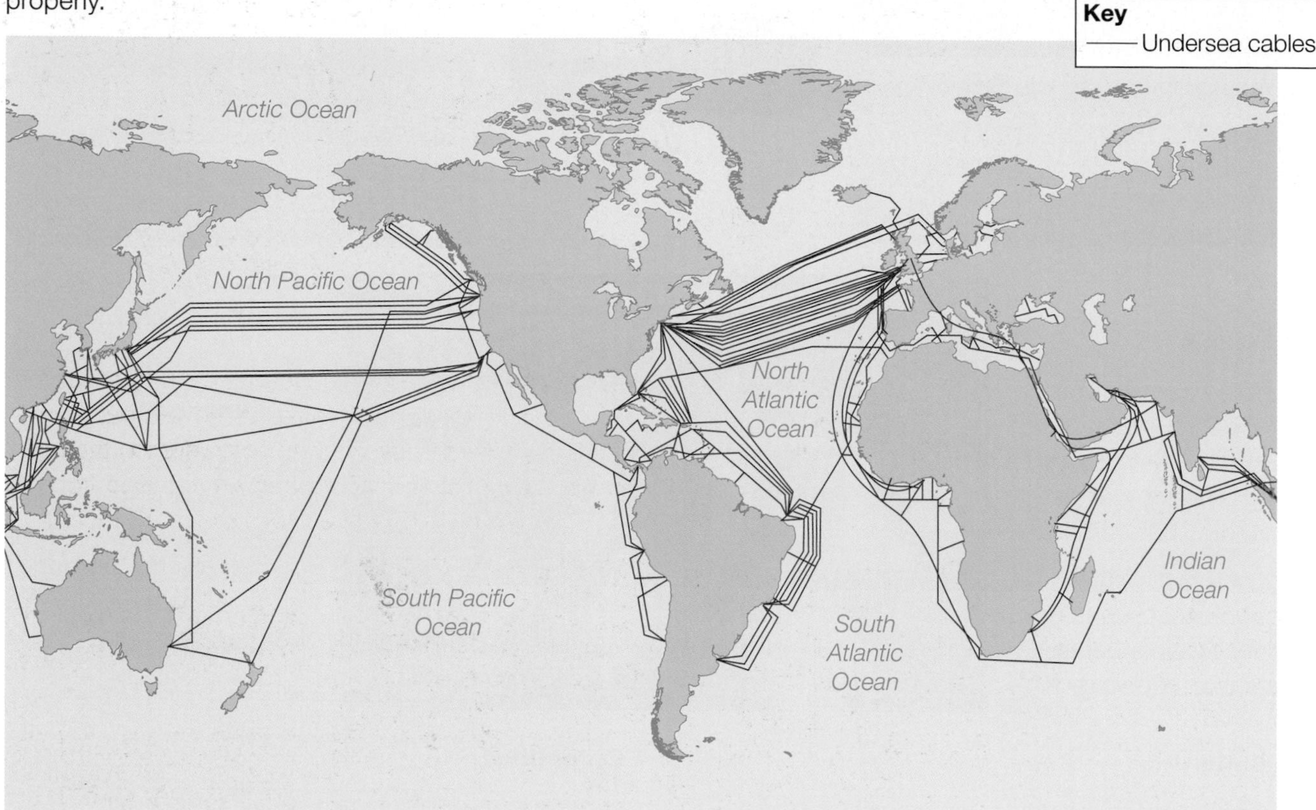

▲ *International communications via the Internet go through submerged cables deep under the world's seas and oceans*

As you saw on page 251, several different factors have affected the process of globalisation. But it's the revolution in **information and communications technology (ICT)** – through the use of computers, mobiles and other digital media – which has really helped to speed up the process.

Our global communications go through the undersea cables shown on the map above. These expensive fibre-optic cables have been laid at huge cost across the world, in order to send information, conversations and messages from one country to another. For example, the cables Flag-Europe-Asia and Sea-Me-We 4 are two of the most vital information pathways between Europe and Asia. They are responsible for 75% of all communications in the Middle East and South Asia. See-Me-We 3 links Europe, the East and Australia.

Advances in ICT have also allowed:

- the development of local industrialised regions with strong links to the rest of the world, such as Motorsport Valley in the UK (see the box opposite)
- the transfer of some key business services abroad, such as call centres.

Motorsport Valley

In 2010, 8 out of 12 Formula 1 (F1) teams were based in the UK – in an area known as Motorsport Valley. Motorsport is hugely important to the UK's economy, because it contributes over £6 billion a year. 4500 companies are based in Motorsport Valley. They employ 38 500 people, including 25 000 engineers.

Motorsport Valley is a 'technology cluster'. Many companies there work closely with each other, and also with nearby universities, research centres and industrial consultancies. The area is a **global centre** for engineering solutions and the production of all aspects of high-performance cars. Members of the Motorsport Industry Association export more than 60% of their products and services, so good global communication links are vital for companies in this area.

▲ *The Red Bull Racing Formula 1 team is based at Milton Keynes in Motorsport Valley*

Call centres

The people in the photo are working in a **call centre** in India. They answer telephone queries from overseas customers – many of them calling from the UK. Their computers provide them with the information they need to answer those queries.

Banks and insurance companies were some of the first businesses to set up call centres in the UK, but many large companies now have them. However, in the mid-1990s, many of those call centres were moved from the UK to South Asia – especially to India. But why?

The rapid development of IT has led to fast and easy international communications, with low telephone and Internet costs.

India has links with the UK as a result of the Commonwealth.

English language skills are good in India.

IT is well taught in the Indian education system.

India produces 3 million graduates a year. That's a lot of well-educated people.

Wage costs are much lower in India. A call-centre worker in India might earn £2500 a year, compared with £18-20 000 in the UK.

YOUR QUESTIONS

1. Explain these terms: ICT, call centre.
2. Give four reasons why so many call centres were moved from the UK to countries like India.
3. Explain the connection between the cutting of an undersea cable off the Egyptian coast and British Airways. Use these words in your answer: globalisation, ICT, call centre.
4. Locate Motorsport Valley on a blank map of the UK. Add boxes to your map which give information about: its transport links, IT communications, the companies located there, links with nearby universities and research centres, the products and services offered by the its companies.

   Hint: You will need to do some research for this question. Begin by searching the Internet for websites connected with Motorsport Valley.

12.3 » Global manufacturing

On this spread you'll find out about transnational corporations, and look at Nokia in more detail.

Transnational corporations

The chances are that your mobile is a Nokia. Nokia is the world's largest mobile phone manufacturer. Forty years ago, not many people had even heard of Nokia, but now it's a household name and a big **transnational corporation**, or **TNC**.

TNCs are multinational companies. They're so big that they often have their headquarters and main office in one country, and smaller offices and factories in poorer countries – where workers are cheaper to employ. More than 90% of the 500 largest companies in the world are based in richer countries (more than 450 of them in North America, Europe and Japan). But the tide might now be turning. Between 2000 and 2007, the proportion of the world's top 1000 companies based in poorer countries increased from 5% to 19%. This trend is called 'Globalisation 2.0'.

▼ *GDP (2009 estimates) for selected countries*

Country	GDP in 2009 (billion US$)
USA	14 140
China	8748
India	3570
UK	2128
Poland	689
Chile	242
Kenya	63
Honduras	33
Jamaica	24
Haiti	12

The top five TNCs

The top five TNCs are hugely wealthy. Some of them make more money in a year than many poorer countries produce in GDP, as the tables show.

Rank	Company	Type of business	Revenue in 2009 (billion US$)	Headquarters
1	Walmart	Retail	413.8	USA
2	Exxon Mobil	Oil and gas	310.5	USA
3	Shell	Oil and gas	278.1	Netherlands and UK
4	BP	Oil and gas	246.1	UK
5	Saudi Aramco	Oil and gas	216.0	Saudi Arabia

◀ *The top five TNCs in 2009*

TNCs – good or bad?

Some people are in favour of TNCs, and some are against them, as you can see below.

Nokia

- Nokia started in 1865 as a wood mill on the Nokia River in Finland (from which it gets its name) – making wood pulp for paper. Another company, the Finnish Rubber Works, opened factories making rubber tyres. It also used Nokia as its brand name. Soon after the First World War, the Finnish Rubber Works took over Nokia Wood Mill and Finnish Cable Works, which produced telephone cables.
- In 1967, the three companies formally merged to form the Nokia Corporation. By then, they made products ranging from paper, car tyres and Wellington boots, to telephone cables and computers!
- Then, in 1984, Nokia bought a company, Salora Oy, which made car phones. Soon after this, it launched one of the world's first mobile phones – weighing 5kg!
- Finally, in 1987, Nokia introduced the first hand-held mobile. The company then decided to sell all of its other operations, so that it could just concentrate on making mobiles. Now Nokia controls 40% of the global mobile market.

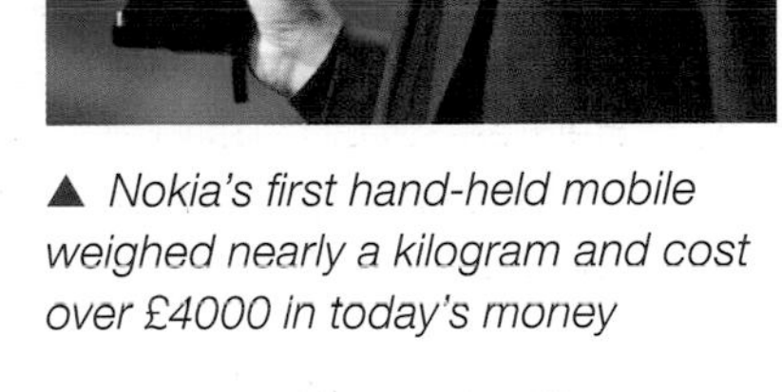

▲ *Nokia's first hand-held mobile weighed nearly a kilogram and cost over £4000 in today's money*

Today, Nokia is a large and rapidly growing TNC:

- In 2009, it was the world's 85th largest company (up from 119th in 2007), selling phones in over 150 countries.
- It employs over 123 000 people in 120 countries – the main ones are shown below.
- It is still based in Finland – making it by far the largest Finnish company.

▼ *Nokia operates around the world*

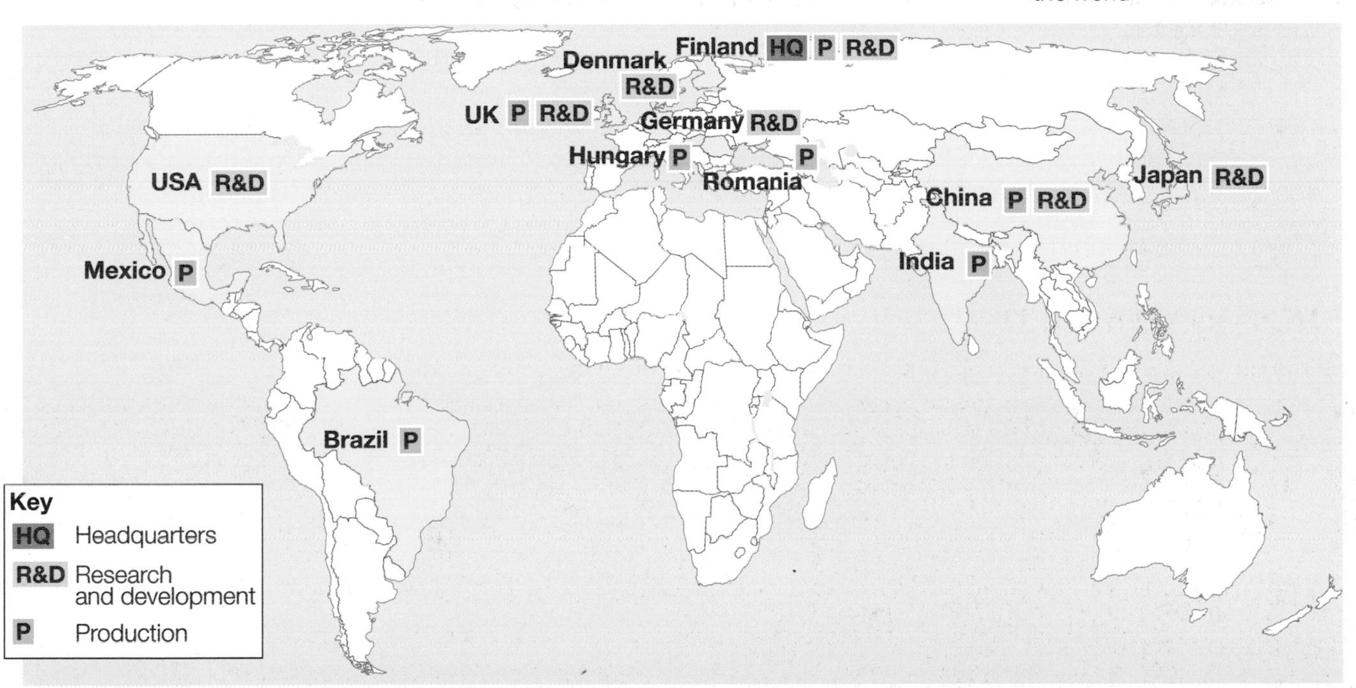

YOUR QUESTIONS

1 Write a definition for TNC, and give three examples of TNCs.

2 Draw up a table with two columns headed 'Advantages of TNCs' and 'Disadvantages of TNCs'. Complete your table using the text on the photo opposite.

3 a Look at the map above. Describe the distribution of Nokia's operations. Mention the headquarters, research and development, and production operations in your answer.

b Suggest reasons for the different distribution of the 3 operations.

4 Update the information about Nokia. You could use the company website www.nokia.com or try other websites like http://en.wikipedia.org/wiki/Nokia

12.4 » The changing world of manufacturing

On this spread you'll find out how manufacturing has declined in some parts of the world and increased elsewhere.

China rules

Look at the map. It shows global clothing exports – and the big green shape on the right is east Asia, which includes China. China exports more clothes than any other country in the world, but it hasn't always been the world's main manufacturing centre.

In the mid-nineteenth century, it was Britain that was seen as the 'workshop of the world'. It produced more than half of the world's iron, coal and cotton cloth – and employed vast numbers of people in those industries. But the box below shows how things have changed.

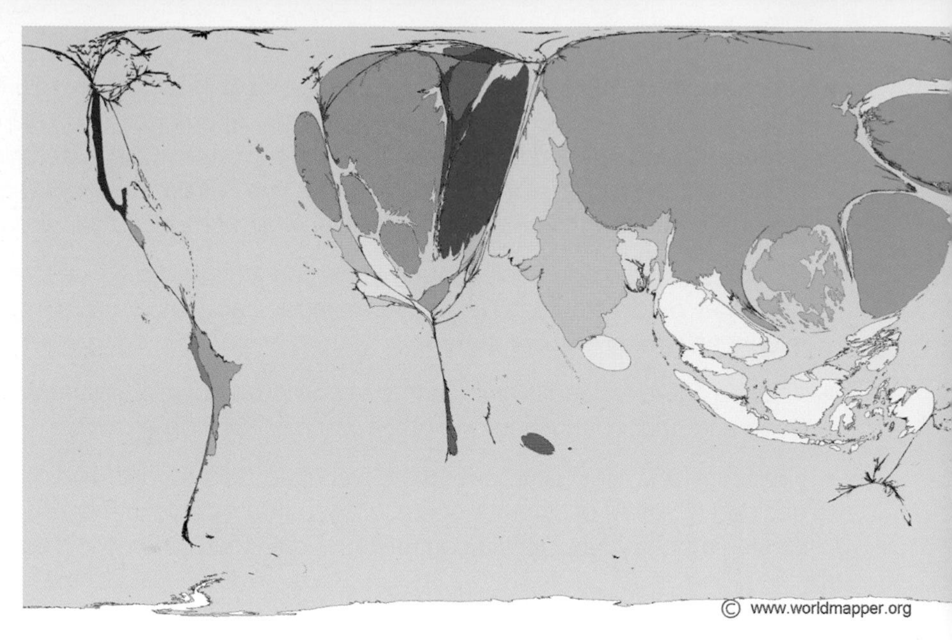

▲ *Global clothing exports. On this map, countries have been drawn in proportion to the volume of clothing they export, so Asia is much larger than either the Americas or Africa.*

Changing industry in the UK

Before 1800, most people in the UK worked in farming or related activities (the primary sector). But the Industrial Revolution of the nineteenth century changed all that. Lots of people moved to the towns or cities for work – making steel, ships or textiles (the secondary sector). Others worked in mines.

Then, in the late twentieth and early twenty-first centuries, it all changed again. There was a big shift to jobs in **service industries** (the tertiary sector) – and then to **quaternary industries**. Between 2000 and 2006, the percentage of people working in service industries increased from 70% to 80%. However, in 2010, the new Coalition government decided that it wanted to 'rebalance the economy' by rebuilding the UK's manufacturing base and relying less on service industries.

▼ *Changing employment structure in the UK*

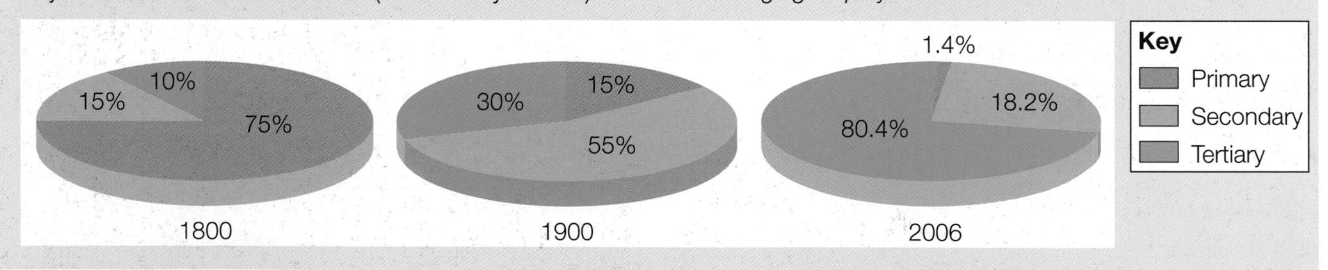

Deindustrialisation

For decades, the UK has been experiencing **deindustrialisation**. This is the name given to the decline in manufacturing (secondary) industry and the growth in tertiary and quaternary industry. In the UK this happened because:

- machines began to replace people in many manufacturing industries, e.g. robots were increasingly used on car production lines
- other countries, such as China, were able to produce goods more cheaply
- British goods were too expensive, due to low productivity (not enough being produced), a lack of investment, and high interest rates and wages.

Industrialisation

The 'Asian Tigers' form part of a group of **Newly Industrialised Countries (NICs)**, which have developed large manufacturing industries very quickly and seen their exports and GDPs grow rapidly. This process is called **industrialisation**. The NICs include Hong Kong, Singapore, Thailand, South Korea and Taiwan, as well as Brazil, Mexico, Argentina, India and China. The graphs show how three of the East Asian economies have grown.

The Asian NICs were able to develop very quickly because:

- their workers were prepared to work for long hours for little pay to begin with – so their goods could be produced cheaply and competitively
- national governments there introduced long-term industrial planning and tried to attract new industry with financial incentives
- these Asian countries are located close to the main shipping lanes – and ships are the cheapest way of transporting bulky goods long distances
- their manufactured goods were targeted at global rather than local markets.

The economic growth of the NICs continued throughout the 1980s and 1990s – when it was slowing down elsewhere, and world manufacturing was declining.

▼ *Economic growth in East Asia*

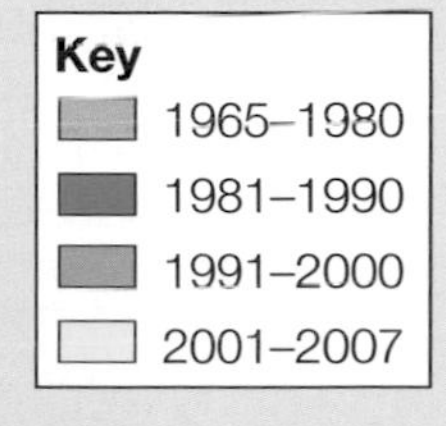

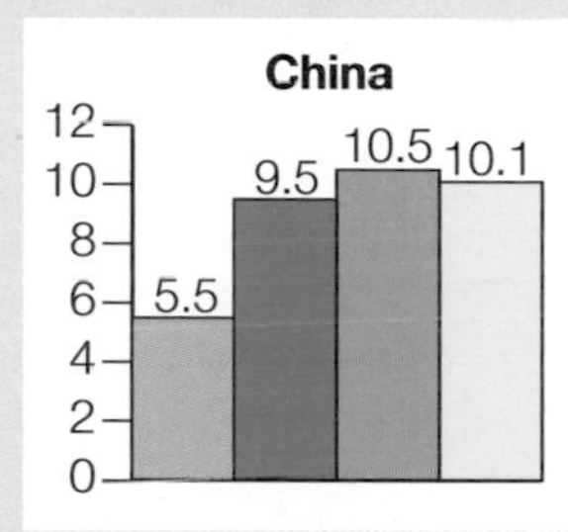

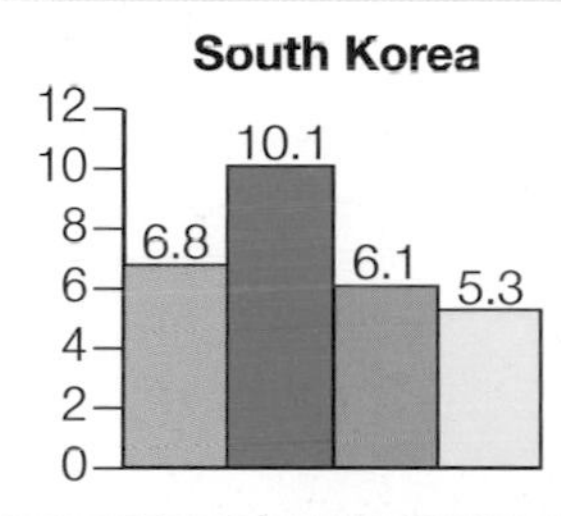

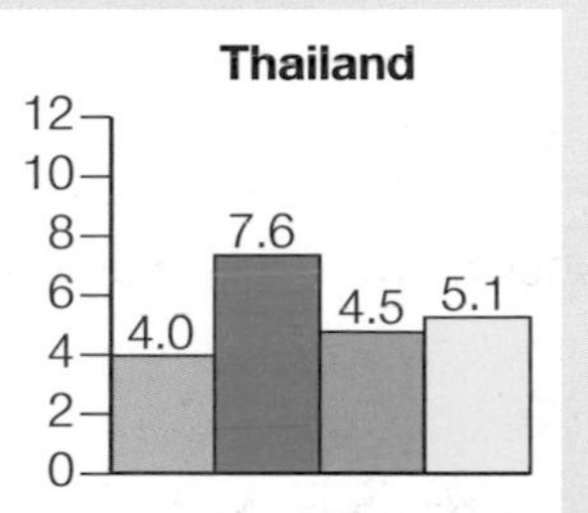

Factors affecting manufacturing

A number of factors can contribute to the decline of manufacturing in one place and its increase somewhere else.

Legislation

Legislation, or laws, in both the UK and the EU affect things like:

- the number of hours that people can work
- the minimum amount of money that workers should be paid.

Strikes

- Strikes in the 1970s in the UK badly affected many manufacturing industries, and they lost customers to overseas competitors as a result.
- In 2010, companies in China – like Toyota – were hit by strikes as workers demanded higher pay.

Health and safety regulations

These exist in countries like the UK to protect workers and keep them safe, but they can also add to a company's costs. In other countries, these regulations may not exist at all – so companies can take short cuts to keep their costs down and productivity high.

Tax incentives and tax-free zones

- In Assisted Areas in England, businesses can get tax breaks, subsidies and grants, which make it cheaper for them to operate there.
- Tax-free zones (see page 259 on China's Export Processing Zones) attract businesses looking to reduce costs.

YOUR QUESTIONS

1. What is the difference between industrialisation and deindustrialisation?
2. Describe why, and how, the UK has experienced deindustrialisation.
3. **a** Label the countries listed in the text on industrialisation on a blank map of the world.
   b Annotate the map to describe why and how these economies have grown (use figures in your answer).

12.5 » China – the new economic giant

On this spread you'll learn how China has become an economic giant.

Made in China

You'll probably never know exactly where your mobile was made. Companies like Nokia (page 255) don't reveal which factory an individual phone comes from. But there's a good chance that it was made in China – which makes half of all mobiles. And also that it could have come from the Longhua Science and Technology Park in Shenzhen (see below) – one of the world's biggest factories. Longhua produces high-tech gadgets for some of the best-known global companies, including Nokia, Apple and Sony.

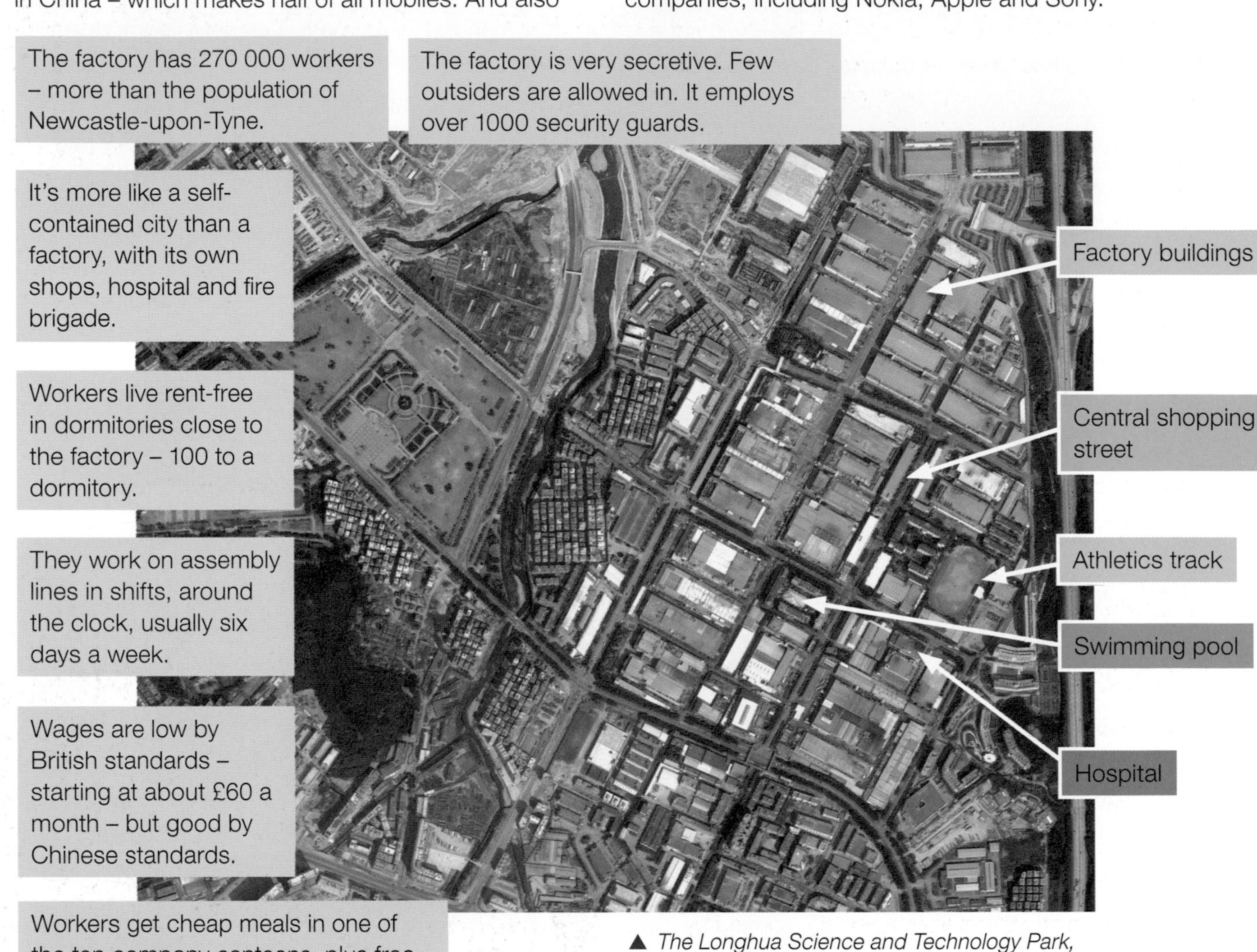

▲ *The Longhua Science and Technology Park, viewed from above*

How did Shenzhen grow?

Shenzhen was a small fishing village until 30 years ago. Today it's one of the world's fastest-growing cities.

- In 1979, the Chinese government made Shenzhen China's first **Special Economic Zone (SEZ)**.
- Shenzhen was chosen because of its location in south-east China, close to Hong Kong (which was then a prosperous British colony – taken back into Chinese control in 1997).
- The SEZ offered tax incentives for foreign companies to build new factories there.
- The foreign companies were also attracted by China's low wages, long working hours and lack of strict workplace regulations (e.g. health and safety).
- Millions of people flocked to Shenzhen from other parts of China to find better-paid work. Its population is now 9 million and growing.

Changing China

After China's communist government took power in 1949, it kept the country isolated from the rest of the world. The government planned the economy centrally, and goods were only produced for China's own people. No private wealth was allowed.

However, in recent decades, China has undergone massive economic changes. Since the early 1980s, China's economy has doubled in size every eight years – so it's no longer a poor country.

After 1986, China's communist government developed an 'Open-Door Policy' to encourage overseas investment. In the 1990s, it also allowed Chinese individuals to become personally wealthy – by producing goods and services without government interference (although most of China's largest companies are still either totally or partly State-owned). But, despite these new economic freedoms, China's political and social organisation has stayed tightly controlled.

Transnational corporations now invest in China to take advantage of the low cost of labour and the benefits of the Special Economic Zones. The SEZs also include **Export Processing Zones**. Now, foreign-owned companies, or those in partnership with Chinese companies (like BMW), produce half of China's exports. 60% of the increase in world trade since 2004 has been as a result of China's industrialisation.

▲ *China has formed several partnerships with foreign companies, and this factory in Shenyang – a partnership with BMW – has already starting producing cars*

Export Processing Zones are a type of Special Economic Zone where businesses can import raw materials, make them into finished products and then export them without paying any duties or tariffs.

The downside

Unfortunately, China's economic growth hasn't all been good news:

- China has 16 of the 20 most air-polluted cities in the world (see right).
- 30% of China suffers from acid rain, caused by coal-fired power stations.
- 70% of China's rivers and lakes are polluted.
- 20% of China's population (about 268 million people) live on less than $1 a day.
- Child labour was used by some factories making souvenirs and toys for the 2008 Beijing Olympics.

▼ *Air pollution in Shanghai, China*

YOUR QUESTIONS

1. Write definitions for these terms: Special Economic Zone, Export Processing Zone.
2. Write a newspaper article of no more than 150 words about how Shenzhen grew from a fishing village to an industrial city.
3. Work with a partner to produce a PowerPoint presentation with the title 'The Changing World of Manufacturing'. Use information from Spread 12.4, as well as this spread.

12.6 » Increasing demands for energy

On this spread you'll explore why demands for energy are rising, and what impacts this will have.

2010 – The diary of a disaster

20 April – The Deepwater Horizon oil rig exploded in the Gulf of Mexico, killing 11 people and damaging the oil well on the seabed.

30 April – Crude oil started to wash ashore in Louisiana, USA. The amount of oil leaking from the damaged oil well was estimated at anything between 5000 and 70 000 barrels a day.

15 July – The oil leak was finally stopped after a cap sealed the well.

▲ *The 2010 Gulf of Mexico oil leak was the biggest environmental disaster in American history*

There were far-reaching consequences from the 2010 Gulf of Mexico oil disaster – for people (whose livelihoods were affected), the environment (see the photo), and the oil industry. BP – which operated the affected oil rig – saw its reputation and market value drop dramatically, and it also had to put billions of dollars aside to pay compensation for the damage caused). However, this disaster was just one of the results of our increasing demands for energy (see the boxes opposite).

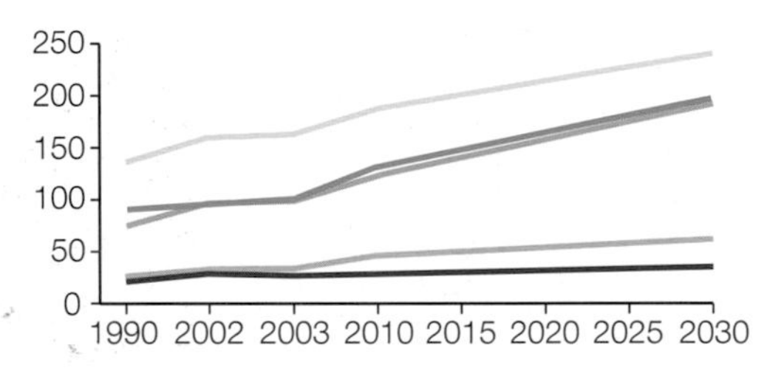

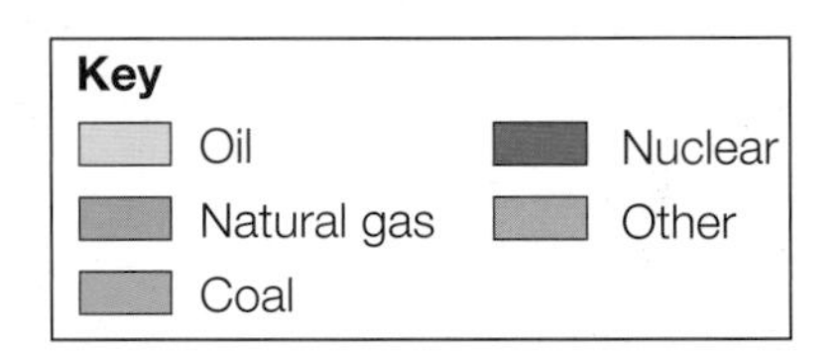

▲ *Worldwide energy use (in quadrillion BTU)*

How much energy do we use?

The graph shows the amount of energy that the world is using now, and is expected to use up to 2030. In 2008, the total amount of oil used around the world was 80-85 million barrels a day. The table shows that just eight countries were using half of this total.

Demand is rising

The world is already using vast quantities of energy, and demand is still rising:

- By 2010, global oil consumption was 86.6 million barrels a day. China was using 10% of this, and India 3%.
- By 2030, global energy demand is expected to have grown by 50%. The demand from developing countries is expected to rise by 85%, compared with 19% in industrialised countries.

Why is demand rising?

- Global population growth. In 2009, the world's population was 6.79 billion. By 2050, it's predicted to rise to 9 billion. The simple fact is that all those extra people will use more energy.
- The growing economies of China and India need more and more energy for their industries to operate and expand.
- China and India between them contain 40% of the world's population – but their average energy use per person is low. If their energy use per person increased to the USA's level, demand for energy would be off the scale.

Country	Oil consumption
USA	20.7
China	7.2
Japan	5.2
Russia	2.9
Germany	2.7
India	2.6
Brazil	2.2
UK	1.8

▲ *Oil use in 2008 (million barrels a day)*

- China currently relies heavily on coal for electricity generation, but it also accounts for a third of the global growth in demand for oil. The biggest reason for the increasing demand for oil is for transport, especially cars.
- Increasing wealth in countries like China and India means that there's more demand for electrical goods (requiring energy to make them and operate them) – and also cars to replace traditional bicycles.

▲ *In the 1990s, this Beijing street would have been filled with bicycles, but now the car is king*

The impacts of increasing energy use

Our increasing use of energy is having a range of impacts, or costs.

Environmental impacts

- Greater use of fossil fuels (coal, oil, gas) is increasing carbon dioxide emissions and driving climate change. This is leading to changing weather patterns (e.g. more floods, droughts and hurricanes), as well as more-permanent environmental changes (e.g. desertification, rising sea levels, melting glaciers and ice sheets, and damage to fragile ecosystems like coral reefs).
- Increasing demand for oil means that we have to look for it in areas that are difficult to get at (like the deep waters of the Gulf of Mexico), or that are environmentally sensitive (like the Arctic).

◀ *Severe weather events are expected to become more frequent, widespread and intense as climate changes – causing economic, environmental and social impacts*

Economic impacts

- Climate change will cost money. The economic costs of disasters like hurricanes are doubling every decade. Rising temperatures will also lead to a reduction in crop yields – a further cost.
- Oil-rich countries spend huge amounts of money to increase the amount of oil they produce. Saudi Arabia has spent $50 billion.
- As easily accessible energy reserves are used up, more and more money has to be spent exploring remote areas for new reserves, e.g. the Arctic or off the Falkland Islands in the South Atlantic.

Social and political impacts

- The droughts and floods caused by climate change can lead to famine, disease and homelessness, which can take years to recover from.
- Some countries rely heavily on others for their energy supplies, which can create problems if they fall out. For example, much of Europe depends on a gas pipeline from Russia that runs through Ukraine. But, in both 2006 and 2008, the pipeline was turned off after disagreements between Russia and Ukraine.
- In some countries, like the USA, political parties are funded by oil and gas companies, which means that these companies can influence government policies.

YOUR QUESTIONS

1 a Use the graph opposite to describe **(i)** how the amount of energy we use is likely to change over the coming decades, and **(ii)** which energy sources will be most affected.

b Explain which parts of the world the increasing demands for energy are likely to come from.

2 Work in pairs to create a spider diagram to show why demand for energy is rising.

3 Hold a class debate. The topic is 'The costs of increasing energy use are too high'.

12.7 » Sustainable development – 1

On this spread you'll find out how sustainable development can be achieved through the use of renewable energy.

Different types of energy

The different types of energy can be classified, or grouped, as:

- **non-renewable**, or finite, e.g. oil and gas. Continued use of these energy sources will eventually use them up, and they can't then be replaced.
- **renewable and sustainable.** These sources can be used again and again, e.g. solar, wind and wave power. This group also includes biogas, which is made from organic material like cow dung.

The two diagrams on the right show the current sources of the world's energy, plus how quickly those sources are increasing in use. Solar power is the biggest growth area, but from a very low starting point (see below).

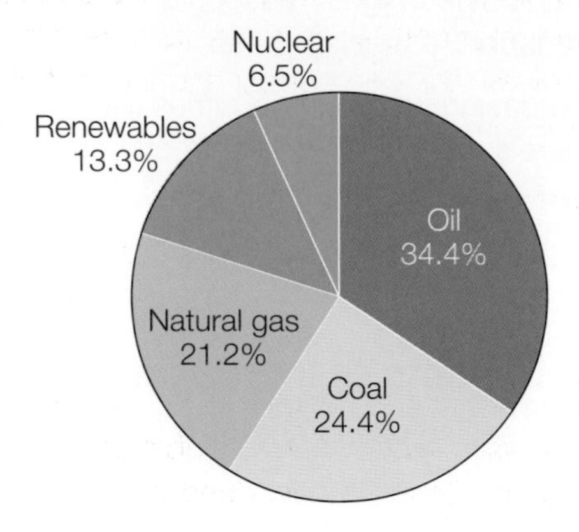

▲ *A breakdown of the world's energy sources*

▼ *Average annual growth rates for different energy sources, 2001-2006*

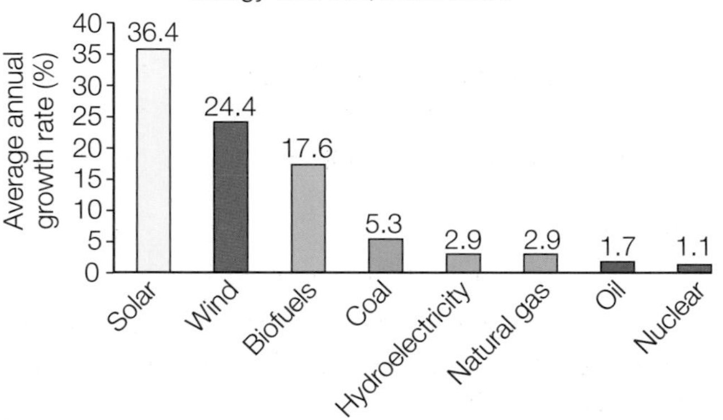

Is our use of energy sustainable?

The simple answer is No. On Spread 12.6, we saw how fast the demand for energy is expected to rise. This is going to cause a real problem. Although the pie chart shows that renewables as a whole provide 13.3% of our global energy needs, the amount provided by solar power, for example, is tiny – only 0.04%. And oil, which currently provides 34.4% of our energy needs, is eventually going to run out. So, what can be done to help?

Turning to renewable energy

India and energy fact file

- Population: 1.16 billion
- Access to electricity: urban areas = 93% of people; rural areas = 50%.
- About 400 million Indians don't have any access to electricity at all.

Energy source	%
Coal, peat	40.8
Combustible renewables and waste (e.g. firewood and cow dung)	27.2
Oil	23.7
Natural gas	5.6
Hydroelectric	1.8
Nuclear	0.7
Other renewables	0.2

▲ *The sources of India's energy in 2007*

▼ *Rural areas like this village in northern India traditionally rely on firewood as a source of energy*

Development project

ASTRA stands for the Application of Science and Technology in Rural Areas, and has been a major development project in India. Its researchers visited villages like the one opposite to find out about rural people's lives.

Problems

- They found that most rural families spent hours doing routine things like collecting fuel and water, and preparing and cooking food.
- The most commonly used fuel was wood. But, with an increasing population, suitable firewood was becoming scarcer. So dried cow dung was often burnt instead.
- Women and girls did most of the domestic work, which left little time for helping in the fields, or for school.

The solution

Cow dung! Instead of just burning the dung like wood, it can be used to produce **biogas**. This gas can then be used for cooking and powering electricity generators.

Benefits

Biogas has brought a range of benefits:

- Women and children gain 2 hours a day, because they don't have to collect firewood. So children have more time to go to school.
- About 80% of families use this time to earn extra money.
- The slurry that's left after fermentation is used to increase crop yields, because it's high in nutrients.
- Cattle are kept in compounds to make collecting the dung easier. So they don't graze in the forest and eat all the vegetation.
- Many villagers use the biogas to power electricity generators. These generators can be used to pump water up from boreholes for domestic use and to irrigate crops, as well as to generate electricity for light.

◀ *Burning wood and cow dung on traditional stoves affects women's health – causing lung and eye problems.*

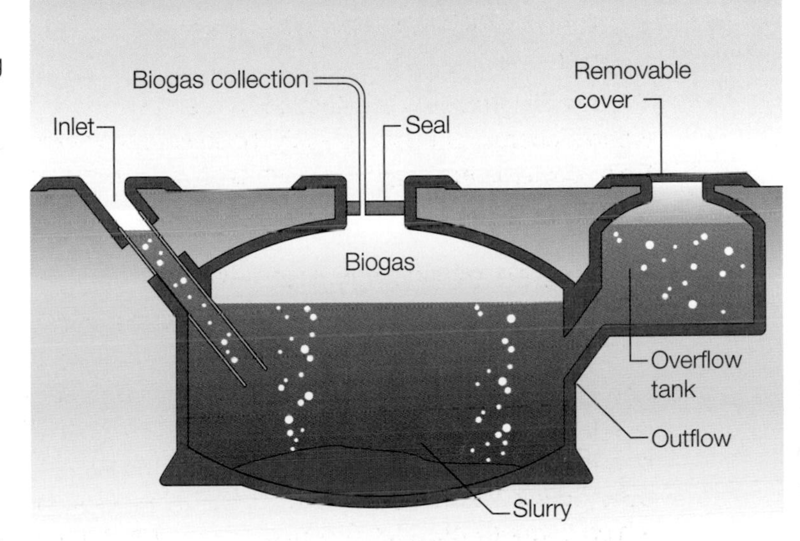

▲ *A biogas plant. The dung is collected up and fed into a lined pit that forms part of a biogas plant. The pit is then sealed by a metal dome, leaving the dung to ferment and produce methane gas, which is then piped into people's homes.*

Sustainable development

The Indian biogas plants are an example of **sustainable development**. They meet the needs of people today without compromising the ability of future generations to meet their own needs. Across India, 4 million cow-dung biogas plants have been built so far, and 200 000 permanent jobs created.

YOUR QUESTIONS

1 Describe the difference between renewable and non-renewable sources of energy, and give three examples of each.

2 a Use the figures in the table opposite to draw a pie chart of India's energy sources.

Hint: First, add up all of the percentages for the different types of renewables, to create a total renewables percentage.

b Compare your Indian pie chart with the one showing the world's energy sources.

Hint: 'Compare' means identify the similarities and differences between the two pie charts.

3 Do you agree with the view that our use of energy is not sustainable? Explain your answer.

On this spread you'll explore how the costs of globalisation can be reduced to ensure that the environment is protected, and that there are enough resources left for future generations.

The costs of globalisation

Globalisation has led to:

- increasing demand for goods
- increasing consumption of resources, such as energy and water
- the creation of lots of waste
- pollution of air, sea and water
- global warming and climate change, due to the burning of fossil fuels.

Reducing the costs – local level

Reduce, re-use, recycle

- Landfill sites in the UK are running out of space. For example, Londoners alone produce 3.4 million tonnes of rubbish a year. Incineration (burning) reduces waste by 75% in weight and 90% in volume. But incineration and landfill both release greenhouse gases into the atmosphere.
- Recycling is cleaner, greener – and provides new raw materials. But collecting, sorting, and processing recyclable waste still uses more energy than simply generating less waste in the first place!
- Composting produces humus that improves soils.

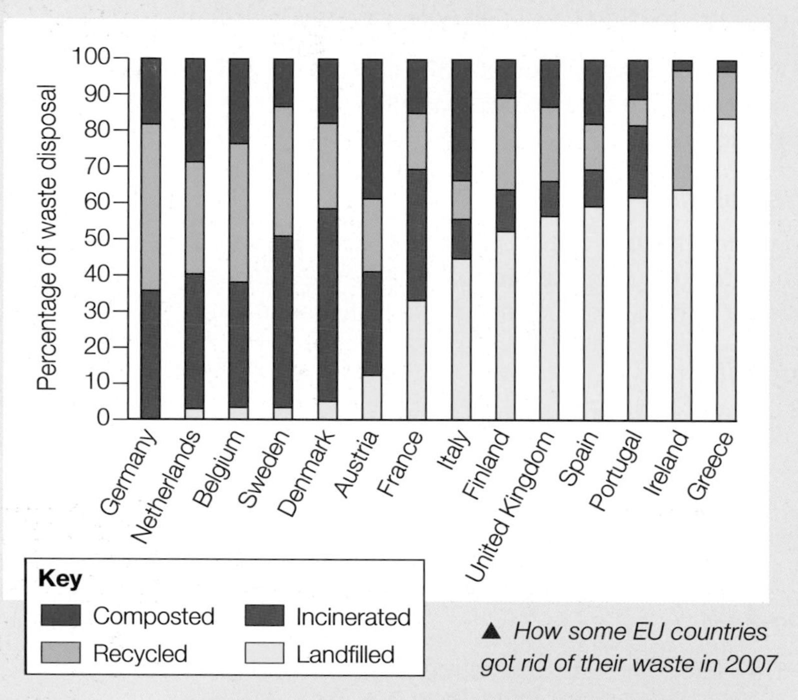

▲ *How some EU countries got rid of their waste in 2007*

At least 80% of our rubbish in the UK could be reused, recycled or composted. The graph shows how the UK's waste disposal in 2007 compared with other EU countries. For instance, Germany used no landfill disposal at all and recycled or composted about 65% of its waste, while Greece put most of its waste into landfill sites and only recycled or composted about 17%.

In 2007, the UK was composting or recycling about 35% of its waste. Under EU rules, the recycling rate in member states has to rise to 50% by 2020. This target has been adopted in England and Northern Ireland, but the devolved governments of Scotland and Wales have set ambitious targets of recycling 70% of their waste by 2025.

Recycling energy and waste – CHP

Electricity power stations waste 65% of the heat they generate, but **Combined Heat and Power** plants (**CHP**) can be so efficient that they only waste 5%. CHP captures and recycles the waste heat produced in electricity generation.

CHP plants:

- still often use fossil fuels, but – because they're so efficient – they cut emissions overall
- can use different fuels in the same boiler, e.g. straw and wood pellets alongside fossil fuels
- can also burn waste and convert it into energy.

Reducing the costs – international level

The Kyoto Protocol and Copenhagen Accord

The Kyoto Protocol was a global agreement made in 1997 to reduce greenhouse gas emissions. The targets for reducing emissions varied between different groups of countries. In 2009, world leaders met at Copenhagen to try to agree a new deal on climate change, because the Kyoto Protocol's targets were due to expire in 2012. The result of that meeting was the Copenhagen Accord, which:

- recognised the need to limit the global temperature
- promised to give billions of dollars of aid to developing countries to help them deal with the impacts of climate change.

Many people were disappointed that the deal wasn't legally binding, and that it didn't include targets to cut greenhouse gas emissions. Instead, countries were given a deadline to spell out their plans to cut emissions. 55 countries met the deadline. Some examples of the targets put forward include:

▲ *Changing the way we produce electricity – like this solar power station in California – is one way of reducing carbon emissions*

- The USA – to reduce emissions by 17% below 2005 levels by 2020.
- The EU as a whole – to reduce emissions to 20% below 1990 levels by 2020.
- Brazil – to reduce emissions growth by 36-39% below 'business-as-usual' levels by 2020.

Carbon credits

This is a system with the aim of reducing greenhouse gas emissions. Companies buy credits that allow them to emit a certain amount of carbon. The idea is that the cost of buying the credits will encourage them to produce fewer carbon emissions. Carbon credits can also be bought or sold if companies produce more, or fewer, emissions than they planned.

▶ *People or organisations can also volunteer to offset the carbon dioxide pollution they create – for example, by paying for trees to be planted. But offsetting has been criticised, because it doesn't encourage people to reduce their emissions.*

YOUR QUESTIONS

1. Explain these terms in your own words: CHP and carbon credits.
2. In small groups, discuss the advantages and disadvantages of the ways in which the costs of globalisation mentioned on this spread can be reduced. Produce a table listing the main advantages and disadvantages.
3. Work with a partner to draw up a spider diagram of all the things that you and your families could do to reduce your use of resources.
4. How effective do you think the Copenhagen Accord will be in reducing carbon emissions? Explain your answer.

On this spread you'll learn about some of the impacts of our increasing demand for food.

Food crisis

In 2008, a TV news programme reported that global food prices had almost doubled since 2005, e.g. the price of rice (a staple food) had gone up by 70%. People all over the world were facing a food crisis. Why?

- Extreme weather had destroyed crops across the world, e.g. grain harvests in 2007-8 were down 60% in Australia and 10% in China. A reduced supply of food meant higher prices in the shops.
- The growing global population was demanding more and more food.
- The diets of many people in countries like China and India were changing as they became wealthier. They were eating more meat instead of grains, which meant that cereal crops had to be fed to animals to produce meat, instead of directly to people.
- Less land was available to grow food crops, because of drought, deforestation, climate change and urbanisation. Also, crops like maize were increasingly being used to produce biofuel instead of food.
- High oil costs had forced up the price of things like fertiliser, food processing and transport – and these costs were added to the price of the food.

Increasing demand for food has a range of impacts – environmental, political, social and economic. Some of the impacts can be positive, but they often aren't.

▲ *People in a number of countries, including Haiti, rioted over rising food prices in 2008. Haitians usually spent 50-80% of their income on food. But, as prices rose in 2008, they were forced to spend almost all of their income just buying food.*

your planet
It takes 8 kilograms of grain to produce 1 kilogram of beef.

Environmental impacts

In poorer countries

The increasing need for food in some of the world's poorest countries has forced people to try to farm **marginal land** (land that's not really suitable for farming, but will just about do). The reasons why people are driven to farm marginal land include rising population (so more food is needed) and wars, which often drive people from their land into marginal areas. But farming these marginal areas has many negative environmental impacts:

- Growing crops uses up valuable nutrients, and poor soil becomes even less fertile. Eventually nothing will grow.
- Overgrazing by farm animals reduces the amount of vegetation, leaving soil exposed.
- Without vegetation there's nothing to hold the soil together, so it's blown or washed away.

▼ *Taking farm animals out to graze in Mali in the Sahel region (below the Sahara desert). This is very marginal land.*

Closer to home

Today's supermarkets can provide almost anything we want – whether it's in season or not. So, if you want strawberries in the winter or sprouts in the summer, that's no problem. But the food we eat has often travelled huge distances to reach us (by plane or ship). These distances are referred to as **food miles** (or kilometres), see the map. Flying food to the UK from all over the world adds a lot to our **carbon footprints**.

Carbon footprints are a measure of our carbon emissions. Aircraft emit a lot of carbon high up in the atmosphere, so they add a lot to our carbon footprints.

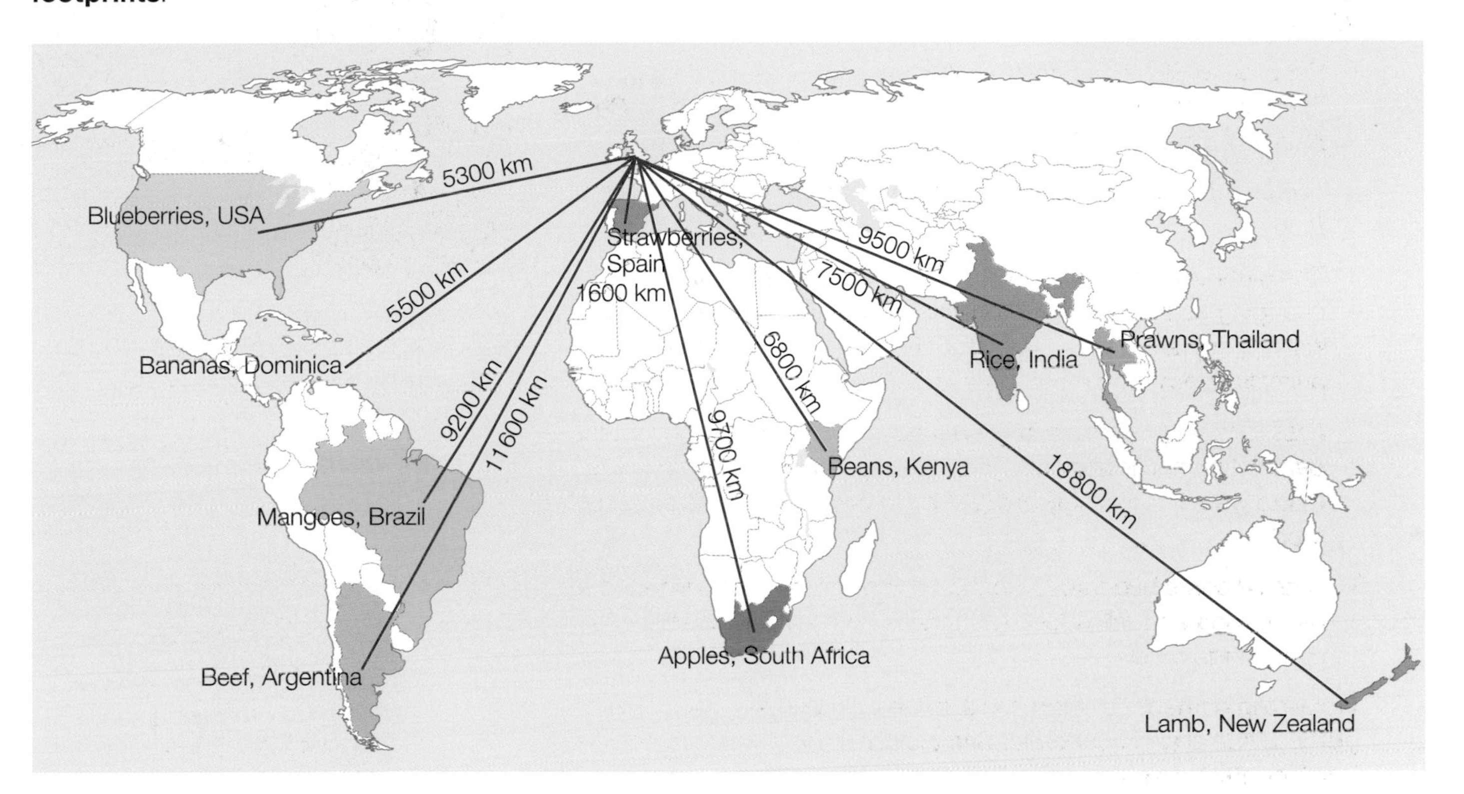

▲ *Transporting food long distances uses more oil and makes food more expensive*

Imported or home grown?

Take apples. The map above shows that in the UK you can buy apples from South Africa, which is a lot of food miles for a crop we can grow in the UK. Not only that, but imported apples have to be treated with chemicals – and waxed – to improve their appearance and make them last longer. So, is home grown better?

Apples grown in the UK in modern orchards are often treated with high levels of pesticides. But that's not all. Even though apples don't grow all year round, they can still be stored for up to 10 months using a combination of special gases and refrigeration. So the Bramley apple you buy in the spring could have been picked the previous year.

Locally produced food

The Campaign to Protect Rural England (CPRE) is just one of a number of organisations campaigning to encourage the use of locally produced food. They've been working with the government and supermarkets to encourage an increasing demand for, and supply of, locally produced food.

YOUR QUESTIONS

1 Explain to your partner what these terms have to do with increasing demands for food: marginal land, food miles, carbon footprint.

2 List four reasons for the food crisis identified in 2008.

3 Work with a partner to create a mind map of the environmental impacts of our increasing demands for food.

4 a Find out about the food that grows in your region. Do a Google search for British regional food or try your own region, e.g. south-west England.

b How could you persuade people to buy more locally produced food?

On this spread you'll find out about some of the political, social and economic impacts of our increasing demand for food.

Political impacts – battle for the Nile

Simon Kitra's front lawn in Uganda opens onto the world's longest river – the Nile. The water slips gently past on its journey to the Mediterranean. When Simon listens to his radio, all he hears is anger and disagreement about how the Nile's water should be used.

Negotiations and disagreement

For a decade, the countries located in the Nile Basin have been negotiating about how to share – and protect – the river at a time of changing climate and rising population. The talks eventually broke down in 2010. On one side are Egypt and Sudan, which rely heavily on the Nile's water because of their dry desert climates. On the other side are the remaining countries, which actually supply most of the Nile's water (particularly Ethiopia, which supplies 85%).

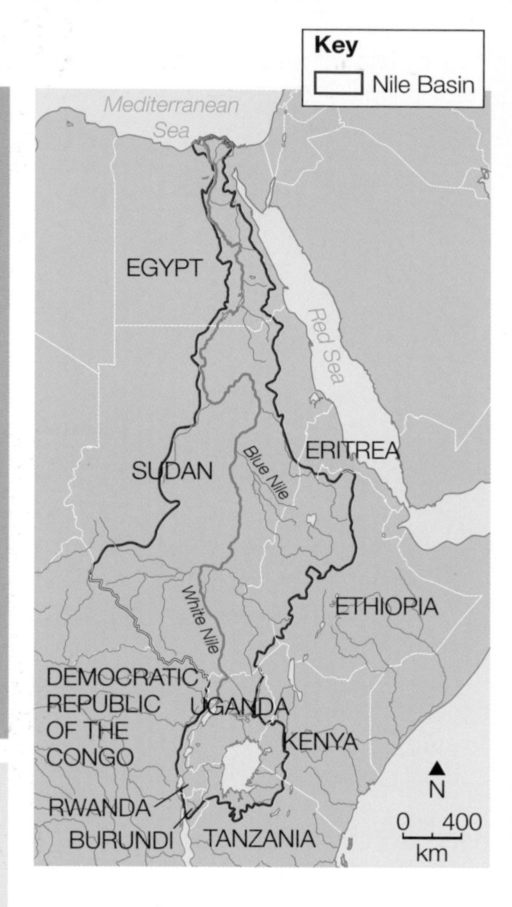

Uganda

- The Ugandan population is expected to triple by 2050, to 97 million – raising demand for food and water.
- The government's priority is to build dams to produce electricity, which will restrict the flow of water to the downstream countries of Sudan and Egypt.

Conflict

Henriette Ndombe is the director of the Nile Basin Initiative. She said that the breakdown of the Nile talks in 2010 was very serious, and could lead to a regional water conflict.

Sudan

- The Blue Nile and the White Nile meet in Sudan.
- Southern Sudan has swamps and rainforests, but a large part of the north is the Nubian Desert.
- Sudan is facing the problems of desertification and a falling water table.
- The Sudanese government wants to expand the use of irrigation to increase the food supply, which will mean taking more water from the Nile.

Egypt

- Egypt's population was 79 million in 2010, but is expected to rise to 122 million by 2050 – leading to a massive increase in demand for water and food.
- The country relies on the Nile for 90% of its water supply.
- The Nile's water is used to irrigate the farmland on either side of the river – allowing crops to be grown for domestic use and for export (to meet the growing global demand).

Ethiopia

- Ethiopia's population was 85 million in 2010, but a high population growth rate of 3.2% means that the population is expected to reach 150 million by 2050 – leading to greater demands for both water and food.
- With the pressure of its growing population, Ethiopia wants to keep more of the Nile's water for its own needs. This is likely to lead to serious disputes with Sudan and Egypt, if they receive less water as a result.
- Northern Ethiopia has a successful agricultural economy, which relies on irrigation to grow crops like coffee for export, as well as food for local people.
- The Ethiopian government wants to build big dams to create hydroelectric power, which can then be exported to neighbouring countries to generate much-needed income.

Social and economic impacts

Kenya

Subsistence farmers grow food to feed themselves and their families, with very little left over to sell. However, globalisation and the growing global demand for food has led more and more subsistence farmers to change over to farming cash crops instead. For example, if you look at where the green beans in your supermarket have come from, the chances are that they're from Kenya.

Growing cash crops should put more money into farmers' pockets, so that they can improve their families' quality of life. But there are downsides:

- Overseas supermarkets want reliable quality and supplies. What happens if there's a drought? The supermarket might change its supplier.
- If overseas sales fall for any reason (like supermarkets finding a cheaper supplier in another country), Kenyan farmers' incomes will also fall.
- If their incomes fall, they might not be able to afford enough food to feed their families – and, because they are now growing cash crops instead of subsistence crops, they can't rely on feeding themselves anymore.
- Growing cash crops requires good water supplies (not always reliable in Africa), and the use of fertilisers and pesticides (which are expensive and can lead to debt).

▲ *In 2010, The UK's Department for International Development set up a fund which gave a grant to Waitrose to encourage it to stock Kenyan beans. This should guarantee a regular income for Kenyan farmers.*

India

India has a big problem – how to feed its huge and growing population of 1.2 billion.

In the 1960s and 70s, farmers in the Punjab (India's most important agricultural region) began growing high-yielding varieties of wheat, rice and cotton, instead of their traditional mixture of crops. The new seeds produced larger yields than previous crops, but needed high inputs of fertiliser and water. As a result of this change to normal farming practice, there are now serious problems in the Punjab, and many other parts of India:

- In order to provide their new crops with enough water, farmers had to drill wells. But as the water was pumped out, the water table fell and the wells had to be drilled deeper and deeper. And that meant that farmers had to buy more-expensive and more-powerful water pumps.
- The high-yielding crops also drained the soil of nutrients, so farmers now have to use three times as much fertiliser as they used to, in order to produce the same amount of crop.

The high cost of the water pumps and the fertiliser has forced many Indian farmers into debt – and, for many of them, suicide is the only way out. The number of farmer suicides in India is increasing at an alarming rate.

YOUR QUESTIONS

1. Use the information on this spread and Spread 12.9 to design a poster about the impacts of the increasing global demand for food. You should include sections about the environmental, political and social impacts, and distinguish between positive and negative impacts.
2. Should Kenya carry on growing cash crops, or should its farmers go back to growing food for their own people? Explain your point of view.
3. In 1995, the Vice President of the World Bank predicted that in the twenty-first century wars would be fought over water, rather than land or oil. Do you agree with him? Justify your answer.

   Hint: 'Justify' means give evidence to support your statements.

What do you have to know?

This chapter is from **Unit 2 Human Geography Section B** of the AQA A GCSE specification. It is about the growth of tourism, its impacts, and managing tourism for the future. The table shows how the pages in this chapter match the content in the specification.

Specification content	Pages in this chapter
The global growth of tourism, where tourists go and why, and the economic importance of tourism.	p272-275
Tourist areas in the UK need to be managed well. • The economic importance of tourism to the UK. • What affects the number of visitors. • The life cycle of a tourist destination. • Strategies to ensure the success of tourism.	p276-279 Case study of tourism in the Lake District
Mass tourism in tropical areas has advantages and disadvantages. Strategies to maintain its importance, and reduce its impacts.	p280-283 Case study of mass tourism in Jamaica
The development of tourism in extreme environments can have negative impacts. How extreme environments can cope with the development of a tourist industry.	p284-287 Case study of tourism in Antarctica
Sustainability and ecotourism – stewardship and conservation, the benefits of ecotourism, and sustainable development.	p288-291 Case study of ecotourism in the Amazon Rainforest

Your key words

- Domestic destinations, short-haul destinations, long-haul destinations
- External factors
- Sustainable management
- Honeypot town
- All-inclusive hotel
- Mass tourism, charter flights, package holidays
- Economic leakage
- Carbon footprint
- Responsible tourism, ecotourism
- Grey market
- Polar Code
- Stewardship
- Ecolodge
- Sustainable development

Exam help ...

Advice See pages 297-299 for information on how to be successful in your exams.

Practice See page 312 for exam questions on this chapter.

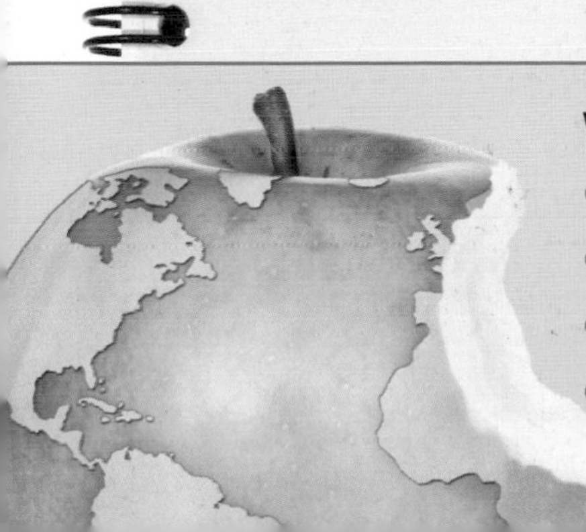

What if ...

- the Lake District was shut to tourists?
- no-one went on holiday?
- the only impact tourism had was a footprint?
- we all went to Antarctica on holiday?

13.1 » Holiday!

On this spread you'll find out about the different sorts of places where people go on holiday.

Where to go?

School's out! It's the end of term – and time for a break. What's your ideal holiday? If it's spills and thrills you're after, then perhaps you'd fancy Tema Park in Las Vegas. Or maybe a spot of extreme mountain biking? Perhaps a city break with some shopping and sightseeing is more your thing?

Tema Park and extreme mountain biking might be too much of an adrenalin rush for you, and might not be your average sort of holiday, but most of us go somewhere to escape. Our holiday destinations tend to be:

- cities, like New York and Paris
- beaches and beach resorts, like Palma Nova in Mallorca
- mountains and ski resorts, like Courchevel in the French Alps
- rural areas, like the Lake District (see pages 276-279).

▲ *Wish you were here?*

◀ *... or here? This is Tema Park, Las Vegas.*

Where do tourists visit, who goes there, and why? The table below gives some answers.

Environment	Examples	Who goes there?	Why?
Cities	London Paris New York Las Vegas Hong Kong	• Young adults (singles and couples) • Recently retired people	• Excitement • Shopping • Nightlife • Sightseeing and culture (art, food, theatre)
Mountains	Lake District Alps Himalayas Rockies Andes	• Sightseers • People who enjoy the natural environment • Mountain bikers, climbers, walkers, skiers	• Natural beauty • Physical challenge • Winter sports
Coasts	Blackpool Jurassic Coast (Dorset and Devon) Palma Nova Miami Beach Maldives	• Water sports enthusiasts • Young people • Young families • Retired people • Walkers	• Sun, sand and sea • Water sports and outdoor activities • Natural environment

For some people, cities and coasts are not enough – they like to go to extremes. Extreme environments are destinations that are very challenging (some people like that!). They can be very:

- cold, such as Antarctica
- hot and humid, such as the Amazon rainforest
- dry, such as the Sahara desert
- steep and rugged, such as the Himalayas.

How far do we go?

Holidaymakers visit destinations all over the world.

- **Domestic destinations** are located in the tourist's own country. So, for British people, they are destinations in the UK.
- **Short-haul destinations** can be reached by an air flight of less than 3 hours. For tourists from the UK, they are places in Europe and around the Mediterranean Sea.
- **Long-haul destinations** are further away and include tropical destinations in countries such as Jamaica, Kenya, and Thailand.

And where do we go?

- British people take 52 million domestic holidays each year, plus 47 million short-haul and 13 million long-haul flights.
- London is a top UK destination – half of all the money tourists spend in the UK is spent there. That's about £19 billion!
- The top overseas destination for British tourists is Spain, which accounts for a third of the 60 million visits that British people make abroad each year. That's nearly twice as many as France (the second most popular overseas destination for the British).
- The USA is Britain's favourite long-haul destination (5% of Britons holidaying abroad go there).

▲ *Sightseeing in New York. From one extreme …*

▲ *… to another. This is Antarctica.*

YOUR QUESTIONS

1 a What is a domestic destination?

b What is the difference between short-haul and long-haul destinations?

2 Choose one of the holiday destinations shown in the photos. Write a postcard to a friend to tell them what you think it's like.

3 Where have your classmates been on holiday?

a Conduct a class survey and record your results on a world map.

b What is the most popular destination? Why do you think that is?

4 These places are all tourist destinations: Aviemore, Benidorm, Blackpool, Klosters, London, Phuket, Rome, Rio de Janeiro.

a Where are they? (Use an atlas to help you.)

b Which are domestic, short-haul and long-haul destinations?

c Which are city, ski and beach destinations?

d Find out one thing you could do, or place you could visit, for each destination.

your planet

For $100 million you can book a seat on a space mission with Space Adventures Limited. How about that for a holiday?

your planet

Several tourists have already been to the International Space Station (350 km up).

13.2 » Global tourism is growing

On this spread you'll find out why tourism is growing, and how important it is in different parts of the world.

How much has tourism grown?

More and more people have been travelling – on holiday, on business, and to visit friends and relatives. Globally, tourism more than doubled in just 18 years – from 438 million tourists in 1990 to 922 million in 2008.

Why has tourism grown?

Tourism has grown, and people have been able to travel more, because they have:

- more leisure time – longer weekends and longer holidays
- more money – higher wages and more income available to spend on things like holidays
- cheaper travel costs – reduced airfares due to the growth of budget airlines, such as easyJet and Ryanair
- better health, especially in later life, so more older adults and pensioners now have adventurous holidays (they're often referred to as the 'grey market', spending 'grey pounds')
- the Internet, which has helped people to search for cheaper travel deals themselves, rather than use travel agencies.

Who goes where?

Tourists are people who travel to other places and stay there for a short time. Just over half of them travel for leisure, recreation and holidays. Most of the remainder are either business tourists (about 15% of the total), who travel for work, or those who travel for reasons like visiting friends and relatives, on religious pilgrimages, and for health reasons.

So, where does everyone go? The ten most popular countries in 2008 are listed in the table.

Rank	Country	Millions of tourists
1	France	79.3
2	USA	58.0
3	Spain	57.3
4	China	53.0
5	Italy	42.7
6	UK	30.2
7	Ukraine	25.4
8	Turkey	25.0
9	Germany	24.9
10	Mexico	22.6

your planet

In 2008, the UK was ranked seventh in the world in terms of the money it earned from tourism. It was behind the USA, Spain, France, Italy, China, and Germany.

Where tourism really matters

Tourism matters a lot to many of the world's poorer countries. It can make a big difference to their economies, by bringing jobs and income to places where it's badly needed. Money from tourism can also pay for big **infrastructure** projects, like new roads and bridges and improved water supplies and energy systems. All of these things can help a country to develop.

The percentage of a country's **GDP** provided by tourism shows how important tourists are to the national economy. GDP stands for gross domestic product. It is the total value of all the goods and services provided by a country in a year. GDP is a measure of how well off a country is.

The Caribbean

The islands of the Caribbean include some of the world's poorest countries, like Haiti. Some islands, like St Lucia, rely heavily on tourism – it accounts for 37% of their GDP. However, in Haiti, tourism provides only about 7% of GDP.

Europe

The economic importance of tourism varies across Europe. Some countries receive a lot of tourists, so tourism is economically important to them, e.g. it provides more than 16% of Spain's GDP, but less than 6% of Romania's.

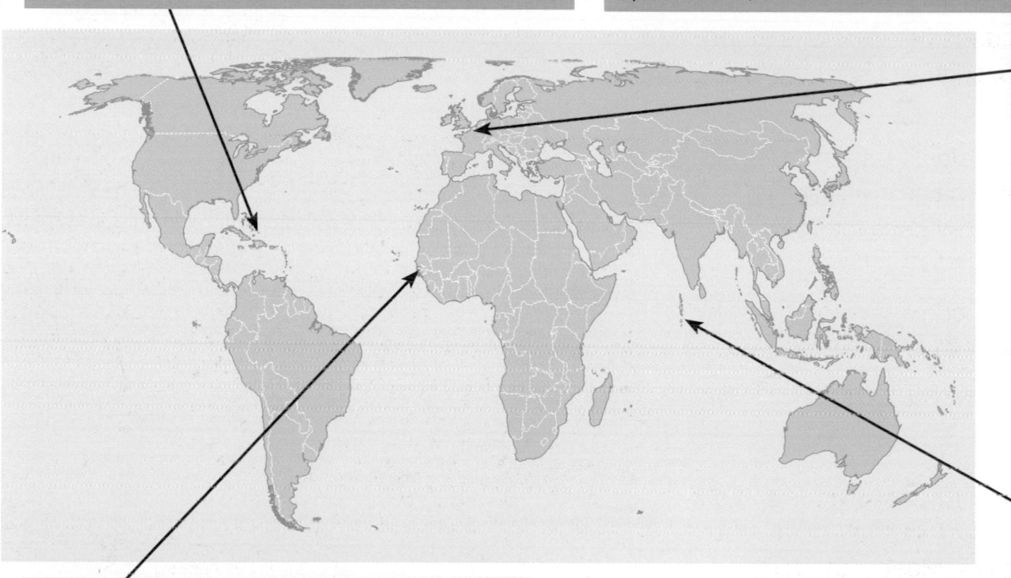

Gambia

Tourism provides 17% of Gambia's GDP. That's very high for Africa, where tourism often accounts for less than 10% of GDP.

Maldives

Tourism is very important to the Maldives, making up almost 30% of its GDP.

▲ *The economic importance of tourism in different parts of the world*

YOUR QUESTIONS

1 What do these terms mean: infrastructure, GDP?

2 a How much did tourism grow between 1990 and 2008?

b Give two reasons why tourism grew so much.

3 Look at the table of the ten most popular countries for tourists in 2008. Choose one of them and prepare a PowerPoint presentation to explain why it's so popular with tourists. You'll need to do some research to find out more about it.

4 Why do the governments of some of the world's poorer countries want tourism to grow even more?

Hint: Think about the benefits that tourism can bring.

5 Almost 30% of the economy of the Maldives depends on tourism. Suggest possible reasons why some of the people who live there are pleased to have so much tourism, while others are not so sure.

On this spread you'll find out why the Lake District is a popular UK tourist destination.

Why is the Lake District so popular?

The Lake District National Park is in Cumbria. It's one of 15 National Parks in the UK, and is the third most popular (after the Peak District and the Yorkshire Dales). Over 8 million tourists a year visit the Lake District, and below are some reasons why it's so popular.

your planet

Lake Windermere changes depth, depending on the weather. In January 2005, it rose by a metre overnight (that's equal to 17 thousand million litres of water) because of heavy rain!

Scenery

- The Lake District is a region of beautiful, mountain scenery.
- Among the hills (fells) are numerous lakes.
- The highest mountain is Scafell Pike (977 m) and Windermere is England's longest lake (17 km).

Activities

- Outdoor activities include walking, climbing, sailing, fishing and canoeing.
- Other activities include cruising on the lakes and visiting picturesque towns, such as Keswick and Ambleside.

Heritage

Famous writers, including William Wordsworth and Beatrix Potter, lived in the Lake District. Their homes are now visitor attractions. Wordsworth's guide to the Lake District began to attract tourists there when it was published in 1810.

Transport links

- Tourists travelling to the Lake District from Manchester use the M6.
- The West Coast Main Line railway between London and Glasgow passes close to Kendal, and goes on to Penrith and Carlisle.

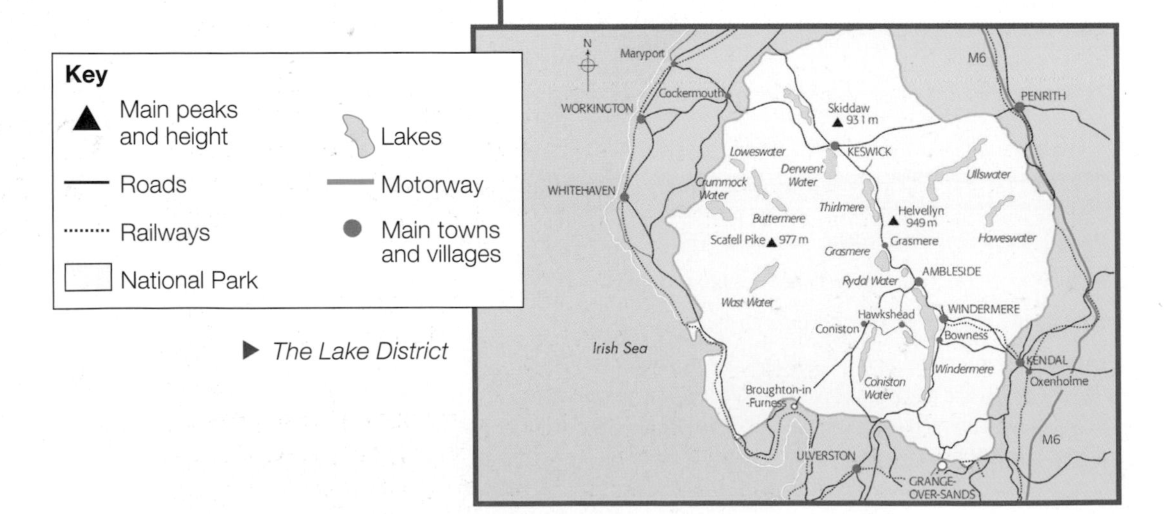

▶ *The Lake District*

How important is tourism?

Tourism is very important to the British economy. More than 1.3 million people (that's almost 1 in 20 of the total working population) work in tourism in the UK. Over 8% of the UK's GDP depends on it.

In the Lake District:

- tourism employs 20 000 people full time – and another 35 000 in seasonal jobs.
- visitors spend more than £600 million per year – nearly £80 each. The pie chart shows where the money goes.
- Tourism keeps services like shops, post offices and buses busy.
- Tourists create a demand for food, helping to keep local farmers in business. Farmers also make money by turning barns into holiday apartments and using fields as campsites.
- Money from car park charges, and taxes paid by tourism businesses, help local councils to provide better services for local people, such as better roads, leisure centres, and play facilities for children.

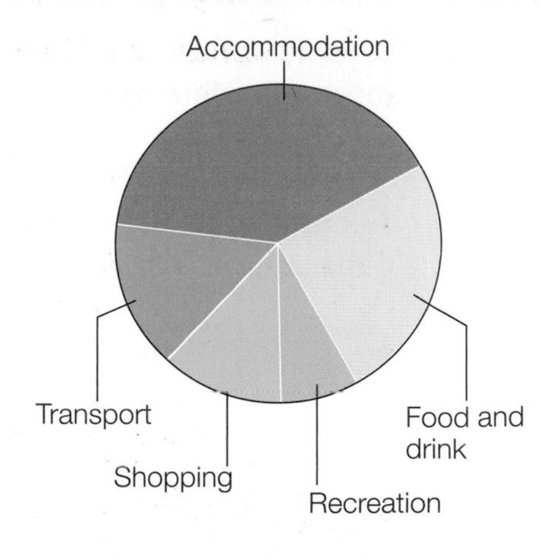

▲ *How tourists to the Lake District spend their money*

What's the problem with lots of tourists?

Tourists bring benefits to the Lake District. But they also bring problems:

- Tourist cars and coaches cause traffic congestion in towns with old, narrow streets. Over 80% of tourists use cars and less than 5% use public transport.
- Footpaths are worn away when too many people use them. The ground becomes hard and bare as plants are trampled underfoot.
- Many tourism jobs in the Lake District are seasonal, so a lot of people are unemployed in the winter. Many are also low-paid and part-time.
- Properties for sale are bought up as holiday homes for tourists, which pushes prices up, so that local people can't afford to buy their own homes.
- 15% of homes in the Lake District are second homes. This has a disastrous effect on businesses like pubs and local shops, which might be forced to close – destroying communities.

▲ *Too many walkers and mountain bikers can cause huge damage*

What affects visitor numbers to the UK?

Many overseas tourists come to the UK. However, their numbers vary from year to year, depending on **external factors**, such as:

- **exchange rates**. Overseas tourists have to change their own currencies into pounds when they visit the UK. The exchange rate for their currency decides how many pounds they will get for their money. If the exchange rate means more pounds, more overseas visitors are likely to come to the UK, because it will seem cheap.
- **security**. Terrorism is a big issue for many people. When terrorist attacks occur in the UK, they are reported around the world and can put people off coming here, because they think it's too dangerous.
- **the state of the global economy**. When the global economy is doing well, more people travel and visitor numbers go up. When it isn't, people don't spend as much on foreign holidays and visitor numbers go down.

YOUR QUESTIONS

1 Draw two spider diagrams to show both the benefits and the problems that tourism brings to the Lake District.

2 Design a poster to attract visitors to the Lake District. Focus on its scenery, activities or heritage.

Hint: You might want to do some more research to add more information to your poster. Try this website to begin with: www.lakedistrict.gov.uk

3 An average visitor to the Lake District spends about £80.

- **a** What do they spend their money on?
- **b** Suggest some jobs likely to be created by this spending.

On this spread you'll find out how the Lake District could cope with large numbers of tourists and still make sure that tourism is successful.

What's the Lake District's future?

As the table shows, the number of visitors to the Lake District has generally been rising since 2000. But what will happen in the future?

Year	2000	2002	2004	2006	2008
Million tourists	7.7	8.0	8.2	8.1	8.3

▲ *Visitors to the Lake District*

The life cycle of a tourist destination

The diagram below is a model of the life cycle of a tourist destination. It shows how the numbers of tourists visiting a destination can be expected to change over many years.

Tourism to the Lake District grew slowly following the publication of William Wordsworth's *A Guide to the Lakes* in 1810. In the 1840s and 1850s, railways reached Kendal, Coniston and Windermere. People from big cities like Manchester were now able to reach what had previously been a wild and remote area – and the number of tourists quickly grew. When the Lake District became a national park in 1951, and the M6 motorway was built through Cumbria in the late 1960s and early 1970s, tourism continued to increase.

So, what will happen in the future? It's likely that the Lake District will continue to attract tourists, because:

- more people from the UK are staying in this country for their holidays
- the Lake District is easy to travel to for a day visit, a short break, or a full holiday.

That means that the Lake District will have to cope with increasing numbers of visitors, both from the UK and abroad. The next page details some different ideas about how this could be achieved.

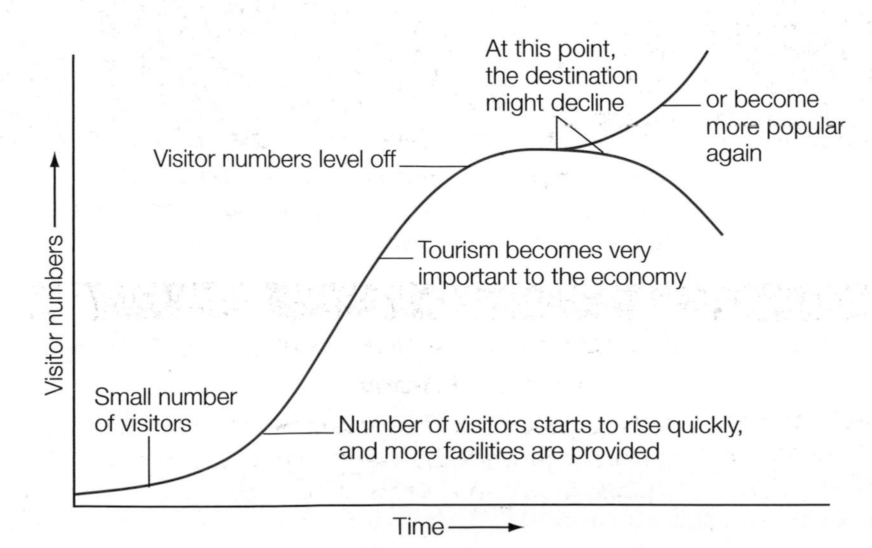

◀ *How tourist destinations change over time*

How can tourism still be successful?

Two organisations, working for the future success of tourism in the Lake District are the Lake District National Park Authority (LDNPA) and Cumbria Tourism. Together they have a 'Vision for the National Park in 2030'. They plan to:

- promote the Lake District around the world to bring in more tourists and increase the economic benefits of tourism
- work with businesses to improve the quality of visitor accommodation, attractions and facilities
- persuade visitors to stay longer, so they spend more money
- encourage young people to enjoy the Lake District, so they return later with their own children.

By doing these things, the LDNPA and Cumbria Tourism want to make sure that tourism in the future is managed successfully by bringing economic benefits to the area without damaging its special environment and heritage. They want tourism to grow and succeed **sustainably**.

How could the Lake District cope in the future?

Some ideas for coping with large numbers of tourists	Would it work?
A National Park entry charge Vehicles could be charged to enter the National Park. This happens in the USA and Australia.	It would probably reduce the number of cars and the congestion they cause. It would also raise money to help maintain the National Park. ***But*** it could discourage visitors – meaning less money for tourism businesses.
Limit visitor numbers *Either* limit the number of visitors allowed into the Park, *or* stop mountain biking and large groups of walkers using some footpaths.	It would reduce the environmental impact of tourism. ***But:*** • it is difficult to monitor • it limits people's freedom.
Repair worn-out footpaths This could be done using local stone and other natural materials.	It would keep eroded and popular footpaths open, and stop even more damage. ***But:*** • it's expensive work, unless it's done by volunteers • stone footpaths look odd.
Build bypasses around congested towns Ambleside is badly affected by traffic congestion (see below). A bypass road could be built around the town.	It would reduce traffic jams and pollution in crowded town centres. ***But*** Ambleside is in a narrow valley, so building a new road around it would be difficult and expensive.
Improve public transport • Operate more buses and trains. • Make the railway along the west coast of the Lake District twin track instead of single track.	Making the railway twin track would cut the journey time, so tourists are more likely to use it. ***But:*** • operating more buses and trains may need subsidies at first • doubling the railway tracks would be very expensive.

▲ *Traffic congestion in Ambleside – a* ***honeypot*** *town in the Lake District* *(see page 282)*

▲ *Encouraging more tourists to visit the Lake District in winter could help to tackle seasonal unemployment. But it can be very cold and snowy at that time of year!*

YOUR QUESTIONS

1 How would you persuade more tourists to visit the Lake District in winter?

2 How well would the following ideas for coping with the impact of large numbers of tourists work in the Lake District? Explain your answers.
- Park-and-ride schemes
- Hotels running minibuses for their customers
- Having fewer car parks in towns

3 What do you think should be done about the Lake District? Produce a PowerPoint presentation to explain your favourite ideas for coping with increasing numbers of tourists there.

13.5 » Jamaica – totally tropical tourism

On this spread you'll find out what mass tourism is, and about the economic effects of tourism on Jamaica.

What's Jamaica like?

What's the connection between Bob Marley and Usain Bolt? Both came from Jamaica – a tropical Caribbean island. Not only has Jamaica been home to some famous people, but it's also one of the Caribbean's top tourist destinations. One reason for that is its hot, tropical climate. At sea level it's hot all year, and there's plenty of sunshine too (at least 7 hours a day). But, like all places with a tropical climate, Jamaica does have rain – as the climate graph shows.

▲ *Sun, sea and sand – who wouldn't want to go to Jamaica?*

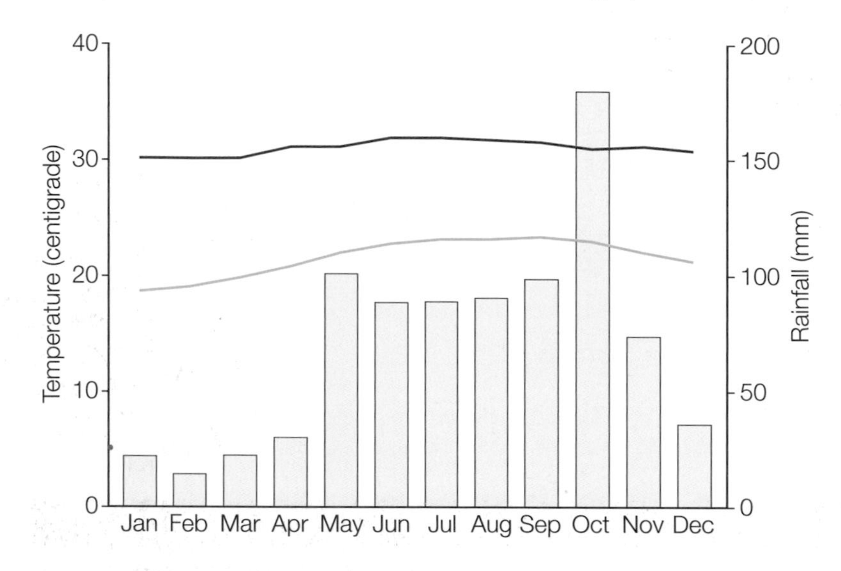

Key
—— Average daily temperature (max)
—— Average daily temperature (min)

◀ *A climate graph for Montego Bay, Jamaica*

Jamaica and tourism

Kingston, on the south coast, is Jamaica's capital, but most tourists stay on the north coast – in resorts such as Ocho Rios, Montego Bay and Negril. Here the beautiful sandy beaches have been developed into tourist resorts. Many of the hotels are **all-inclusive**.

All-inclusive hotels provide tourists with accommodation, meals, entertainments, drinks and activities for one all-inclusive price. The hotels are set in their own grounds, usually with private beaches and swimming pools. Apart from the people who work there, the only people allowed into the hotels are their customers. Many never venture out of the hotel.

What is mass tourism?

Mass tourism is when large numbers of tourists visit the same destination. Holiday companies arrange special flights, called **charter flights**, to transport them. Many holidays like this include flights, airport transfers and accommodation (plus some meals) as a package, so they're called **package holidays**.

Most mass-tourism package holidays are to short-haul destinations, such as resorts in Spain. But long-haul package holidays to tropical destinations like Jamaica have become more popular since the 1980s. Nearly 1.8 million tourists visited Jamaica in 2008, compared with 0.6 million in 1982 – and only 0.3 million in 1966.

Because of the large numbers of tourists involved, mass tourism can have major effects (both good and bad) on tourist destinations and the people who live in them.

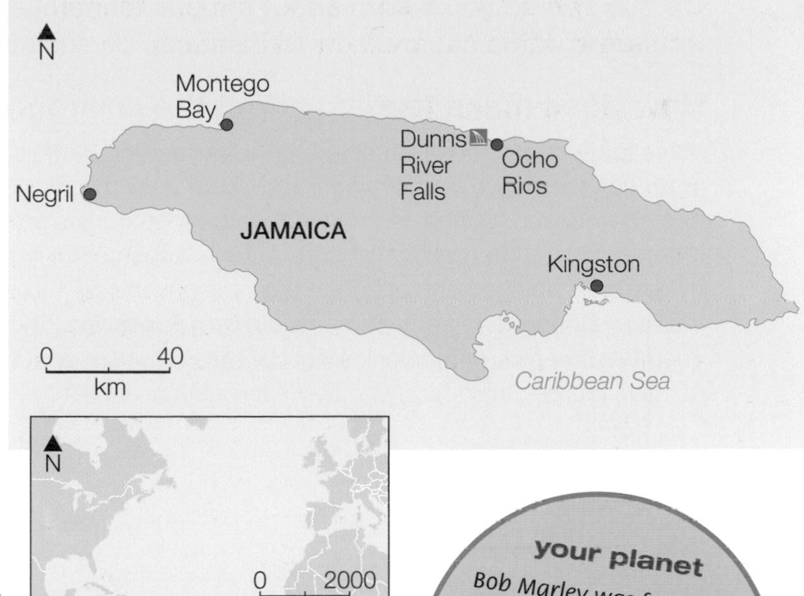

your planet

Bob Marley was famous for his Jamaican reggae music. Now Levi Roots is famous for his Reggae Reggae sauce. He's from Jamaica and loves to cook food the Jamaican way.

How does mass tourism affect Jamaica's economy?

Tourism brings in a lot of money for Jamaica – about 20% of Jamaica's GDP in 2009. In 2008, tourists spent nearly $2 billion there.

Positive economic effects	Negative economic effects
• The money spent by tourists makes tourism businesses, like hotels, profitable. • Those tourism businesses employ many local Jamaican staff. • The Jamaican tourism workers spend their wages in other Jamaican businesses, which in turn become more profitable and employ more local staff. • The taxes paid to the Jamaican government by businesses, workers and tourists provide money which helps Jamaica to develop. • Jamaicans learn skills in the tourism industry that can be used in other parts of the economy. • Many tourism jobs pay well by Jamaican standards. • Tourist resorts and the people who live there become richer.	• Many tourism businesses are owned by foreign companies, so most of the profits end up abroad. This is called **economic leakage**. • Some tourism staff are foreigners. They also send their wages home. This is economic leakage too. • Economic leakage also means less tax revenue for the government to develop Jamaica. • Jobs in tourism are often seasonal. • Some skilled Jamaicans leave to work abroad for more money. • Tourist destinations attract Jamaicans from poor inland areas, where businesses lose out. • Tourist jobs and money are concentrated in the resorts, so inequalities with other parts of the country increase.

YOUR QUESTIONS

1 Explain these terms: all-inclusive, mass tourism, charter flights, package holidays, economic leakage.

2 a Use the climate graph to describe Jamaica's climate.

b Explain why Jamaica's climate attracts British holidaymakers.

3 Work as a group of four. Each adopt one of the following roles:

- Myron – a Jamaican hotel receptionist
- Larry – a Jamaican hotel manager
- Leon – a Jamaican shop owner in an inland town
- Tom – a tour guide from Canada

You have been asked to contribute to an online discussion forum about whether more tourism would be good for Jamaica's economy. What will you say? Write a 2-minute podcast of your views.

On this spread you'll learn about the environmental effects of mass tourism in Jamaica, and how tourism can be sustainable.

How does mass tourism affect the environment?

Mass tourism has mostly affected the areas around Jamaica's north coast resorts, east of Negril and as far as Ocho Rios. They've become built-up, congested and polluted. Most tourists arrive at Montego Bay's international airport. Their environmental impact on Jamaica begins here, although their journey this far has already made a difference to each person's **carbon footprint**. The diagram illustrates the environmental impacts of mass tourism on a taxi minibus journey from Montego Bay international airport to Negril.

Carbon footprint

A carbon footprint is a measure of the amount of carbon someone's lifestyle adds to the atmosphere, and travel is part of that. A tourist visiting Jamaica from the UK will have travelled to a UK airport and then flown across the Atlantic Ocean. Both parts of the journey will have emitted carbon dioxide and other greenhouse gases into the atmosphere.

▼ *Tourism's environmental trail*

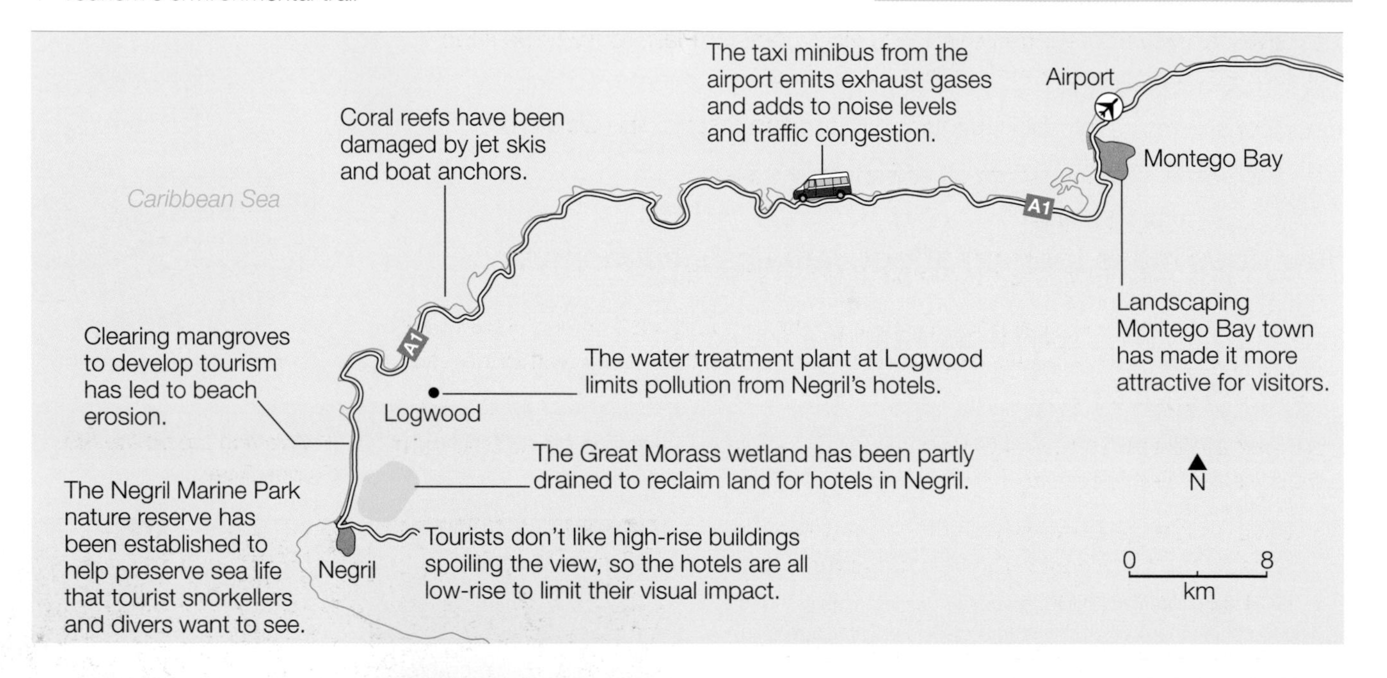

▶ ***Honeypots** are a consequence of mass tourism. People swarm like bees to beautiful attractions like Dunn's River Falls in Jamaica. Do you think they spoil it? Some people do.*

How can tourism grow and be sustainable?

Jamaica is a relatively poor country – its GDP is only about US$8000 per person per year, compared with the UK's US$36 000. Tourism accounts for 45% of the money Jamaica earns from abroad (almost $2 billion in 2008). So, increasing tourism could help to raise the standard of living of Jamaica's people.

However, mass tourism also has some negative effects, as you've already seen. Many Jamaicans have come to dislike it. They don't see how it helps them or the country. So, there is a puzzle for the government to solve – how can they develop tourism but avoid its negative effects?

They need to find sustainable solutions – ways for tourists to visit Jamaica without damaging its future and the future of its people.

What's the difference between responsible tourism and ecotourism?

Both are about tourism helping local people, while not harming the environment. But, with ecotourism, people visit places because of their natural environment, whereas responsible tourism can happen anywhere – a city or a holiday resort, for example.

What's been tried?

The Jamaican government has been following a Master Plan, to try to develop sustainable tourism. Its three main ideas have been to:

- limit the development of mass tourism to existing resorts, like Ocho Rios
- spread small-scale tourism to other parts of the island
- involve local people more.

As part of its Master Plan, the government has encouraged:

- community tourism – local people running small-scale guesthouses. This helps to bring tourists to less-developed towns, such as Port Antonio, without mass tourism's negative effects.
- **responsible tourism**, which involves local people and aims to do as little harm as possible. For example:
  - Local guides take visitors to off the-beaten-track attractions, such as the Rio Grande River (pictured on the right).
  - Tourists are encouraged to buy local food and crafts from Jamaican traders.
  - Smaller inland hotels employ local staff and use locally grown food.
- tourists and local people to get in touch with each other through the Jamaican Tourist Board's 'Meet-the-People' website initiative. This helps both visitors and locals to understand each other, which is an important benefit of tourism.
- educating tourists and locals about how to avoid negative environmental effects.

▼ *A rafting trip on the Rio Grande River*

YOUR QUESTIONS

1. What do these terms mean: carbon footprint, honeypots, responsible tourism?
2. Use the information opposite to draw up a table showing the positive and negative effects on the environment of a holiday trip to Negril.
3. Design a poster to promote sustainable tourism in Jamaica.
4. **a** Explain why the Jamaican government thinks community tourism is a good idea.
   b Suggest why some people favour more mass tourism.

On this spread you'll find out why more tourists are visiting extreme environments, like Antarctica.

What are extreme environments?

Extreme environments are places where people find it very difficult to live. They can't farm there. There are no cities. They're wild and inhospitable. They're places like mountains, deserts, rainforests – and Antarctica.

What's extreme about Antarctica?

- **Where it is.** Antarctica is centred around the South Pole, so you can't get more extreme than that! It's surrounded by the Antarctic (sometimes called the Southern) Ocean.
- **Its size.** Antarctica is a continent. It has an area of 5 million square miles (8 million km^2). That's one and a half times the size of the USA!
- **The emptiness.** No-one lived there at all until 1897. Hardly anyone lives there now, except scientists living in research stations. There are about 50 research stations dotted about. In summer, the largest – McMurdo – is home to about 1000 people, which drops to 200 or so in the winter.
- **The cold.** The temperature is generally below freezing. Incredibly cold temperatures have been recorded inland, such as -60°C! On the coast, temperatures can sink to -30°C, but it can also warm up in summer – sometimes as high as freezing point (0°C)!
- **The wilderness.** There are hardly any people and hardly any buildings (outside the research stations). The natural, largely white, landscape is home to wildlife like penguins, especially along the icy coastline.

your planet

Penguins may be birds but they can't fly. It's the males that act as 'mum' and look after the eggs. They keep them warm by balancing them on their feet and covering them with their belly flap.

Why are more people going to Antarctica?

In 1992, 6700 tourists visited Antarctica. By 2009, that figure had jumped to 45 000 – more than 6 times as many! But it's not just Antarctica that's proving popular. As remote parts of the world become more accessible, people are becoming more adventurous and the demand for adventure holidays in extreme environments is increasing. Below are some of the reasons why more people are visiting Antarctica.

The growing popularity of ecotourism – people want to visit wild places because of the attraction of their natural environments and unique wildlife, while causing as little environmental impact as possible.

The wildness of the environment – Antarctica is one of the few unspoilt places left on Earth.

The comfort factor – nowadays, tourists to Antarctica don't have to be superfit and intrepid, like the explorers of the past. They can marvel at the amazing scenery and incredible wildlife from the comfort of their cruise ships, as they journey along the Antarctic coastline – occasionally going ashore in small boats to get closer to nature.

Financial factors – there is a growing market of younger single people with high incomes, plus a **grey market** of older people (often recently retired), who can afford to pay for tours to extreme environments like Antarctica.

Ease of access – more tour operators are running more trips to satisfy the growing demand

What do tourists to Antarctica do?

As well as sightseeing from cruise ships and landing to camp onshore, tourists:

- fly over the ice in light aircraft and helicopters
- climb rock and ice faces
- cruise inlets in small boats
- visit scientific research stations
- hike
- scuba dive under the ice (see right)
- kayak
- explore the shallow sea bed in underwater vehicles.

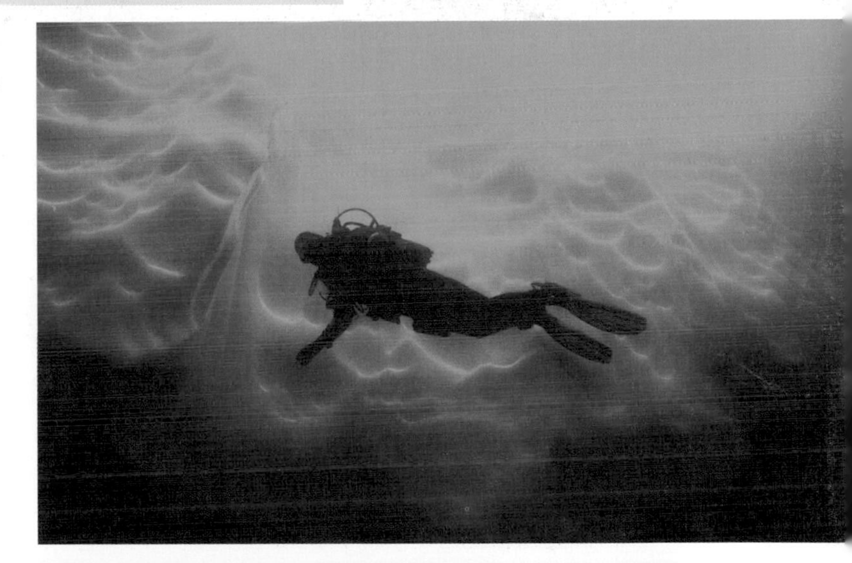

YOUR QUESTIONS

1. What is meant by these terms: ecotourism, the grey market.
2. Conduct a class survey.
   a. How many people think they would enjoy a holiday in Antarctica? Why?
   b. Who would enjoy another extreme environment? Where?
3. Write a newspaper article of about 200 words to explain the appeal of Antarctica for tourists. The article needs to explain why people would want to go to such a cold place. Include what's good about the landscape and the wildlife, as well as how tourists get about and what they do when they're there.

   Hint: When you're given a word limit, stick to it. 200 words isn't a lot, so you need to think carefully about what you want to include.
4. Use the responsible travel website www.responsibletravel.com to investigate a holiday in a different extreme environment. Find out where you can go (search by destination), and what you can do there.

On this spread you'll learn how Antarctica is coping with tourism.

Going down ...

Antarctica 2007. The Canadian cruise ship *Explorer* hit submerged ice off Antarctica and sank. All 154 passengers were rescued, but oil from the ship leaked into the sea, causing pollution.

What impact does tourism have on Antarctica?

Most tourists to Antarctica go on cruise ships like *Explorer*. Cruising is the easiest way to get around, to see the spectacular frozen landscape – and to meet the wildlife. However:

- when cruise ships come too close to the shore, they can disturb wildlife. This can affect feeding, breeding and the rearing of young.
- ships can also have accidents – especially in the treacherous, icy waters that surround Antarctica's often rocky shoreline – just like *Explorer*. Wrecked ships leak oil, polluting the sea and beaches and threatening the birds, animals and plants that live there.

▲ *Tourists landing on shore can threaten and disturb the wildlife they've come to see*

At the moment, tourism's impact on Antarctica is limited. This is because tourism there is internationally controlled and carefully monitored (plus very expensive; a 7-day trip to Antarctica costs about £25 000). Tourists spend most of their time on board their cruise ships and don't venture far inland.

However, as you've already seen, tourism to Antarctica is becoming more popular, so in future it is likely that its impact will grow. Extreme environments – whether it's Antarctica or the Andes – are often fragile and can easily be damaged by people. In Antarctica, more tourist ships means that the likelihood of a serious pollution incident in the future is increased.

What's being done to help Antarctica?

Extreme environments need protection. A number of things are being done to protect Antarctica.

The Treaty of Antarctica

This has been in force since 1961. Nearly 50 countries are signed up to protect Antarctica from mining, drilling for oil, pollution and war – at least until 2048. These countries now want to discuss ways to regulate tourism.

Research

There are over 50 research stations in Antarctica. Scientists observe, record and measure the weather, plants and animals, the ice and its movement, the rock beneath the ice, the sea, and the Earth's magnetic field. The more people understand Antarctica, the more likely it is that it can be protected for the future. Tourists can visit some of the research stations and learn more about how important it is to protect the environment.

Tourism

The IAATO (International Association of Antarctica Tour Operators) was set up in 1991. It has guidelines on things like the number of people allowed on shore, activities and wildlife watching. Tour operators and tourists are not allowed to leave anything behind – no rubbish of any sort. Cruise ships carry their used (or grey) water back to port. They don't dump it in the sea around Antarctica.

In 2010, the British government suggested to the other countries who signed the Treaty of Antarctica that a new agreement, covering tourism to Antarctica, was needed. This would limit the number of tourists and where they could go, plus also ban any hotel building.

The point of such an agreement would be to allow some visitors to enjoy Antarctica without spoiling it for the future – in other words to manage tourism sustainably.

Other measures being put in place to protect Antarctica

After 2011, ships won't be allowed to use heavy fuel oil – potentially the most polluting in the event of an accident.

From 2013, the new **Polar Code** will limit the number and size of ships visiting Antarctica. Ships carrying more than 500 people won't be allowed to land anyone, and only 100 tourists will be allowed ashore at any given time.

YOUR QUESTIONS

1. What are: The Treaty of Antarctica, The Polar Code?
2. Design a leaflet to advise tourists landing on Antarctica about how to 'leave nothing but footprints and take nothing but photographs'.
3. Draw a concept map, using the information on pages 284-287, to show the impact that tourism might have on Antarctica in 20 years.

On this spread you'll find out about the need for stewardship and conservation, and about ecotourism in the rainforest.

Why should we look after the rainforest?

The Amazon rainforest is hot, steaming and teeming with life. It has a lush appearance, but appearances can be deceptive. It's a fragile environment and it needs looking after.

Over the last 50 years the Amazon rainforest has changed. People have moved in and cleared thousands of square kilometres every year for timber, farming, mining and road building. The result of this clearance is that 20% of the rainforest has now been destroyed. But does that really matter? Look at the photo and text boxes and see what you think.

Indigenous peoples

Amerindians have lived in the rainforest for centuries. However, increasing contact with the outsiders now moving into the forest, means that they risk catching diseases which they have no resistance to. These diseases can be fatal. Clearing the rainforest also means that the Amerindians lose their homes and way of life.

Global warming

The living rainforest absorbs a lot of carbon dioxide from the atmosphere. However, cutting down the trees and burning them to clear the land, results in carbon dioxide being added to the atmosphere instead – which, in turn, increases global warming.

Amazon rainforest

Tourism

Ecotourists like to visit the rainforest and meet its wildlife and people. They bring money to the places they visit. If the forest is cleared, they won't want to come and communities will suffer financially.

Flooding

Clearing the rainforest trees reduces interception and means that, when it rains, the fertile topsoil is washed into the rivers and lost. The rivers also silt up as a result, which means that they are more likely to flood and damage farms, homes and businesses downstream.

Ecosystem

Clearing the rainforest damages the fragile ecosystem there. Animals and plants lose their habitats and many species become extinct. We also suffer as a result, because Amazon rainforest species have provided the source of many ingredients used in modern medicines. Increasing rainforest clearance means that potential future medicines might be lost before they're discovered.

The need for stewardship and conservation

The rainforest is a fragile environment. If we're going to keep it for future generations, we need to look after it. **Stewardship** means caring for the environment of a place as though it was our own. It means carefully looking after the plants, animals and people who live there.

Stewardship is an important way of conserving the environment. It helps to protect it from harm, so it continues to thrive into the future. Ecotourism can play a big role here. It's a really good way of helping local people while looking after the environment.

Ecotourism in the Amazon rainforest

Ecotourism usually involves small-scale tourism. The Yachana ecolodge in the Amazon rainforest in Ecuador is one example of an ecotourism development. The **ecolodge** is a guesthouse where a small number of ecotourists can stay. It's basically a small environmentally friendly hotel that is surrounded by nature.

The Yachana ecolodge is next to the Napo River – a tributary of the Amazon – close to the village of Mondaña. It is set in its own, protected, 1200-hectare section of rainforest, which is home to thousands of species of tropical plants and animals. Every room has a view of the river, safe drinking water and a private bathroom with a hot shower. Its dining room serves the guests meals made from locally grown food.

Most of the people who work at Yachana are local. They have jobs in the kitchen, dining room, garden – and help to look after the guests and their bedrooms. The lodge also employs Amerindian guides to show guests the forest environment and its creatures, how local people live, and how they use plants for medicines.

The Yachana ecolodge offers a range of ecotourism activities. They involve visiting the natural environment in small groups and causing as little harm as possible to the area and to the local people. The activities help tourists to better understand the environment and the lives of local people. They include:

- rainforest hiking
- birdwatching
- swimming in the Napo River
- canoeing
- photography
- visiting the local village
- learning to make traditional 'mokaua' pottery
- taking part in a traditional ceremony
- visiting a nearby biological research station

▲ *The Yachana ecolodge*

▲ *The Amazon Basin*

▼ *A traditional cleansing ceremony being performed on an ecotourist by an Amerindian at the Yachana ecolodge*

YOUR QUESTIONS

1 a What is an ecolodge?

b Explain, in your own words, what stewardship is.

2 Imagine that you're staying at the Yachana ecolodge. Write a blog entry describing what you've done and explaining why your holiday is an example of ecotourism.

3 Make a wall-display to explain the importance of looking after the Amazon rainforest.

4 Hold a group discussion about whether all tourism should be ecotourism.

On this spread you'll explore how ecotourism can benefit the environment, the local economy and the people of the Amazon rainforest, and how it can also help places develop sustainably.

your planet
The Amazon is 6400 km long but it doesn't have a single bridge crossing it!

What's good about ecotourism?

Ecotourism benefits the environment and the local economy of the places ecotourists visit, together with the lives of the local people.

The environment

- Ecotourism means that the environment the ecotourists are visiting will be looked after. The trees are not cut down, but conserved for the future, because the forest is now an important attraction and economic asset.
- Because ecotourism is small-scale, ecotourists travel in small groups. This means that they consume few resources, cause little pollution and are less likely to cause physical damage, like trampling vegetation.
- The Amazon rainforest is an important global resource. By absorbing carbon dioxide from the atmosphere, the trees act as a brake on increased global warming. Therefore, anything which promotes their preservation is a real benefit – not just for the local environment but for the world.
- The Yachana ecolodge on page 289 recycles its waste and uses renewable solar power.

▲ *A guided rainforest hike in a small group*

The local economy

- Ecotourism developments, such as the Yachana ecolodge, employ mostly local people. Their wages are then spent in local markets.
- Ecotourism also provides local farmers with two new markets: the tourist developments themselves, like the ecolodge, plus the local people who work for the ecolodge and who don't have time to grow their own food any more, because they're too busy with their tourism jobs.
- Ecotourists like to visit local villages and interact with the people. They pay for extra services and buy souvenir handicrafts from them. This puts more money into the local economy.

▲ *An ecotourist interacting with the local people*

People's lives

- As a result of ecotourism, some local people can now afford consumer goods, like televisions and radios – and motorboats instead of canoes. These change people's traditional lives and can be seen as a negative impact. But many younger people welcome them.
- The extra money in the local economy means that more can be spent on healthcare and education – leading to higher literacy levels and life expectancy.
- Because more people are better off now, as a result of ecotourism, fewer feel the need to migrate (move away) to cities in search of work. Many migrants are younger adults, who leave an older population behind. One result of less migration from the Amazon is a better balance of people from different age groups – with more younger, fitter people to earn money, grow food and look after the elderly.

▲ *The local people's lives are changing*

How does ecotourism help sustainable development?

Ecotourism, like the holidays provided by the Yachana ecolodge, helps to contribute to the **sustainable development** of the area. This means that ecotourism:

- values and conserves the natural environment
- improves the well-being of local people (their standard of living and quality of life).

The flowchart shows how ecotourism helps with the Amazon rainforest's sustainable development.

Ecotourism gives the living rainforest an economic value. Previously it had to be destroyed to make money. The trees were chopped down for timber, and large areas were cleared for farming that didn't last because of damage to the soil through erosion and the loss of nutrients. These developments were not sustainable – they were short-term and destructive.

However, ecotourism is sustainable. Local people, businesses and government are now keen to conserve the rainforest environment, because that's what attracts tourists. They can see that keeping the rainforest intact is the key to their future long-term prosperity.

YOUR QUESTIONS

1. Name three ways in which ecotourism benefits the Amazon rainforest.
2. Make a PowerPoint presentation to explain how ecotourism helps sustainable development.
3. Draw and complete a mind map like the one below on a piece of A3 paper. Make use of information from both Spreads 13.9 and 13.10.

Sustainable development

The environment and traditional ways are sustained.

Local people, local government and local businesses value nature and tradition more.

The forest and its wildlife become economic resources – and so do the traditions of its local people.

Ecotourism brings money to the area.

Ecotourists visit the Amazon rainforest because of its natural environment.

▲ *Learning traditional skills at the Yachana ecolodge – how to use a blowpipe for hunting*

Local fieldwork investigation

- You will carry out a fieldwork investigation based on a question, or **hypothesis**.
- For the investigation you will collect **primary data** by doing fieldwork, and then you will write a report, of up to 2000 words. This will be worth 25% of your total GCSE grade.

An **hypothesis** is a prediction you can prove or disprove by investigation.

Primary data is information you find out for yourself. (Secondary data is information you obtain from other sources, like books or websites.)

Stages in your fieldwork task

There are five main stages:

1 Introduction – in class before you go.

- Decide on an hypothesis and suggest what you expect to find out.
- Say how the hypothesis is linked to what you have studied.
- Find out some relevant background information of the place you investigate.

2 Method and data collection – prepare in class, then collect data outside.

- Decide on a range of fieldwork techniques. Describe how to carry them out, and give reasons for choosing them.
- Collect the data!

3 Data presentation – in class when you get back.

- Use a range of ways to present the data, such as maps, graphs, drawings and photographs.

4 Interpretation and analysis – in class at the end.

- Describe key findings shown by the data.
- Explain what the findings mean.

5 Evaluation and conclusion – in class at the end.

- Go back to your hypothesis, and draw a conclusion (decide whether it was right or wrong.)
- Evaluate (consider) how successful and useful the investigation was.
- Identify any limitations of the investigation, and suggest how it could be improved or extended.

Fieldwork tips

Do as much thinking and planning as you can before you go out on your fieldwork trip. It can save you a lot of trouble later on.

On your fieldwork trip, work as a team. You may need to share data, so the data you collect might help the rest of the group, and their data might help you.

During the write up of your task, you are not allowed to take any of your work out of the class or bring anything in. But that doesn't stop you planning and thinking at home – bring your ideas to class in your head!

Your introduction

The hypothesis

- You will be given a fieldwork task statement that will be quite general such as, *Investigate the success of inner city revitalisation.*
- You need to adapt the fieldwork task to a particular place, and give the investigation some focus: *Has the revitalisation of London's Docklands been successful?* The **aim** of your investigation is to discover the answer to this question.
- Suggest what you expect to find out. This then becomes your basic hypothesis. *The regeneration of London's Docklands has been very successful* or *The regeneration of London's Docklands has not been at all successful.*
- Give reasons to explain your hypothesis. Make use of the geography you have studied so far, and try to identify some key processes and geographical ideas that helped you form the hypothesis.

Setting the scene – location and background

Research the area you are going to be investigating. This is your **secondary data**.

- Where is the area located?
- What is the area like?
- What issues are affecting the area and why?
- What processes are affecting the area, and with what effect?

Methods

Now you need to plan how you are going to collect your primary data – what you need to find out, and how you intend to do it. Here are some ideas for our example investigation.

Method	Purpose
Land use map for your study area	Show what kinds of land use exist now, and how they may have changed – can you find an historic map from before the revitalisation?
Age-of-buildings map for your study area	Identify the most recent buildings which have resulted from the revitalisation.
Environmental Quality Survey (EQS) at different places	Compare the environmental quality (building quality, noise, open space) in areas that have been revitalised with areas that haven't.
Questionnaire to find out people's ideas about the success of revitalisation	Find out whether people think they have benefited from revitalisation or not, and see whether their ideas vary with age or gender.
Field sketches	Give your own impressions of the areas you visit – the annotations are more important than artistic quality!
Photographs	Take photographs of different areas to be used alongside the EQS, to give visual reference.

Data presentation

Once you have been out and collected all of your primary data, you need to bring it all together, and present it clearly. This means that you can easily analyse the results. Choose varied ways of presenting data.

- If you are dealing with statistics, produce graphs.
- If you are comparing statistics from different places, draw a graph for each place, and position them around a map, linked to their area.
- Find ways of linking data. Try presenting the results of an Environmental Quality Survey (EQS) on the same page as an annotated photograph of the area.
- You can use ICT to present your results, but you can also draw some by hand.

Graphs

Choose the right kind of graph for the data you want to display. Take a look at the table on the right, which will give you lots of ideas.

Type of graph	When to use it	What it looks like
Bar chart	For comparing categories e.g. how many types of building are retail, housing, office etc.	
Line graph	Showing continuous measurement e.g. changes over time.	
Mirror graph	Like age-sex diagrams (or population pyramids), these are good for comparing two categories (like gender) or categories of data collected in two different places.	
Pie chart	When you want to compare proportions or percentages instead of raw numbers.	
Scatter graph	When you want to show a relationship e.g. between gradient of slope and velocity of a river.	

Maps

Choose the right kind of map for the information you want to show. The table below gives some examples.

Type of map	When to use it	What it looks like
Choropleth map	• Where there are categories e.g. building heights grouped by number of storeys.	Shaded areas with a key. Uses categories which range from light shades of a colour for low values to dark shades of the same colour for high values.
Dot map	• To show distribution e.g. particular kinds of shop in a city or town.	Dots displayed on a map to indicate an occurrence. The dots might be coloured e.g. a red dot represents a café, a blue one shows an antiques shop.
Isoline map	• Rainfall map with areas between isolines coloured in. • Building heights within a city centre. • Pedestrian counts.	A map showing lines which join values of the same amount e.g. rainfall, or temperature. Often, the areas in between each line are shaded in like choropleth maps.

Annotated photos, maps and graphs

On each of your photos, maps and graphs, you should try and **annotate** what you can actually see on them. (Annotate means to add a piece of information to some text, or an illustration, to describe or explain a particular aspect of it.)

On a set of data you can describe what you see, for example, *Place X has a high percentage of offices*.

You can also compare what you see, for example, *Place X has the highest percentage of offices, but a low percentage of shops*.

On a photo you can describe what you see, for example, *Place X is a pedestrianised area, with a high percentage of shops*.

You can also add information that you have gathered from elsewhere, like from your EQS (Environmental Quality Survey). For example, *There is a lot of traffic noise from the nearby main road, but this precinct is traffic-free*.

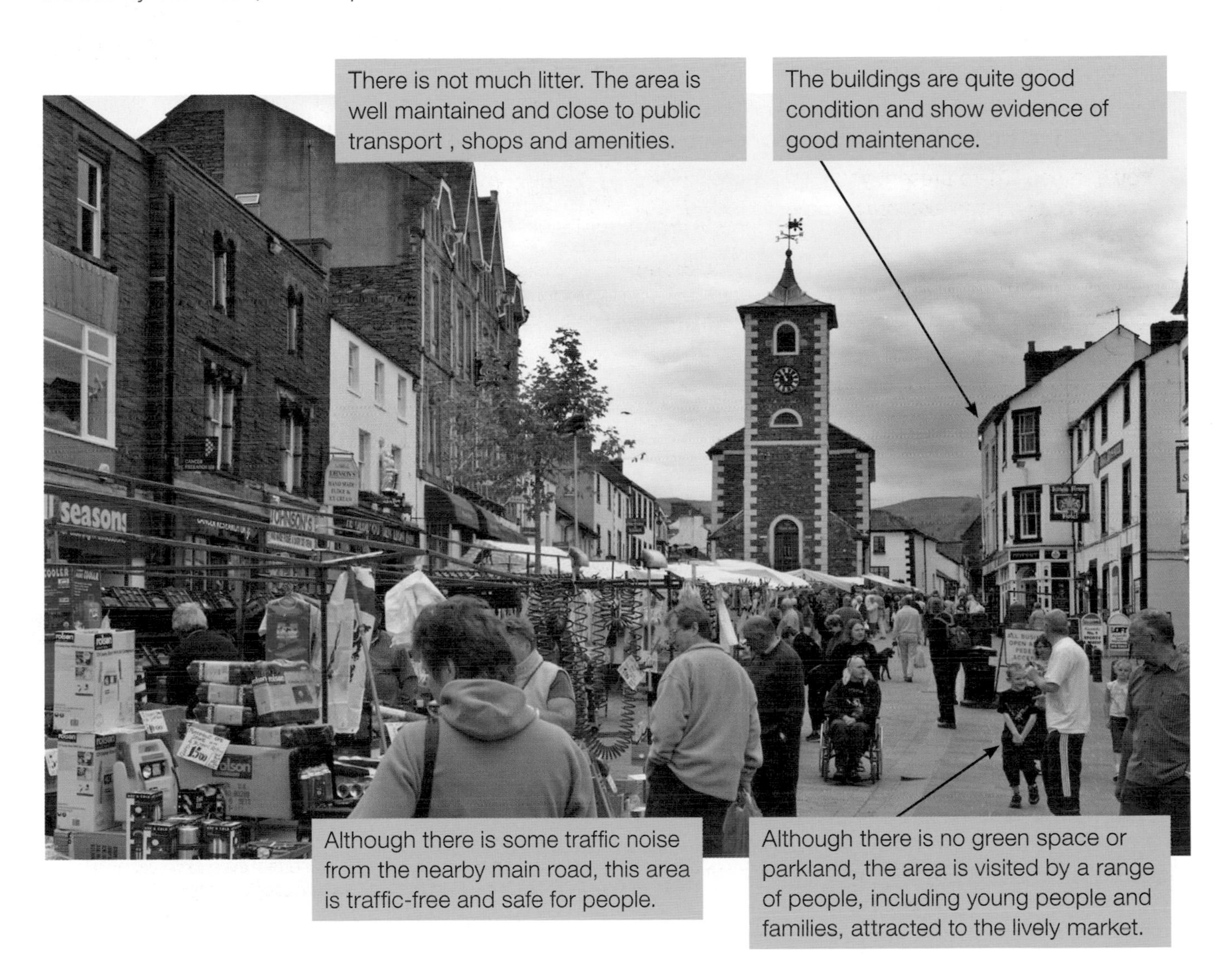

Description, interpretation and analysis

This part is all about drawing ideas out of your results, and understanding what they mean. You will then use these ideas to form a **conclusion** about your hypothesis.

- **Describe** results that you have not already covered in your annotations.
- **Compare** places, for example, *Place Y has the highest pedestrian count of all the places sampled.*
- **Explain likely reasons** for what you see. For example, *Place Y has the highest pedestrian count of all the places sampled because it is near to large stores such as Debenhams and HMV.*
- **Explain possible reasons** for what you see. *For example, Place Z has a low pedestrian count because there are a lot of derelict* (empty and neglected) *buildings in the area.*

Your conclusion

This is where you return to your main hypothesis. Your conclusion should take this statement, and try to decide whether or not it is true.

- Your hypothesis might be correct, incorrect, or partly correct.
- You need to decide on the basis of evidence – from your results and analysis – not just from a general impression.

If you are not sure, it is OK to say so. But you must provide plenty of evidence that supports each side of the hypothesis, to demonstrate why you cannot decide.

Your evaluation

This is where you decide on the **validity** of your investigation.

- Is what you have found out reliable?
- Can your data be trusted?
- Did the study go well?

There are three main areas you can think about: your **methods**, the reliability your **conclusions**, and the **relevance** of the investigation.

The box on the right gives some questions you can ask yourself.

Analysis – where do I start?

1 Start with general points.

Environmental quality tends to be lower in areas that have not been revitalized, than in areas that have been.

2 Move onto more specific points to show differences between places.

Canary Wharf has the highest percentage of people in professional and managerial jobs (78%) whereas Shadwell has the lowest (16%). However, Shadwell has the greatest percentage of people in work, with only 5% unemployed.

3 Then compare data within places.

Shadwell may have the lowest unemployment, at 5%, but it also has the highest percentage of people in part-time work (21%).

Evaluation – questions to ask

Did your data collection methods work well? Did some work better than others?

Would any of the results be different at another time of day/of the week/of the year?

Are there any **anomalies** (strange results) that you can't explain?

Are your conclusions valid, even though you only took a small sample?

Is your study area unique, or would you get similar results in other places?

Who might be interested in your investigation and why? How might they use it?

How could you extend your investigation? Which parts would you like to develop further?

How to be successful in exams

How are your exam papers structured?

You will be assessed on three Units:

- Unit 1 – this is Physical geography (chapters 1 – 7 in this book).
- Unit 2 – this is Human geography (chapters 8 – 13 in this book).
- Unit 3 – this is your Local fieldwork investigation (which might also be called **controlled assessment**. You can find advice about this on pages 292 – 296 of this book).

Your Unit 3 work will have to be done under **controlled conditions**, which means you won't always be able to talk to your friends, or ask your teacher for advice. This will feel very much like an exam.

But you only take actual exams for Units 1 and 2, which is where these next pages might help! Each exam will last 1 ½ hours, and each will be worth 37.5% (a bit over a ⅓) of your total GCSE Geography mark.

Unit 1 Physical geography	*Unit 2 Human geography*
Section A	**Section A**
The Restless Earth	Population Change
Rocks, Resources and Scenery	Changing Urban Environments
Challenge of Weather and Climate	Changing Rural Environments
Living World	
Section B	**Section B**
Water on the Land	The Development Gap
Ice on the Land	Globalisation
The Coastal Zone	Tourism

In both exams you only answer three questions. You have to choose one from Section A and one from Section B, but then your third one can come from either section.

Don't answer too many questions! If you answer more than three, you will run out of time. And make sure you answer at least one from each section. You don't want to lose marks by answering the wrong questions.

How to choose the right question

When you're under pressure in your exam, the most important thing you can do is to choose the right question for *you*. Look at the suggestions below and take time at the start of your exam to select the best questions.

Make a decision based on your strengths and knowledge. Don't be put off by what your friends have said before the exam.

Don't pick a question just because it has a great photo, or a cartoon that looks interesting. How difficult is the whole question?

Only choose questions about a topic that you have studied! Another question might look easier, but that is only because you have not learnt the details.

Read all of the questions carefully. And then choose which ones to answer. You might want to pencil a few notes for each question as you choose it, to jog your memory when you come to answer it.

If you realise you have chosen the wrong question, be confident and change. If you then find yourself running out of time, start writing bullet points. This is better than not writing an answer at all.

What is the question really asking?

Before your exam, become familiar with the common command words. These are the key to understanding what the question really wants you to do. Look at the table below to see what some command words are, and what they really mean.

Command word	What it means	Example
Account for	Explain the reasons for.	Account for the presence of volcanoes along constructive plate boundaries.
Compare	Identify similarities and differences.	Compare the responses to flooding in an LEDC and an MEDC.
Define	Give a clear meaning.	Define 'natural increase'.
Describe	Say what something is like; identify trends.	Describe what the graph shows.
Explain	Gives reasons for why something happens.	Explain how fold mountains form.
How far/To what extent	Give both sides of an argument.	How far were the floods due to human causes?
Justify	Give evidence to support your answer.	Which management option do you think is most suitable? Justify your choice.
List	Just state the answers or options. No explanation or description is needed.	List four push factors for international migration.
Outline	You need to describe and explain, but mostly describe.	Outline the trend in population growth.

Interpreting the question

It is very easy to look at a question, and think you know what it's asking. But a question that looks familiar might actually be slightly different to what you were expecting. Give yourself time to properly understand the question. Try thinking about these four areas:

Command words

Identify the command word, and be certain of what it is asking you.

Theme or topic

This is the subject – the particular part of your geography course – that the question is asking about.

Focus

This is the area that the theme has been narrowed down to.

Case studies

A number of questions might be looking for examples of specific places you have studied.

Take a look at this example question, and see how it can be broken down into the four key areas.

Using a named example, explain how quarries can reduce their impact on the environment **either** during **or** after extraction. *(6 marks)*

Command words

'Explain' – you need to do more than simply list some methods; give reasons for how these methods help.

Theme or topic

This question is about quarrying, which is part of the Rocks, resources and scenery topic.

Focus

The examiner is asking particularly about how the quarries are managed, to reduce their environmental impact. You are being asked to talk about methods used while the quarry is active, or to talk about methods used once the quarry is no longer required. You don't need to do both.

Case studies

The question says 'Using a named example'. This means that you need to talk about a particular place that you have studied. The example from this book that you could use is the Cotswold Water Park (pages 46-47).

Why don't you practise this technique on some of the exam-style questions in the next section of this book?

Exam-style questions

1 The restless Earth

1 (a) What do you understand by the term 'supervolcano'? *(2 marks)*

1 (b) What are the correct terms (names) for these two layers in the Earth's structure?

(i) The layer which is at the Earth's surface.

(ii) The layer which is at the centre of the Earth.

(iii) A layer made of solid rock. (There is more than one – just choose one.)

(iv) A layer made of molten rock. (There is more than one – just choose one.)

(4 marks)

1 (c) Study **Figure 1**, which shows a map of the Earth's plate boundaries.

(i) Name the two plates labelled **A** and **B** in **Figure 1**. *(2 marks)*

(ii) What type of plate boundary (**constructive** or **destructive**) lies between Plate **A** and Plate **B**? *(1 mark)*

(iii) Name one major kind of landform often created at the boundaries between plates like **A** and **B**. *(1 mark)*

(iv) Explain how the landform you named in part (iii) above is formed. You may draw a labelled diagram to help you answer this question. *(4 marks)*

1 (d) Study **Figure 2**.

(i) Name the volcanic features labelled **A**, **B**, **C**, and **D** in **Figure 2**. *(4 marks)*

(ii) Describe three ways in which feature **A** can be harmful to people and their activities. *(3 marks)*

1 (e) In what ways can fold mountains be of economic benefit to people? *(4 marks)*

TOTAL MARKS: 25

Figure 1

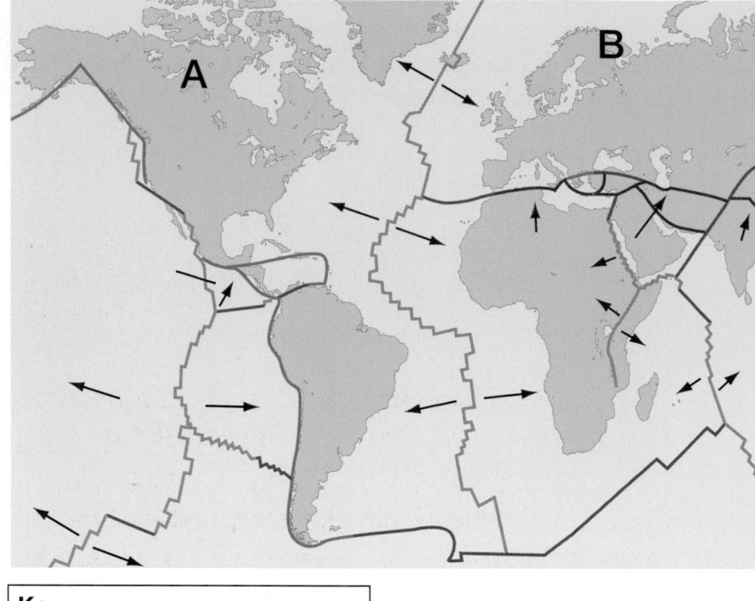

Key
← Direction of plate movement

Figure 2

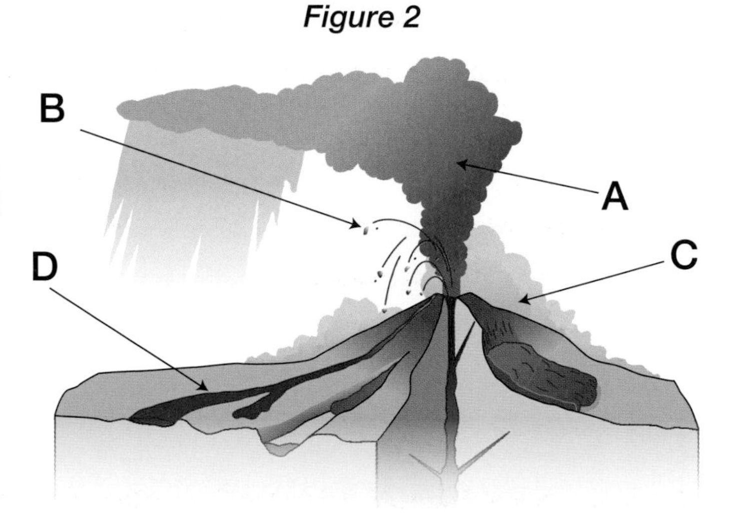

2 Rocks, resources and scenery

2 (a) (i) Copy **Figure 1**.

Figure 1

Rock type	Age of rock type (in millions of years)	Description of this rock type	Rock location letter from Figure 2
	2 - 199	Permeable in some areas; impermeable in others	
	about 280	Very hard, upland volcanic rock	
	299 – 359	Hard, upland rock containing lines of weakness called joints and bedding planes	

(ii) Write **Igneous** in **one** box and **Sedimentary** in **two other** boxes in the Rock type column of **Figure 1**. *(2 marks)*

(iii) Use **Figure 2**. Identify the locations of the three rock types and add the correct letters to the table. *(2 marks)*

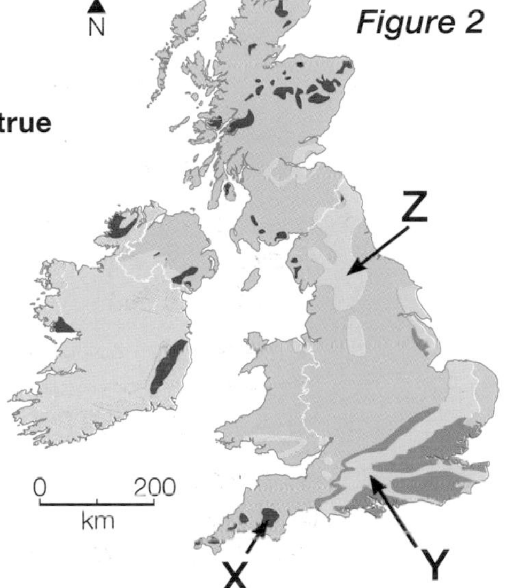

Figure 2

2 (b) Look at these statements about rock types in the British Isles. Answer **true** or **false** for each one. *(3 marks)*

(i) Carboniferous limestone is much younger than chalk or clay.

(ii) Chalk was formed 145-65 million years ago.

(iii) Clay was deposited during the Jurassic Period.

2 (c) Think about these three landscape features:

- Steep-sided, narrow gorges
- Tors in moorland areas
- Vales and downs

Which one is associated with each of the following rock types?

(i) Carboniferous limestone

(ii) Chalk and clay

(iii) Granite

(3 marks)

Figure 3

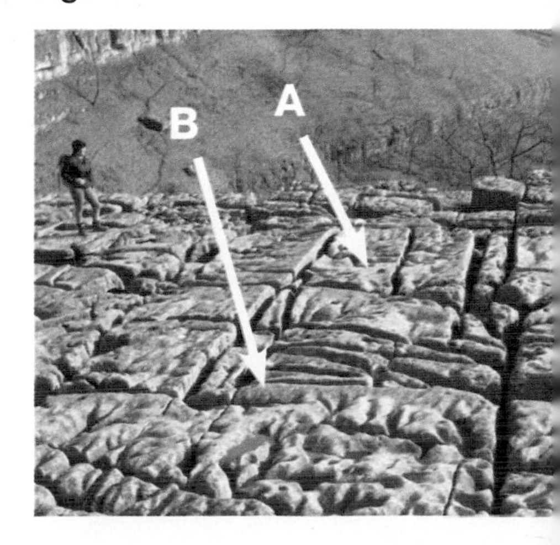

2 (d) (i) Look at **Figure 3**. Name the Carboniferous limestone landscape features at **A** and **B**. *(2 marks)*

(ii) What name is given to areas of exposed Carboniferous limestone like in **Figure 3**? *(1 mark)*

(iii) Suggest **two** types of recreational activity which are appropriate to Carboniferous limestone features and landscapes. *(2 marks)*

2 (e) Describe how the process known as 'freeze-thaw action' leads to the formation of scree on upland hillsides. *(4 marks)*

2 (f) 'Quarrying in rural, upland areas creates more problems than benefits.'

Say whether you **agree** or **disagree** with this statement. Explain your reasons for making this decision, supporting your answer with named examples. *(6 marks)*

TOTAL MARKS: 25

3 Challenge of weather and climate

3 (a) What do you understand by each of these terms?

(i) Weather *(2 marks)*

(ii) Climate *(2 marks)*

3 (b) **Figure 1** shows a cross-section from west to east across Northern Britain.

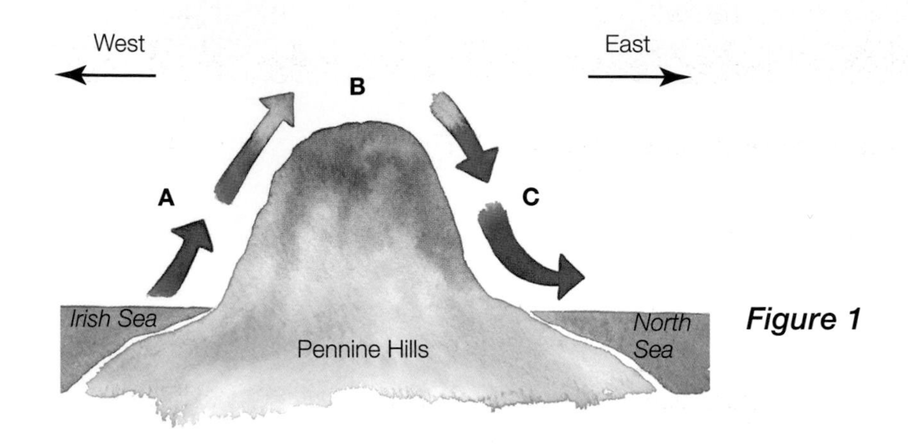

Figure 1

(i) Give the letter of the area which is likely to experience the heaviest rainfall. *(1 mark)*

(ii) What happens to the **temperature** of the air after it has passed over the hills? *(1 mark)*

(iii) What 'type' of rainfall occurs on the western side of the hills in **Figure 1**? *(1 mark)*

(iv) What is the name given to the area marked C on this diagram? *(2 marks)*

3 (c) (i) Anticyclones are areas where the air descends (sinks). Does this mean they are areas of **high** or **low** air pressure? *(1 mark)*

(ii) Explain why night-time temperatures during winter anticyclone conditions are usually very low. You may use a fully-labelled diagram to help you answer this question. *(4 marks)*

3 (d) (i) List **three** different causes of climate change. *(3 marks)*

(ii) Describe the possible consequences of climate change. *(8 marks)*

TOTAL MARKS: 25

4 Living world

4 (a) (i) Copy **Figure 1**.

Figure 1

Key term	Meaning of this key term
	Essential plant foods in the soil
	Feed on dead and waste material, and make things rot, so that they can be recycled and used again
	Feed on plants – or each other
	Plants which use sunlight, water and goodness from the soil to create their own food

(ii) Place these four words next to their correct definitions in **Figure 1**. *(3 marks)*

- Consumers
- Decomposers
- Nutrients
- Producers

4 (b) Hot desert regions are economically valuable in many ways. Some of these ways are:

- Agriculture
- Mineral extraction
- Tourism

Choose two of these activities and suggest why they are important to hot desert regions. *(4 marks)*

4 (c) (i) What is meant by the term 'sustainable'? *(2 marks)*

(ii) Explain how a deciduous woodland may be managed sustainably. *(4 marks)*

4 (d) Using **Figure 2** (a climate graph for a tropical rainforest area), as well as your own knowledge, describe and give reasons for the global distribution of tropical rainforests. *(6 marks)*

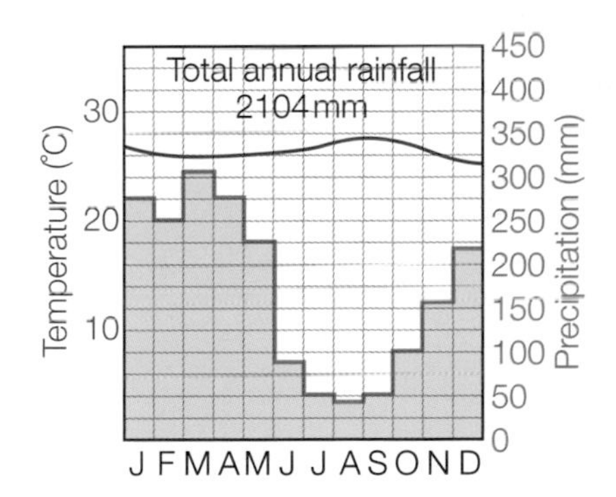

Figure 2

4 (e) Explain how both local and global environmental changes may result from large-scale deforestation in tropical rainforest areas. *(6 marks)*

TOTAL MARKS: 25

5 Water on the land

5 (a) Name **two** ways in which a river transports its load of material. *(2 marks)*

5 (b) Describe the process known as 'attrition'. *(2 marks)*

5 (c) Look at the Ordnance Survey map extract on page 98 of this book. *(3 marks)*

(i) Name the river valley feature at each of these grid reference positions.

- 943759
- 946775
- 942764

(ii) Compare the valley cross-profiles in squares 9575 and 9475. Use diagrams to help you answer this question. *(4 marks)*

5 (d) Copy **Figure 1** and label it to show the key features of a typical waterfall. *(4 marks)*

5 (e) Explain how **any two** activities by people can increase the risk of river flooding. *(4 marks)*

Figure 1

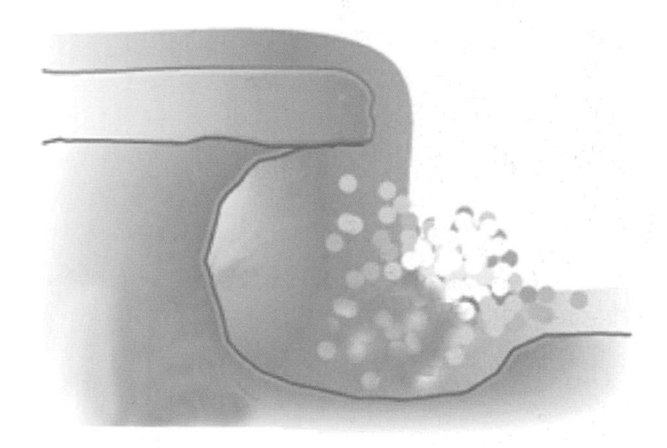

5 (f) Describe some of the advantages and disadvantages of different 'hard engineering' methods designed to reduce river flooding. *(6 marks)*

TOTAL MARKS: 25

6 Ice on the land

6 (a) Decide whether each of these statements is **true** or **false**.

(i) Most drumlins are found in lowland areas or valley bottoms.

(ii) Drumlins are features of deposition as well as transportation.

(iii) Drumlins often reach more than 300 m (984 ft) in height.

(iv) Most drumlins are formed of boulder clay, which makes them suitable for grazing sheep and cattle.

(4 marks)

6 (b) 'Ablation' and 'accumulation' are processes which affect an area's 'glacial budget'. What does each process involve? *(4 marks)*

6 (c) **Figure 1** shows a typical, upland glaciated area.

Figure 1

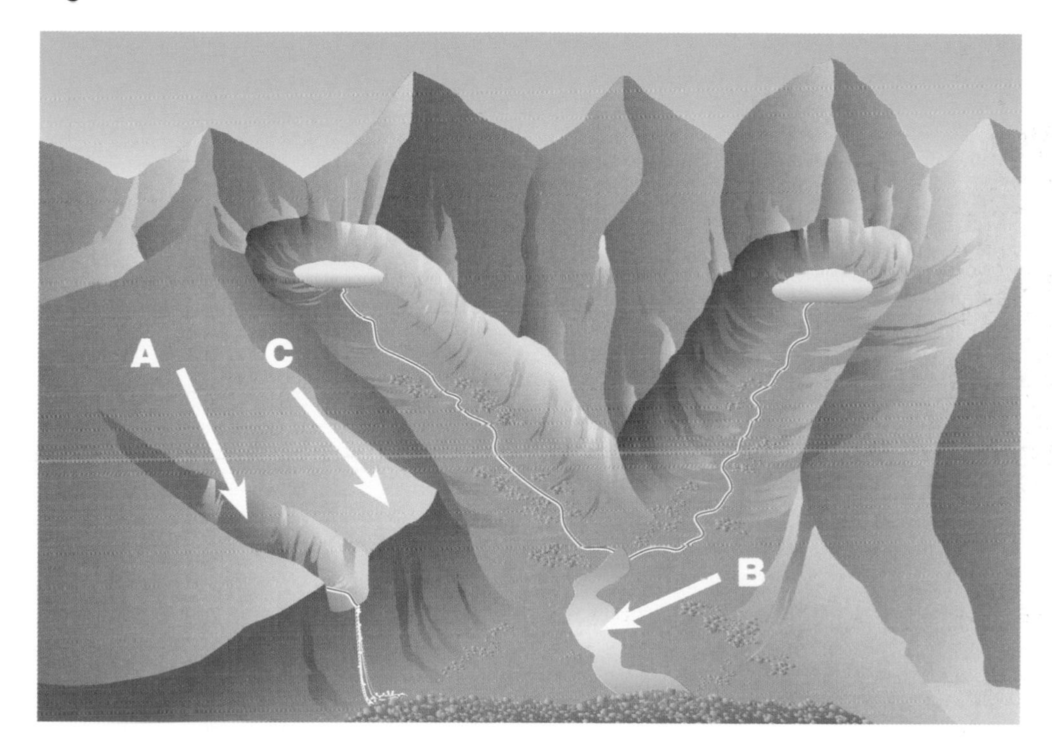

(i) Name the features labelled **A**, **B** and **C** in **Figure 1**. *(3 marks)*

(ii) Explain how **any one** of features **A**, **B** and **C** was formed. Include a fully-annotated diagram in your answer. *(6 marks)*

6 (d) Name an upland, glaciated region you have studied. Describe ways in which your chosen area is economically important, and suggest how this might endanger the natural environment. *(8 marks)*

TOTAL MARKS: 25

7 The coastal zone

7 (a) Write down the meanings of 'backwash' and 'fetch'. *(2 marks)*

7 (b) Copy and complete the paragraph below, with the help of **Figure 1**. You have a clue for the first gap.

Figure 1

Headlands form where there is much ________ **(harder/softer)** rock than in the area surrounding them. Where there are cracks in a headland's rock, ________ are likely to form. If two of these are adjacent and erode so far back that they meet, the rock above them forms a ________. When this collapses, part of the headland becomes surrounded by water. This tall, isolated pillar of rock is called a ________. Over thousands of years, headlands retreat inland, leaving behind flat, rocky areas on the shore called a ________ . *(5 marks)*

7 (c) Look at the Ordnance Survey map extract on page 143.

(i) What **types** of coastal landform are walkers most likely to see from these grid reference positions?

- 031769
- 040786
- 055825

(3 marks)

(ii) Explain how the structures along the shoreline in square 0379 reduce the need for beach replenishment. *(4 marks)*

7 (d) (i) Name one area you have studied where cliff collapse is a serious problem. *(1 mark)*

(ii) Explain why cliff collapse occurs so often in this area. You may use a labelled sketch to help you answer this question. *(4 marks)*

(iii) Describe the main social and economic impacts of cliff collapse on the local population in that area. *(6 marks)*

TOTAL MARKS: 25

8 Population change

8 (a) (i) Complete **Figure 1** using the following information: *(2 marks)*

- Females: 40-49 years old = 3.9 million people
- Males: 50-59 years old = 5.5 million people

(ii) What is the name for the type of model shown in **Figure 2**? *(1 mark)*

Figure 1

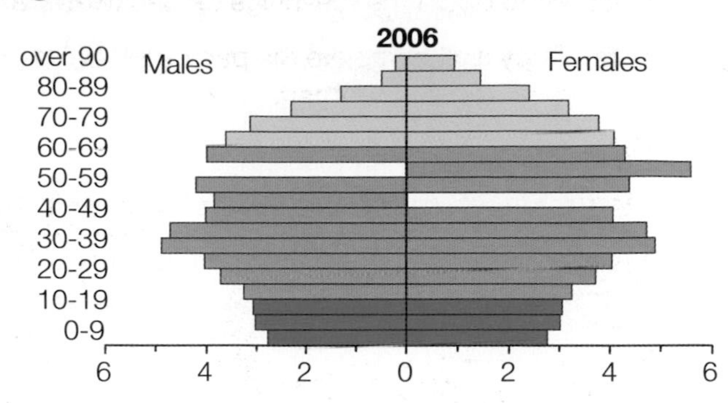

Figure 2

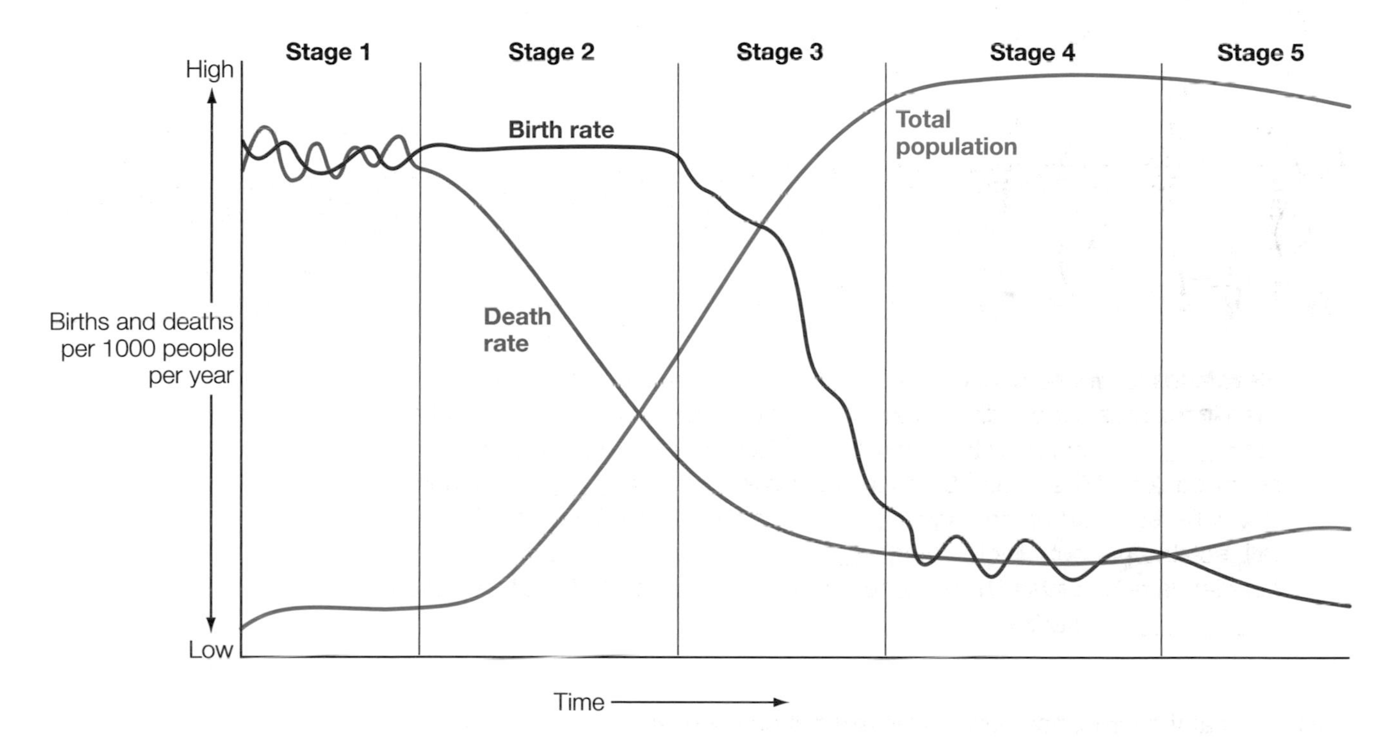

(iii) Which stage of this model is the pyramid in **Figure 1** most likely to relate to? *(1 mark)*

(iv) Give **two** reasons for your answer to part (iii). *(2 marks)*

(v) Explain why the rate of population change can be both small and changeable in Stages 4 and 5 of the model. *(4 marks)*

8 (b) Outline some of the benefits and issues experienced by communities with ageing populations. *(4 marks)*

8 (c) Describe some of the social and economic problems which China's 'one-child policy' is now causing. *(4 marks)*

8 (d) (i) Explain what is meant by the term 'economic migrant'. *(1 mark)*

(ii) What are the likely benefits of economic migration across Europe? Name host and source countries where appropriate. *(6 marks)*

TOTAL MARKS: 25

9 Changing urban environments

9 (a) What do the following two terms mean? *(2 marks)*

(i) Urbanisation

(ii) Settlement functions

9 (b) The most expensive land is usually to be found in the commercial heart of a city. Explain why this happens. *(2 marks)*

9 (c) With the help of **Figure1**, explain why the central areas of many British towns and cities have been in decline in recent years. *(4 marks)*

Figure 1

9 (d) Explain the statement: 'Ethnic groups often cluster within major cities throughout the world.' *(4 marks)*

9 (e) Explain how each of the following control measures can reduce traffic-related problems in urban areas. *(3 marks)*

(i) Congestion charges

(ii) Park-and-ride facilities

(iii) Integrated transport policies

9 (f) Effective waste disposal is just one way of making settlements more sustainable. In what **other** ways can their sustainability be increased? *(4 marks)*

9 (g) Describe how a squatter settlement you have studied has been redeveloped to the benefit of its inhabitants. *(6 marks)*

TOTAL MARKS: 25

10 Changing rural environments

10 (a) Copy **Figure 1**, which shows some changes due to the development of 'agribusiness'. Add one tick to each row to show an effect that this change has had. *(4 marks)*

10 (b) Give **two** reasons why so many hedgerows have been removed by farmers. *(2 marks)*

10 (c) (i) Look at **Figure 2**. It shows the opinions of different villagers about more people coming to live in their village. Pick **three** people, and give one reason each to explain why they have that opinion. *(3 marks)*

Figure 1

	Eutrophication	Habitat loss	Size of workforce decreases	Soil fertility changes
Hedge removal				
Increase in size of farm machinery				
Use of chemical fertilisers				

Figure 2

I don't mind people coming here – as long as they live in the village all year round!

Shop keeper

It might make my job more interesting!

Village policeman

We're better off than most village schools – they've been under threat for years!

Primary school teacher

Most village incomers are well-off – and enjoy going out in the evening!

Taxi driver

(ii) Describe ways in which life in rural communities can be made more sustainable. Your answer may refer to communities you have studied. *(4 marks)*

10 (d) Give reasons why the rate of soil erosion has increased in recent years in many tropical and sub-tropical rural areas. *(4 marks)*

10 (e) Answer these questions using **Figure 3** and your own knowledge:

(i) What makes this retail park's location attractive to developers? *(4 marks)*

(ii) Why are retail parks like the one in **Figure 3** attractive to shoppers? *(4 marks)*

TOTAL MARKS: 25

Figure 3

11 The development gap

11 (a) (i) What do the letters HDI stand for? *(1 mark)*

(ii) List **any three** kinds of information which are used to calculate HDI averages. *(3 marks)*

11 (b) What do you understand by the term 'Fair Trade'? *(2 marks)*

11 (c) List **three** different ways in which countries can provide aid for others less rich than themselves. *(3 marks)*

11 (d) (i) Copy **Figure 1**. Plot the information below onto the graph. *(3 marks)*

- $5230 : 28 deaths per 1000 live births
- $13730 : 25 deaths per 1000 live births
- $28440 : 5 deaths per 1000 live births.

Figure 1

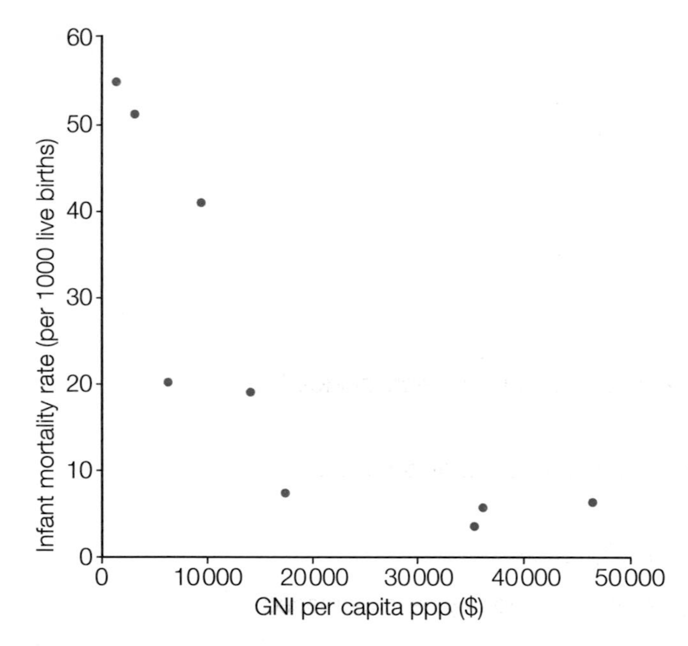

(ii) Add a best-fit line to your graph. *(1 mark)*

(iii) What do you think your best-fit line shows about the link between a country's wealth and its infant mortality rate? *(2 marks)*

11 (e) Suggest ways in which people's 'quality of life' may be put at risk by **either** environmental factors **or** political issues. *(4 marks)*

11 (f) Explain why access to water is a significant factor in people's 'quality of life'. *(6 marks)*

TOTAL MARKS: 25

12 Globalisation

12 (a) (i) Coal is a non-renewable source of energy. Name two other non-renewable energy sources. *(2 marks)*

(ii) **Figure 1** shows how Britain's sources of energy changed between 1970 and 2007. Copy and complete the graph to show that in 2007 Britain used 17% coal and 75% other non-renewable sources. *(1 mark)*

Figure 1

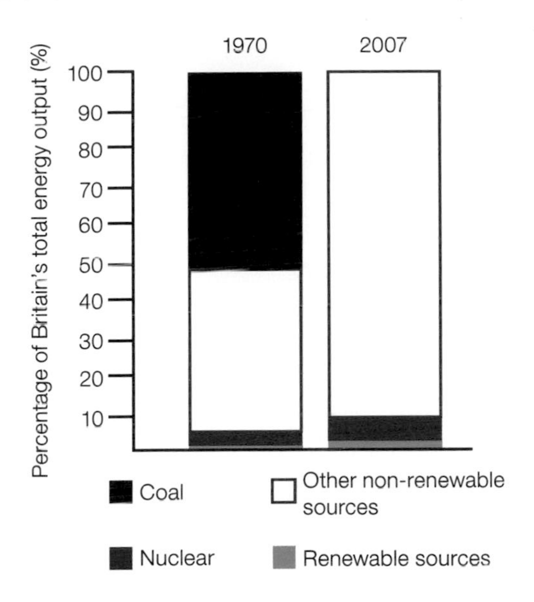

(iii) Name one source of renewable energy which might have been included in this graph. *(1 mark)*

(iv) Describe **three** environmental benefits of using renewable sources of energy. *(3 marks)*

12 (b) (i) What does the term 'globalisation' mean? *(2 marks)*

(ii) What kinds of inventions and developments in technology have increased globalisation in recent years? *(4 marks)*

12 (c) (i) Most transnational corporations (companies) have their headquarters and research departments in an MEDC (one of the world's richer countries). Suggest three reasons why they choose to do this. *(3 marks)*

(ii) Most of these corporations have large factories in LEDCs. Suggest three reasons for their decision to do this. *(3 marks)*

12 (d) Give reasons to explain China's recent growth into a major industrial and exporting nation. *(6 marks)*

TOTAL MARKS: 25

13 Tourism

13 (a) Explain the meanings of the terms below.

(i) Honeypot site

(ii) Mass tourism

(iii) Short-haul destination *(3 marks)*

13 (b) Suggest **three** reasons why global tourism has more than doubled in less than twenty years. *(3 marks)*

13 (c) (i) Name a UK national park, coastal holiday resort or city which is popular with tourists. Describe the attractions of this location for tourists. *(3 marks)*

(ii) With the help of **Figure 1**, explain why local residents may have issues with large numbers of visitors. *(4 marks)*

Figure 1

13 (d) In what ways can increased tourist activity lead to improvements in a country's infrastructure? *(4 marks)*

13 (e) Describe ways in which tourist activities can harm the natural environment. Refer to named locations where appropriate. *(8 marks)*

TOTAL MARKS: 25

Ordnance Survey Symbols

ROADS AND PATHS

M I or A 6(M) — Motorway
A 35 — Dual carriageway
A31(T) or A 35 — Trunk or main road
B 3074 — Secondary road
Narrow road with passing places
Road under construction
Road generally more than 4 m wide
Road generally less than 4 m wide
Other road, drive or track, fenced and unfenced
Gradient: steeper than 1 in 5; 1 in 7 to 1 in 5
Ferry — Ferry; Ferry P – passenger only
Path

PUBLIC RIGHTS OF WAY

(Not applicable to Scotland)

1:25 000 1:50 000

Footpath
Road used as a public footpath
Bridleway
Byway open to all traffic

RAILWAYS

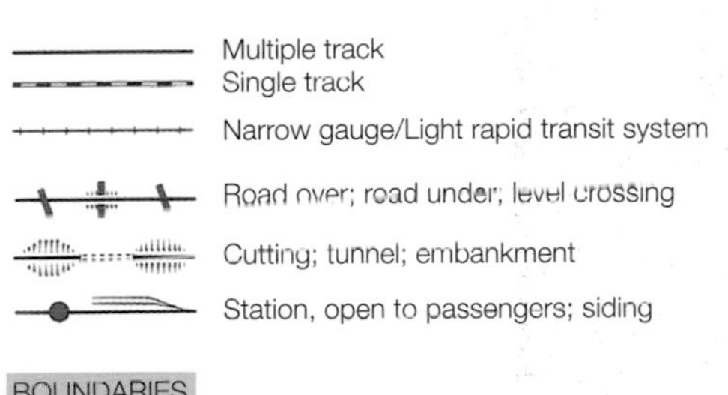

Multiple track
Single track
Narrow gauge/Light rapid transit system
Road over; road under; level crossing
Cutting; tunnel; embankment
Station, open to passengers; siding

BOUNDARIES

National
District
County, Unitary Authority, Metropolitan District or London Borough
National Park

HEIGHTS/ROCK FEATURES

Contour lines

· 144 Spot height to the nearest metre above sea level

outcrop cliff scree

ABBREVIATIONS

P	Post office	PC	Public convenience (rural areas)
PH	Public house	TH	Town Hall, Guildhall or equivalent
MS	Milestone	Sch	School
MP	Milepost	Coll	College
CH	Clubhouse	Mus	Museum
CG	Coastguard	Cemy	Cemetery
Fm	Farm		

ANTIQUITIES

VILLA Roman
Castle Non-Roman
Battlefield (with date)
Tumulus/Tumuli (mound over burial place)

LAND FEATURES

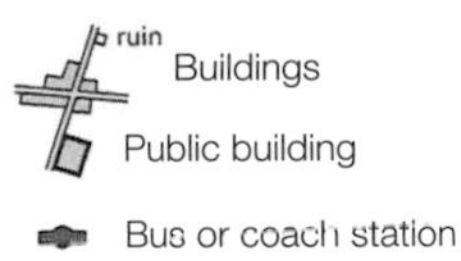

Buildings
Public building
Bus or coach station
Place of Worship — with tower; with spire, minaret or dome; without such additions
Chimney or tower
Glass structure
Heliport
Triangulation pillar
Mast
Wind pump / wind generator
Windmill
Graticule intersection

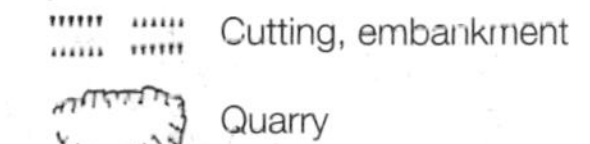

Cutting, embankment
Quarry

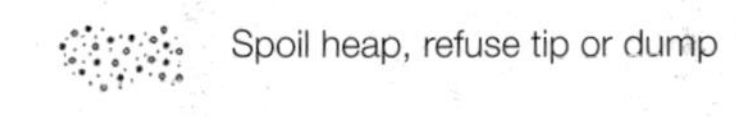

Spoil heap, refuse tip or dump

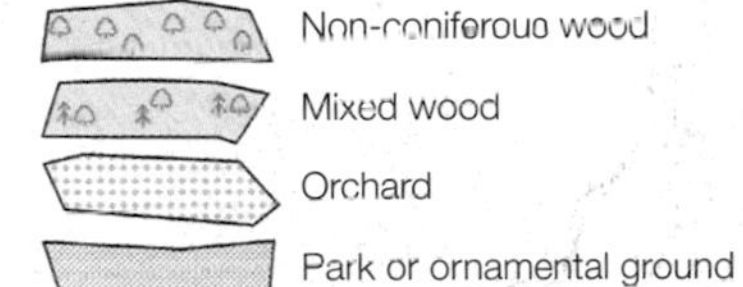

Coniferous wood
Non-coniferous wood
Mixed wood
Orchard
Park or ornamental ground
Forestry Commission access land
National Trust – always open
National Trust, limited access, observe local signs
National Trust for Scotland

TOURIST INFORMATION

Parking
Park & Ride
Visitor centre
Information centre
Telephone
Camp site/ Caravan site
Golf course or links
Viewpoint
PC Public convenience
Picnic site
Pub/s
Museum
Castle/fort
Building of historic interest
Steam railway
English Heritage
Garden
Nature reserve
Water activities
Fishing
Other tourist feature
Moorings (free)
Lectric boat charging point

WATER FEATURES

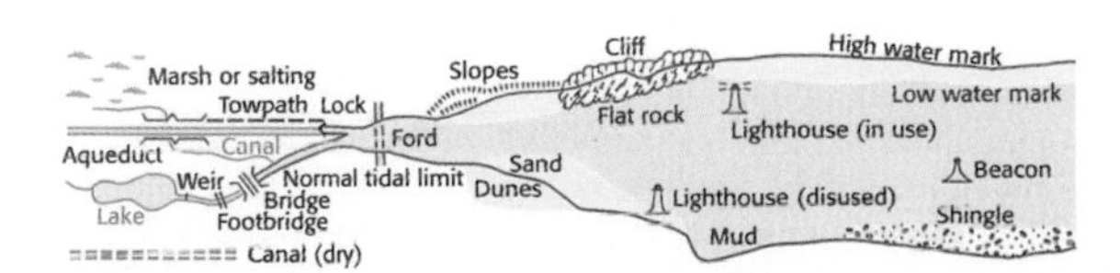

Glossary

abrasion the erosion of a surface caused by pieces of rock carried in rivers and glaciers

aggregate crushed stone, some of which can be used for building roads and making concrete

agribusiness large-scale farm run like a big business, likely to use lots of modern machinery and chemicals

all-inclusive a holiday deal that includes accommodation, meals, drinks and entertainment for one price

alluvium the fine and fertile sediment deposited on wide floodplains

anticyclone an area of high pressure resulting in calm and clear weather

aquifer an underground reservoir of water

arable farming which grows crops

attrition a process where rocks and stones moving along in the water get knocked against each other and are gradually worn away

backwash the movement of water as it drains back down the beach to the sea

biodiversity the varied range of plants and animals (flora and fauna) found in an area

biogas a gas produced by the breakdown of organic matter, such as manure or sewage, in the absence of oxygen. It can be used as a biofuel

biomes large-scale ecosystems, of which the world has eight. These include hot deserts and tropical rainforests

birth rate the number of live babies born in a year for every 1000 people

boulder clay material deposited by a glacier, including huge rocks and fine dust. Sometimes called til

caldera a large ridge forming the edge of a supervolcano, where the original cone has either collapsed or been destroyed in a large eruption

carbon sink a natural ecosystem that has the ability to absorb and store large amounts of carbon, such as a tropical rainforest

Central Business District (CBD) in the middle of MEDC cities, where many businesses and shops are found

climate the average weather pattern measured in one particular area over a period of 30 years

climate change the long term variation in a year's average temperature and rainfall

cold front where cold air pushes in behind warm air, causing the warm air to rise, resulting in clouds and heavy rain

commuter villages small settlements with a large proportion of commuters, meaning much of the population is out during the day

confluence the point where two rivers meet

conservation swaps a deal reached between a country which owes a wealthier country money. The wealthy country cancels the debt as long as the country in debt invests in conservation programmes

conservative a plate margin where the tectonic plates are sliding past each other

constructive a plate margin where the tectonic plates are moving apart

constructive waves waves that push sand and pebbles further up the beach. They have a strong swash and a weak backwash so help to build up the beach

consumers animals which feed on plants (producers) or each other

continental crust the part of the Earth's crust that makes the continents

convectional rainfall during the summer, sunshine heats the ground and causes warm air to rise to a high altitude. The water vapour then cools and forms clouds which can produce heavy rainfall

Coriolis Force where air streams are deflected through the rotation of the Earth on its axis. It can be the cause of storms, including hurricanes

corrie a steep sided hollow on a mountain that has been deepened by the action of ice

death rate the number of people who die in a year for every 1000 people

decomposers fungi and bacteria that feed on dead and waste material, making it break down or rot

demographic transition model a diagram which shows how a country's population changes over time. The model goes through five stages, with the population being stable before rising sharply, then levelling off, and then slowly decreasing.

depopulation when the number of people living in an area falls because residents are moving away. This is often seen in rural villages, when people leave because of poverty and unemployment

depression an area of low pressure resulting in cloudy, wet and windy weather

destructive a plate margin where the tectonic plates collide

destructive waves waves that remove material from the beach. They have a weak swash and a strong backwash, pulling sand and pebbles down the beach when the water retreats

development the use of natural and human resources to achieve a higher standard of living

discharge the amount of water flowing in a river at any one point

diversified when a farmer has branched out into other areas, such as camping, running farm shops and renting converted barns

drumlins small hills made of glacial moraine and formed by glaciers. They can reach 50 metres high, have one steep side and one gently sloping side.

economic migrants people who move away from poorer areas to gain access to better jobs and wages

ecosystem a unit made up of living things and their non-living environment. For example, a pond, a forest, a desert

ecotourism when people visit a place because of its natural environment and cause as little harm to it as possible

elderly dependents the population aged over 65 who may be (or become) dependent on others for care

environment the non-living surrounding of plants and animals that can have an effect on them, such as the climate and soil

epicentre the point on the ground directly above the focus (centre) of an earthquake

erode to wear away rock by the natural processes of rivers, ice, wind and sea

escarpment a landform found in areas of both chalk and clay. Escarpments have a steep scarp slope above a gentle dip slope

ethnic segregation people from different ethnic groups and religions living in separate areas

fairtrade providing better prices, working conditions and terms for farmers and workers in poorer countries

fetch the length of water over which the wind has blown, affecting the size and strength of waves

floodplain a flat area either side of a river which floods when the river overflows

flood warning issued by the Environment Agency when flooding is expected to affect homes and businesses shortly, and people should act immediately

flood watch a warning issued when flooding may affect low-lying land and roads. People should monitor the weather and news reports when they hear this

focus the point of origin of an earthquake

fold mountains a mountain range formed by the collision of a continental plate with either another continental plate or an oceanic plate

food chain the transfer of energy through an ecosystem, showing which animals eat which animals and plants

food miles the distance food travels from the farmer producing it, to the person eating it. The longer the distance, the more carbon dioxide is produced by the journey

food web a complex diagram showing which animals eat which other animals and plants within an ecosystem. The web will contain several food chains and show how they link together and intertwine

freeze-thaw weathering the breaking of rock caused by the cycle of freezing and thawing

globalisation the way business, ideas and lifestyles are spreading more and more easily around the world

global warming the way temperatures around the world are rising. Scientists think we have made this happen by burning too much fossil fuel, releasing carbon dioxide

gorge a steep-sided cut through the landscape formed over thousands of years by a retreating waterfall

green belts areas around the edge of cities with strict planning controls designed to stop houses being built in the countryside

greenhouse effect when gases such as carbon dioxide trap heat around the Earth, leading to global warming

Gross Domestic Product (GDP) the total value of goods and services produced by a country in one year

groundwater flow movement of water through rocks in the ground

groundwater water stored in rocks underground

groynes large wooden fences built on beaches to try and trap sediment and stop it being transported along the coast

hanging valley a high level tributary valley with a sharp fall to the main valley. This is a feature of glacial erosion

hard engineering building structures to deal with natural hazards, such as dams to prevent flooding

hunting and gathering a traditional method of survival involving the hunting of animals and the gathering of food from plants

hurricanes particularly powerful tropical revolving storms, capable of creating a great deal of damage

hydraulic action fast flowing water pushes air into cracks and the force of this causes the channel to break up over time

hydrograph a graph showing how a river responds to a storm, showing the rainfall and discharge over time

ice age a period of time when ice sheets are found on the continents

ice sheet a huge mass of ice covering a vast area of land

igneous rock formed by the cooling of magma or lava

impermeable rocks and soils that do not absorb water

industrialization where a mainly agricultural society develops and begins to depend on manufacturing industries

infiltration movement of water from surface into the soil

informal economy includes jobs where there are no official contracts and no taxes are paid

infrastructure the name used to describe a country's transport links and basic services, such as train routes and hospitals

integrated transport policy a planned system where buses on different routes link with each other and trains to provide a door to door service that rivals car use

intensive farms which have a large input of labour, money or technology

interdependent when countries rely on each other for providing goods and services

irrigated farmland that has water added to it by humans to help crops grow during dry periods

lag time the time between the peak rainfall and the maximum discharge in a river

leached when minerals are slowly washed out of the soil

levees river embankments built as a flood defence

load the amount of material carried by a river

logging cutting down trees for timber to make a range of products, including furniture, houses and paper

long-haul destinations places that are far away from the United Kingdom and take a long time to reach by air, including Jamaica and Kenya

longshore drift the process which material is transported along the coast

meanders bends found along the course of a river

megacities settlements that have over 10 million people living in them

metamorphic rocks created when existing rocks are transformed underground by heat and pressure

moraine all the material moved or carried by a glacier

mudflats flat coastal areas formed when mud is deposited by rivers and coasts

natural decrease (in the population) where the number of deaths is greater than the number of live births

natural increase (in the population) where the number of live births is greater than the number of deaths

non-renewable energy sources that are finite and will eventually run out, such as oil and gas

occluded front where a cold front catches up with a leading warm front and lifts it up; or where a warm front catches up with a leading cold front and slides over it

oceanic crust the part of the Earth's crust which is under the oceans

ocean trench a deep depression on the ocean floor formed at the subduction zones of destructive plate margins

overcultivation the excessive use of farmland, resulting in the soil's nutrients being used up and the land becoming infertile

overgrazing removal of vegetation by grazing animals at a faster rate than the vegetation can regrow

oxbow lake a lake shaped like a crescent that is formed when a river bend is cut off from the main flow and becomes isolated

package holidays vacations that include flights, airport transfers and accommodation in one deal

pastoral farming that rears animals

permeable rocks and soils that absorb water and allow it to drain from the surface into the ground

plates the large pieces of broken crust that cover the Earth

plate margins where two tectonic plates meet

Pleistocene Era a period of time, starting about 2 million years ago and ending 10 000 years ago

plucking when ice freezes onto rock and moves, pulling the rock away

plunge pool the area of deep water at the bottom of a waterfall, formed by hydraulic action and the grinding of rocks and pebbles

pollarding when a tree has its upper branches cut off, leaving the rest of the tree intact. The removed wood is used for building and fuel

pollution the release of contaminates that can cause distress to people, animals and plants

population density the average number of people living in an area

population growth rate the number of people added to, or lost from, the population each year. This can be as a result of natural increase and migration

population pyramid a graph showing the age and sex of a population

primary effects impacts of a natural disaster that happen immediately or soon after the event

producers plants using sunlight, water and nutrients from the soil to produce their own food

pyroclastic flows fast-moving, destructive torrents of hot ash, rocks, gasses and steam from a volcano

rebranding to enhance and positively change the image people have of towns, cities and villages

relief rainfall when winds forces moist air to rise over mountains and it cools, falling as rain

renewable energy sources that are sustainable and can be used again and again, such as solar, wind and wave power

retirement migration the movement of people who have retired to places with a warm climate and open spaces, such as the coast or countryside

ribbon lakes long, narrow lakes that fill the valley floor in some glacial troughs. Examples include Windermere, Wast Water and Coniston Water

Richter Scale a measure of how strong an earthquake is

rural-urban fringe the area where the city and the countryside meet, usually containing large housing and farmland

rural-urban migration the movement of people from the countryside to the cities, normally to escape from poverty and look for work

Saffir-Simpson scale a measure of the strength and intensity of hurricanes

secondary effects impacts of a natural disaster which may occur days, weeks or months after the event

second home a dwelling that is owned by people who live in another home they own. Their extra house may be rented out to tourists

sediment material such as sand and clay that is carried by a river

sedimentary rocks formed by layers of sediment that were deposited under water

severe flood warning issued by the Environment Agency when there is extreme danger imminent, threatening life and housing

shanty towns areas of overcrowded poor housing on the edge of LEDC cities, often built illegally

shock waves the energy waves created by an earthquake

short-haul destinations places that can be reached by a flight of less than 3 hours. For the UK, this means many places in Europe

Site of Special Scientist Interest (SSSI) areas designated to be of importance because of the plants, animals or geology found there

slash and burn a traditional method of farming in the rainforest that sees a small amount of land cleared and farmed for a few years before the people leave it and move on

sliding when large chunks of rock slide down a slope quickly without warning

slumping when cliffs made of clay become saturated and material oozes down towards the sea as mud or debris flow

snow line the height at which snow stays on the ground for most of the winter

soft engineering involves adapting to natural hazards and working with nature to limit damage. For example, planting trees to limit flood risk

solution when minerals are dissolved in water and you can no longer see them

spit a long ridge of sand and shingle, attached to the land at one end and in the open sea at the other

storm surge a rapid rise in the level of the sea caused by low pressure and strong winds

subduction the transformation into magma of a denser tectonic plate as it dives under a less dense plate

subsistence farming where just enough crops are produced to support the farmer and his family

supervolcano a volcano capable of erupting on a much larger scale than a normal volcano

sustainable development meeting the needs of people now and in the future, and limiting harm to the environment

swash the movement of water as it rushes up the beach from the sea

synoptic chart a weather map showing isobars (lines linking areas of equal pressure)

tarn a lake found in a corrie

thalweg the fastest current in a river, forced to the outer bend at a meander

throughflow the movement of water through the soil

tors blocks of granite found at the top of a hill, appearing to balance on top of each other

traction big boulders and stones are rolled and dragged along the river bed

transnational corporations those which operate over several continents and can earn more money than small countries

transport the movement of eroded material by natural processes such as wind, rivers and sea

tributary a smaller river or stream flowing into a larger river

tsunami huge waves caused by an earthquake under the sea

urbanisation the growth of towns and cities and their population

urban sprawl the spreading of towns and cities outwards into the countryside

v-shaped valley a valley with steep sides and a narrow bottom that has been formed by erosion

warm front where warm air rises over cold air, cooling and condensing as it rises to form clouds and rain

waterfall a vertical fall of water where the course of a river is interrupted by a steep drop in the land it is flowing over

water table the level of water in the ground or soil

wave-cut platform area of gently sloping or flat rocks exposed at low tide

weather short term changes in atmospheric conditions such as temperature, wind, sunshine

weathering the breaking down of rock. The three types of weathering are mechanical, chemical and biological

working population those aged between 15 and 64 who are eligible to work. This group is also called the economically active

young dependents the population aged under 15 and dependent on adults to care for them

Index